Technically-Write!

Sixth Edition

Ron Blicq
Lisa Moretto
RGI International

Upper Saddle River, New Jersey
Columbus, Ohio

Editor in Chief: Stephen Helba
Executive Editor: Gary Bauer
Editorial Assistant: Natasha Holden
Production Editor: Louise N. Sette
Design Coordinator: Diane Ernsberger
Cover Designer: Monica Kompter
Cover art: Comstock
Production Manager: Brian Fox
Marketing Manager: Leigh Ann Sims

This book was set by Phyllis Seto. It was printed and bound by Courier Kendallville, Inc. The cover was printed by Phoenix Color Corp.

Pearson Education Ltd.
Pearson Education Singapore Pte. Ltd.
Pearson Education Canada, Ltd.
Pearson Education—Japan

Pearson Education Australia Pty. Limited
Pearson Education North Asia Ltd.
Pearson Educación de Mexico, S.A. de C.V.
Pearson Education Malaysia Pte. Ltd.

10 9 8 7 6 5 4 3 2 1

ISBN 0-13-114878-8

Contents

About the Authors

Ron Blicq and Lisa Moretto are Senior Consultants with RGI International, a consulting company specializing in oral and written communication. They teach workshops, based on the Pyramid Method of Writing presented in this book, to audiences all over the world. In 2001, they opened a second company—RGI Learning Inc.—specifically to deliver their courses on the Web. Their web site is **www.rgilearning.com.**

Ron is Senior Consultant at RGI's Canadian office. He has extensive experience as a technical writer and editor with the Royal Air Force in Britain and CAE Industries Limited in Canada, and taught technical communication at Red River College from 1967 to 1990. Ron has authored five books with Pearson Education and has written and produced six educational video programs, such as *Sharpening Your Business Communication Skills* and *So, You Have to Give a Talk?* He is a Fellow of both the Society for Technical Communication and the Association of Teachers of Technical Writing, and a Life Member of the Institute of Electrical and Electronics Engineers Inc. Ron lives in Winnipeg, Manitoba.

Lisa is Senior Consultant at RGI's United States office. She has experience as an Information Developer for IBM in the US and as a Learning Products Engineer for Hewlett-Packard in the UK. Lisa holds a B.S. in Technical Communication from Clarkson University in New York, and an M.S. in User Interface Design from the London Guildhall University in England. Her specialties include developing online interactive information, designing user interfaces, and writing product documentation. She is a senior member of the Society for Technical Communication and a member of the Institute of Electrical and Electronics Engineers Inc. Lisa lives in Rochester, New York.

(Photo: Mary Lou Stein)

Preface

This book presents all aspects of technical communication that you, as a technician, technologist, engineer, scientist, computer and environmental specialist, or technical manager, are likely to encounter in industry. It contains numerous examples of letters, reports, and proposals, all based on the unique "pyramid" method for structuring information, a technique that has helped countless technical people overcome "writer's block."

With each previous edition of Technically-Write!, changes were introduced to reflect the way technical professionals were currently presenting information in the various industries and in engineering consulting firms. This edition is no exception—in fact we have made more changes than ever before. Here are some of the most noticeable revisions:

- There is a new, opening chapter that traces how, over 100 years ago, the engineering community recognized that more attention needed to be placed on increasing a technical professional's ability to communicate effectively, and how lecturers at universities and colleges responded.
- Chapter 3 has more information on how to plan and write email messages.
- Chapter 4 now includes a personal progress report which helps keep managers informed of the writer's activities.
- There is a comprehensive new chapter (Chapter 7) on how to write informal and semiformal proposals.
- The chapter on writing resumes and attending interviews (Chapter 11) has been enlarged to include new techniques for submitting electronic resumes.
- The glossary has been enlarged to include more computer-related terms.

If you have seen previous editions, you will have noticed that the shape of the book has also changed. The shape will seem slimmer because we have reduced the number of pages by about 7%. We did this partly to help reduce the cost for purchasers, and partly to place some of the assignments and exercises in an instructor's manual and for electronic delivery.

Information about the two companies—H. L. Winman and Associates and Macro Engineering Inc.—has been removed, but many of the model letters and reports, and end-of-chapter assignments, still retain these two companies to provide a logical environment for the documents and exercises.

Along the way, we have very much appreciated the friendly advice and many helpful suggestions from users of the book, both teachers and students, and the advice of reviewers. In particular, we would like to thank the following reviewers: John Roberts (Mohawk College), Lisa Wolanski-McGirr (Keewatin College), George Scott (Seneca College), Alexa Campbell (Red River College), Elizabeth Smyth and Joe Benge (Camosun College). Their ideas have guided us in preparing this sixth edition. We are also celebrating, for it's 32 years since the first edition of *Technically-Write!* was published!

R.B. & L.M.

Supplements

The sixth edition of Technically-Write! is supported by a comprehensive supplements package, which includes the following:

- Instructor's Resource Manual with Transparency Masters — ISBN 0-13-117238-7
- Test Item File — ISBN 0-13-117237-9
- Test GenEQ — ISBN 0-13-117236-0
- Text-Enrichment Website — ISBN 0-13-117234-4
- Instructor Resource CD — ISBN 0-13-117526-2

People as "Communicators"

We are equipped with a highly sophisticated communication system, yet we consistently fail to use it properly. The system comprises a transmitter and receiver combined into a single package controlled by a built-in computer, the brain. It accepts multiple inputs and transmits in three mediums: action, speech, and writing.

We spend many of our waking hours communicating, half the time as a transmitter, half as a receiver. If, as a receiver, we mentally switch off or permit ourselves to change channels while someone else is transmitting, we contribute to information loss. Similarly, if as a transmitter we permit our narrative to become disorganized, unconvincing, or simply uninteresting, we encourage frequency drift. Our listeners detune their receivers and let their computers think about the lunch that's imminent, or wonder if they should rent a video tonight.

As long as a person transmits clearly, efficiently, and persuasively, people receiving the message keep their receivers "locked on" to the transmitting frequency (this applies to all written, visual, and spoken transmissions). Such conditions expedite the transfer of information, or "communication."

In direct contact, in which one person is speaking directly to another, the receiver has the opportunity to ask the transmitter to clarify vaguely presented information. But in more formal speech situations, and in all forms of written and most visual communication, the receiver no longer has this advantage. He or she cannot stop a speaker who mumbles or uses unfamiliar terminology to ask that parts of a talk be repeated or clarified; neither can the receiver easily ask a writer in another city to explain an incoherent passage of a business letter, or the producer of a video program to describe the point the video is trying to make.

The results of failure to communicate efficiently soon become apparent. If people fail to make themselves clear in day-to-day communication, the consequences are likely to differ from those they anticipated, as Cam Collins has discovered to his chagrin.

Cam is a junior electrical engineer at Macro Engineering Inc., and his specialty is high-voltage power generation. When he first read about a recent extra-high-voltage (EHV) DC power conference, he wanted urgently to attend. In a memorandum to Fred Stokes, the company's chief engineer, Cam described the conference in glowing terms that he hoped would convince Fred to approve his request. This is what he wrote:

Cam's request fails to convince

Fred

The EHV conference described in the attached brochure is just the thing we have been looking for. Only last week you and I discussed the shortage of good technical information in this area, and now here is a conference featuring papers on many of the topics we are interested in. The cost is only $228 for registration, which includes a visit to the Freeling Rapids Generating Station. Travel and accommodation will be about $850 extra. I'm informing you of this early so you can make a decision in time for me to arrange flight bookings and accommodation.

Cam

Fred Stokes was equally enthusiastic and wrote back:

Cam

Thanks for informing me of the EHV DC conference. I certainly don't want to miss it. Please make reservations for me as suggested in your memorandum.

Fred

Cam was the victim of his own carelessness: he had failed to communicate clearly that it was *he* who wanted to go to Freeling Rapids!

Elizabeth has a good idea...

Elizabeth Drew, on the other hand, did not realize she had missed a golden opportunity to be first with an innovative computer technique until it was too late to do anything about it. Her story stems from an incident that occurred several years ago, when she was a recently graduated engineer employed by a manufacturer of agricultural machinery. Elizabeth's job was to design modifications to the machinery, and then prepare the change procedure documentation for the production department, service representatives, sales staff, and customers.

"For each modification I had to coordinate three different documents," she explained to us over lunch. "First, there had to be a design change notice to send out to everyone concerned. And then there had to be an 'exploded' isometric drawing showing a clear view of every part, with each part cross-referenced to a parts list. And finally there had to be the parts list itself, with every item labeled fully and accurately."

Elizabeth found that cross-referencing a drawing to its parts list was a tedious, time-consuming task. The isometric drawing of the part was computer generated by the drafting department. The parts list was also keyed into a computer, but by a separate department. However, because the two computer systems were incompatible, cross-referencing had to be done manually.

"And then I hit on a technique for interfacing the two programs," Elizabeth explained. "It was simple, really, and I kept wondering why no one else had thought of it!"

...it was simple and efficient...

Without telling anyone, she modified one of the company's software programs and tested her idea with five different modification kits. "It worked!" she laughed. "And, best of all, I found that cross-referencing could be done in one-tenth of the time."

Elizabeth felt her employer should know about her idea: possibly the company could market the software, or even help her copyright it. So the following day she stopped Mr. Haddon, the Engineering Manager, as they passed in the hallway, and blurted out her suggestion. This is the conversation that ensued:

Elizabeth	Mr. Haddon
Oh! Mr. Haddon! You know how long it takes to do the documentation for a new part...?	
	Yes..s..s..?
The problem is in trying to interface between the graphics computer and the parts list...	
	(*Mr. Haddon appeared to be listening politely, but internally he was growing impatient.*)
...It has to be done by hand, you see...	
	Doesn't the drafting department do all that?
Oh, yes! They do. I was just trying to help them...to speed up their work a bit.	
	You're working for the chief draftsman now?
Oh, no! It was just an idea I had—to modify the software we use...	
	I don't remember issuing you a work order...
No. You didn't. I was doing it on my own... (*She meant she was doing it on her own time.*)	
	You mean the I.T. people asked you to do it?
Well—uh—no. Not exactly...	
	But you have been modifying one of our software programs? Without authority?
(*Reluctantly*) Uh-huh.	
	I thought I had made it quite clear to all the staff: No projects are to be undertaken without my approval! (*His tone was cold and abrupt.*)
I wanted to try...	
	That's final! (*And he turned on his heel and continued down the hall.*)

...but Elizabeth didn't know how to articulate her ideas clearly

They need to focus their messages

Elizabeth's simple suggestion had become lost in a web of misunderstanding. By the time she was through explaining what she had been doing, she had given up trying to offer her idea to the company. And so her idea lay dormant for two years, until a major software company came out with a comparable program. Elizabeth knew then that perhaps there had been market potential for her design.

If Cam Collins and Elizabeth Drew had paused to consider the needs of the people who were to receive their information, they would never have launched precipitously into discourses that omitted essential facts. Cam had only to start his memorandum with a request ("May I have your approval to attend an EHV DC conference next month?"), and Elizabeth with a statement of purpose ("I have designed a software program that can save us hundreds of dollars annually. May I have a few moments to describe it to you?"), to command the attention of their department heads. Both Mr. Stokes and Mr. Haddon could then have much more effectively appraised the information.

Such circumstances occur daily. They are frustrating to those who fail to communicate their ideas, and costly when the consequences are carried into business and industry.

Bill Carr recently devised and installed a monitor unit for the remote control panel at the microwave relay station where he is the resident engineering technologist. As his modification greatly improved operating methods, Janet Reid, Manager of Technical Services at head office, asked him to submit an installation drawing and an accompanying description. Here is part of his description:

> Some difficulty was experienced in finding a suitable location for the monitor unit. Eventually it was mounted on a locally manufactured bracket attached to the left-hand upright of the control panel, as shown on the attached drawing.

Good intentions...

On the strength of Bill's explicit mounting description and detailed list of hardware, Janet instructed project coordinator Phyllis Walters to convert Bill's description into an installation instruction, purchase materials, assemble 21 modification kits, and ship them to the 21 other relay stations in the microwave link.

...resulting in confusion!

Within a week, the 21 resident engineering technologists were reporting to Phyllis that it was impossible to mount the monitor unit as instructed, because of an adjoining control unit. Neither Janet nor Phyllis had remembered that Bill Carr was located at site 22, the last relay station in the microwave link, where there was no need for an additional control unit. Bill had assumed that Janet would be aware that the equipment layout at his station was unique. As he commented afterward: "I was never told *why* I had to describe the modification, or what head office planned to do with my description."

In business and industry we must communicate clearly and understand fully the implications of failing to do so. A poorly worded order that results in the wrong part being supplied to a job site, a weak report that fails to motivate the reader to take the urgent action needed to avert a costly equipment breakdown, and even an inadequate job application that fails to sell an employer on the right person for a prospective job, all increase the cost of doing business. Such mistakes and misunderstandings are wasteful of the country's labor and resources. Many of them can be prevented by more effective communication—communication that is receiver-oriented rather than transmitter-oriented, and that transmits messages using the most expeditious, economical, and efficient means at our command.

Chapter 1
Why Technical People Need to Write Well

Over the past four years we have asked numerous technical professionals: "What is the publication date of the earliest book on technical writing that you own?"

Nearly everyone listed books from the 1960s and 1970s. Yet our research shows that the teaching of technical writing in science and engineering courses began more than one hundred years ago, in 1901, when the Society for the Promotion of Engineering Education (SPEE) published this succinct statement:

> The writing skills of engineering students are deplorable and need to be addressed by engineering colleges.

Technical Communication Overview
http://saulcarliner.home.att.net/idbusiness/historytc.htm
This site includes a brief history of technical communication.

In the early 1900s, technical communication was taught by engineering professors.

These words did not go unheeded. Although technical communication was not part of a technical student's curriculum in those days, and was rarely included even in the range of courses taught by the English department, some engineering and English professors, both in North America and Great Britain, quietly began teaching the importance of good writing as part of other technical courses. After doing this for many years, some of them published books based on the notes they had typed up for their students. The following is a brief history of those texts.

The First Fifty Years

In 1908, T. A. Rickard, an associate of the Royal School of Mines in London, England, published a book titled *A Guide to Technical Writing*.[1] He wrote:

> Conscientious writers try to improve their mode of expression by precision of terms, by careful choice of words, and by the arrangement of them so that they become efficient carriers of thought from one mind to another.

Rickard titled one of his chapters: "A Plea for Greater Simplicity in the Language of Science," having noticed that technical people tended to write in a long-winded way that was not easy for anyone outside their discipline to understand.

In 1922, Karl Owen Thompson, who taught English at Case School of Applied Science in Cleveland, Ohio, published a book titled *Technical*

Exposition.[2] In the introduction to his book he commented on the differences between literary and technical writing:

In today's global community, Thompson would replace "English" with "Language"

> The study of English at a scientific school has a more directly professional application than it has at an academic college. Instead of courses in literature with their cultural purposes, courses are given that prepare the students for the types of reading and writing that will be required of them after they are graduated from college.... English is more than a tool, it is a part of life itself in its many activities.

At the University of Michigan's College of Engineering, J. Raleigh Nelson insisted from 1915 onward that his students write clearly. In 1940 he summed up his thoughts in a book titled *Writing the Technical Report*,[3] in which he wrote:

> In report writing, in particular, there is an increasing demand that the first page or two shall provide a comprehensive idea of the whole report.

This was the first documented reference to what we now refer to as the Executive Summary, which precedes a long report or proposal (see Chapter 6). Reginald Kapp taught electrical engineering at University College in London. Like Nelson, he insisted his students write well. In 1948 he summed up his thoughts in a pocket-sized reference book titled *The Presentation of Technical Information*,[4] in which he particularly drew attention to the importance of identifying the audience before (in those days) putting pen to paper. He wrote:

> You must consider carefully the extent of the reader's knowledge, his range of interests, and...any peculiarities, whatever they may be, that might influence his receptivity for the information you have to impart.

Rickard and Kapp strongly stressed the need to identify the audience before starting to write

Similarly, forty years earlier, T. A. Rickard had written:

> If you describe a stamp-mill to an experienced mill-man, a mining student, or a bishop, you will vary the manner of telling. The most effective will be that which has a sympathetic appreciation of the other fellow's receptiveness. Do not plant carnations in a clay soil, or rice in a sand-heap.[5]

Technical Communication Quarterly
www.attw.org/
Technical Communication Quarterly is the journal of the Association for Teachers of Technical Writing.

(These authors were writing books for technical professionals, who were almost entirely male in the early part of the 20th century. They would write very differently today: for example, T. A. Rickard would probably change *mill-man* to *mill worker* and *other fellow's* to *other person's*.)

A Change in Style

Tyler G. Hicks was a mechanical engineer who taught at Cooper Union School of Engineering. He had written numerous articles and three technical books before turning his attention to engineering writing. In 1959, Hicks wrote *Successful Technical Writing*,[6] a major milepost for books on technical writing because of his refreshing directness and style. Here are three examples:

> Technical writing always pays off. You never lose when you write a good technical piece.... Good writing is a sure road to professional recognition.

> Talk directly to the reader. Bring him into the discussion. Use the personal pronouns "we" and "you," but with discretion.

> Choose verbs that create active impressions to the reader, and steer clear of the passive voice. You thus give life to your style.

Hicks's writing still sits well with today's readers

The five writers discussed here were very conscious that they were preparing their students to take up important roles in the engineering and technical professions. What they had to say to their students then is just as relevant today.

Although writing styles may have changed, the message remains constant

When, as a newly graduated engineer, engineering technician, or computer or environmental specialist, you first become employed in a technical field, you might be surprised to discover that report writing is an integral part of your work. As you advance in your chosen profession, you will also find that you will have to do more and more writing. We hope that *Technically-Write!* helps prepare you for the many situations you encounter.

REFERENCES

1. T. A. Rickard, *A Guide to Technical Writing* (San Francisco: Mining and Scientific Press, 1908) p. 8.
2. Karl Owen Thompson, *Technical Exposition* (New York: Harper & Brothers Publishers, 1922), p. vii.
3. J. Raleigh Nelson, *Writing the Technical Report* (New York: McGraw-Hill Book Company, 1940), p. 39.
4. Reginald O. Kapp, *The Presentation of Technical Information* (London: Constable & Company Ltd., 1948), p. 20. (Reprinted, with slight revisions, and published by the Institute for Scientific and Technical Communicators, UK, 1998.)
5. Rickard, p. 12.
6. Tyler G. Hicks, *Successful Technical Writing* (New York: McGraw-Hill Book Company, Inc., 1959), pp. 1 and 194.

In 2003, the Kapp book was still in print

Chapter 2
A Technical Person's Approach to Writing

Engineering technician Dan Skinner has a report to write on an investigation he completed seven weeks ago. He has made several half-hearted attempts to get started, but never seemed to find the right moment: maybe he was interrupted to resolve a circuit problem, or it was too near lunchtime, or a meeting was called. And now he is up against the wire.

Unless Dan is one of those unusual people who can produce only when under pressure, he is in danger of writing an inadequate, hastily prepared report that does not represent his true abilities. He does not realize that by leaving a writing task until it is too late to do a good job, and then frantically organizing the work, he is probably inhibiting his writing capabilities.

If Dan were to relax a little, instead of worrying that he has to organize himself and his writing task, he would find the physical process of writing a much more pleasant experience. But first he must change his approach.

Every technical person, from student technician to potential scientist to practicing engineer, has the ability to write clearly and logically. But this ability has to be developed. Dan Skinner must first learn some basic planning and writing techniques, then practice using them until he has acquired the skill and confidence that are the trademarks of an effective writer.

Simplifying the Approach

Throughout this book we will be advising you to *tell your readers right away what they most need or want to know*. This means structuring your writing so that the first paragraph (in short documents, the first *sentence*) satisfies their curiosity. Most executives and many technical readers are busy people who only have time to read essential information. By presenting the most important items first, you can help them decide whether they want to read the whole document immediately, put it aside to read later, or pass it along to a specialist in their department.

This reader-oriented style of presentation is known as the "pyramid technique." Imagine every letter, memorandum, or report you write is shaped like a pyramid: there is a small piece of essential information at the top, supported on a broad base of details, facts, and evidence. In most letters and short reports the pyramid has only two parts: a brief **Summary** followed by the **Full Development**, as shown in Figure 2-1(a). In long reports, an additional part—known as the **Essential Details**—is inserted between the Summary and the Full Development, as in Figure 2-1(b).

The writer's pyramid helps you focus your letters and reports

Normally, readers are not aware when a writer has used the pyramid technique. They simply find the letter or report well organized and easy to read. For example, in the opening paragraph of his letter report in Figure 2-2, Wes Hillman summarizes what Tina Mactiere most wants to know (whether the training course was a success and what results were achieved). In the remainder of the letter he fills in background details, states briefly how the course was run, reports on student participation and reaction, and suggests additional topics that could be covered in future courses.

Every document shown in this textbook has been structured using the pyramid technique. The pyramid's application to letters, memorandums, email messages, reports, proposals, instructions, descriptions, and even resumes and oral presentations is described in Chapters 3 through 8, and 10 and 11. For the moment, just remember that using the pyramid is the simplest, fastest, most effective way to plan and write any document, regardless of its length. If Dan Skinner had known about the pyramid technique, he would have found it much easier to get started.

Society for Technical Communication
http://stc.org
With more than 20,000 members worldwide, STC is the largest professional organization serving the technical communication profession. The society's diverse membership includes writers, editors, illustrators, printers, publishers, educators, students, engineers, and scientists employed in a variety of technological fields.

Planning the Writing Task

The word "planning" seems to imply that report writers must start by thoroughly organizing both themselves and their material. We disagree. Organizing too diligently or too early in the writing process inhibits rather

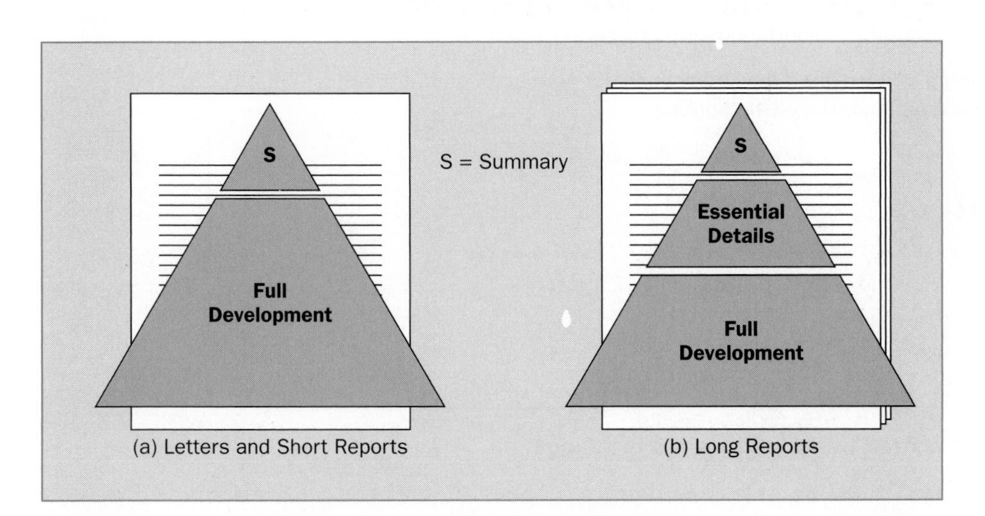

Figure 2-1 The pyramid writing technique.

The Roning Group Inc

Communication Consultants
2002 South Main Drive
Montrose OH 45287

October 16, 2004

Tina R. Mactiere, President
Macro Engineering Inc.
600 Deepdale Drive
Phoenix AZ 85007

Dear Ms. Mactiere,

Results of Pilot Report-Writing Course

The report-writing course we conducted for members of your engineering staff was completed successfully by 14 of the 16 participants. The average mark was 63%.

Summary (main message)

This was a pilot course set up in response to an August 13, 2004, enquiry from Mr. F. Stokes. At his request, we placed most emphasis on providing your staff with practical experience in writing business letters and technical reports. Attendance was voluntary, the 16 participants having been selected at random from 29 applicants.

Full development (all the details)

Best results were achieved by participants who recognized their writing problems before they started the course, and willingly became actively involved in the practical work. A few said they had expected to attend an "information" type of course, and at first were mildly reluctant to take part in the heavy writing program. Our comments on the work done by individual participants are attached.

Course critiques completed by the participants indicate that the course met their needs from a letter- and report-writing viewpoint, but that they felt more emphasis could have been placed on technical proposals and oral reporting. Perhaps such topics could be covered in a short follow-up course.

We enjoyed developing and teaching this pilot course for your staff, and particularly appreciated their enthusiastic participation.

Sincerely,

Wes Hillman

Wesley G. Hillman
Course Leader

enc

Figure 2-2 A letter report written using the pyramid technique.

than accelerates writing. The key is to organize your information in a spontaneous, creative manner, allowing your mind to freewheel through the initial planning stages until you have collected, scrutinized, sorted, grouped, and written the topics into a logical outline that will appeal to the reader.

We recommend that Dan Skinner at first neither make an outline nor take any action that resembles organization. Instead, he should work through seven simple planning stages that are less structured and therefore less confining. These stages are shown in Figure 2-3 and described in detail below.

"Disorganize" the writing task!

1. Gather Information

Dan's first step should be to assemble all the documents, results of tests, photographs, samples, computer data, specifications, and other supporting material that he will need to write his report, or that he will insert into it. He must gather everything he will need now, because later he will not want to interrupt his writing to look for additional facts and figures.

2. Define the Reader

Next, Dan must clearly identify his audience. This is probably the most important part of his planning, for if he does not, he may write an unfocused report that misses its mark. He must conjure up an image of the person or people who will read his report by asking himself six questions:

Pay *primary* attention to the ultimate reader

1. *Who, specifically, is my reader?* If it is someone he knows, his task is simplified. If it is someone he is not acquainted with (such as a customer in an out-of-town firm), he must imagine a persona.
2. *Is he or she a technical person?* Dan needs to know whether he can use or must avoid technical terms.

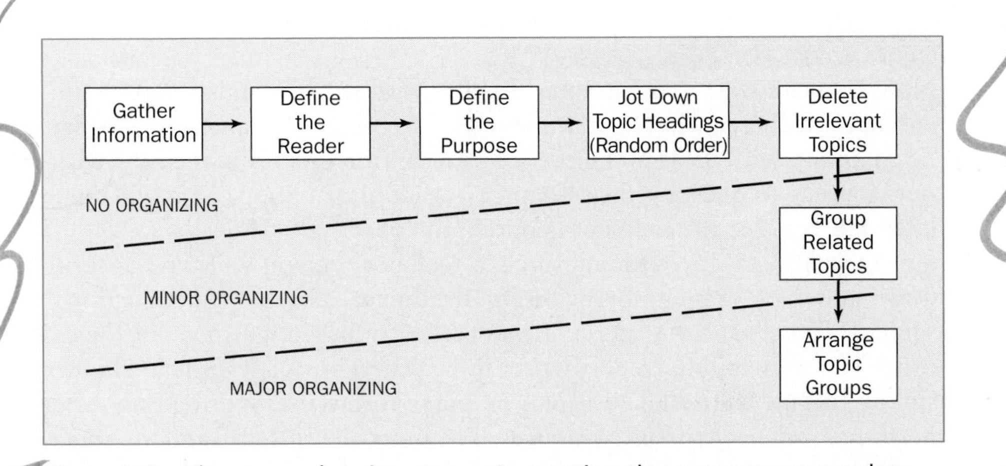

Figure 2-3 The seven planning stages. In practice, these stages can overlap.

3. *How much does the reader know about the subject I will be describing?* This will give Dan a starting point, since he won't need to cover information a reader already knows.
4. *What does the reader want to know or expect to be told?* Dan must be able to anticipate whether the reader will be receptive or hostile to the information he is presenting.
5. *Will more than one person read my report?* If so, Dan must repeat questions 2 through 4 for additional readers.
6. *Who is my* ***primary*** *reader?* The primary reader is the person who will make a decision or take action after reading Dan's report. Often this is the person to whom the report is directed. On occasion it may be one of the secondary readers. For example, a report may be addressed to a department manager, but the person who uses it or does something about it will be an engineer on the manager's staff.

Dan's inability to identify his reader was one of the reasons he had difficulty getting started on his report-writing task.

3. Define the Purpose

Now that he has identified his reader, Dan needs to ask himself one or possibly two more questions:

Decide: Why am I creating this message?

7. *Why am I writing to this person (or these people)?* Dan needs to decide whether his objective is to pass along information (to ***inform*** the reader about something), or to convince the reader to act or react (to ***persuade*** the reader to reply, make a decision, or approve a request).
8. If Dan's purpose is to persuade, then he also needs to ask: *What action do I want the reader to take?* This will help him decide what he wants his email, memo, letter, or report to achieve.

Now Dan is ready to develop a *focused* writing plan.

4. Jot Down Topic Headings

Loosen up: Delve deeply into brainstorming

Now Dan can start making notes. At this third stage he must "loosen up" enough to generate ideas spontaneously. He needs to brainstorm, so that he comes up with ideas and pieces of information quickly and easily, without stopping to question the relevance of that information. That will come later. His role for the moment is purely to collect it.

Normally, at the outlining stage, a technical person will type or write down a set of familiar or arbitrary headings, such as "Introduction," "Initial Tests," and "Material Resources," and arrange them in logical order. But we want our report writer to be different. We want Dan Skinner simply to type the series of topics he plans to discuss, writing only brief headings rather than full sentences. He must do this in random order, making no attempt to force the topics into groups. The topics he knows best will spring readily to mind; those he knows less well may take longer to recall.

When he finishes his initial list, he should scroll up the screen and examine each topic to see if it suggests less obvious topics. As additional topics come to mind he must type them in, still in random order, until he finds he is straining to find new ideas.

Dan must not try to decide whether each topic is relevant during this spontaneous brainstorming session. If he does, he will immediately inhibit his creativity because he will become too logical and organized. He must list all topics, regardless of their importance and eventual position in the final report. At the end of this session Dan's list should look like Figure 2-4.

Let the initial outline develop naturally, loosely

Building OK – needs strengthening
Elevators – too slow, too small
Talk with YoYo – elev mfr (10% discount)
Waiting time too long – 70 sec
Shaft too small
How enlarge shaft?
    Remove stairs?
Talk with fire inspector
Correspondence – other elev mfrs
Talk with Merrywell – Budget $950,000
Sent out questionnaire
Tenants' preferences –
    Express elev   No stop – 2nd flr
    Executive elev   Faster service
    Prestige elev   No stop – ground flr
    Freight elev
Freight elev – takes up too much space
Shaft only 35 × 8 ft (when modified)
Big freight elev – omit basement
Tenants "OK" small freight elev
    (YoYo "C" – 8 ft)
YoYo – has office in Montrose
Basement level has loading dock
Service reputation – YoYo?
    – Others?

Figure 2-4 Initial list of topic headings, typed in random order.

5. Delete Irrelevant Topics

Start grouping your topics into compartments

The fifth stage calls for Dan to print a hard copy to work on, then to examine his list of headings with a critical eye, dividing them into those that bear directly on the subject and those that introduce topics of only marginal interest. His knowledge of the reader—identified in stage 2—will help him decide whether each topic is really necessary, so he can delete irrelevant topics as has been done in Figure 2-5.

6. Group Related Topics

Now start pulling the pieces together

The headings that remain should be grouped into "topic areas" that will be discussed together. Dan can do this by simply coding related topics with

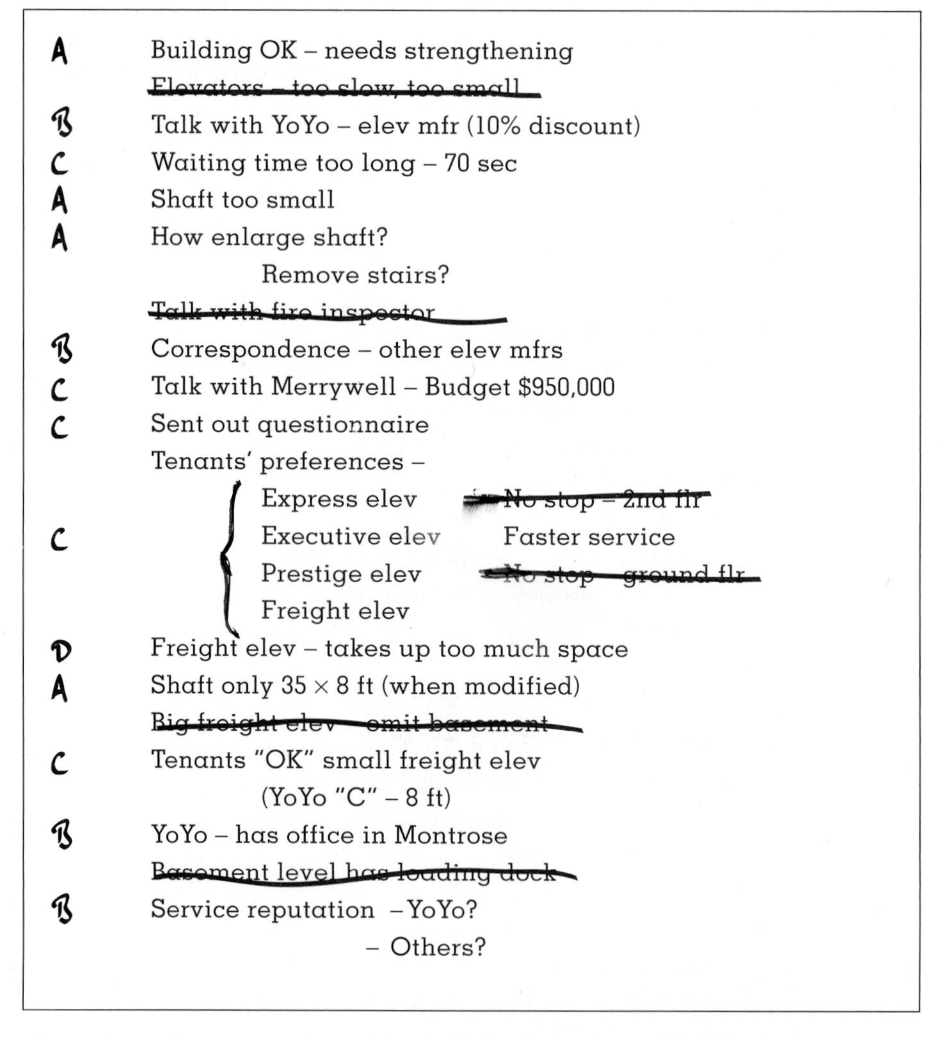

Figure 2-5 The same list of topic headings, but with irrelevant topics deleted and remaining topics coded into subject groups (A–structural implications; B–elevator manufacturers; C–tenants' preferences; D–freight elevator).

the same symbol or letter. In Figure 2-5, letter (A) identifies one group of related topics, letter (B) another group, and so on.

7. Arrange the Topic Groups

At this stage we encourage Dan to take his first major organizational step: to arrange the groups of information in the most suitable order. At the same time he needs to sort out the order of the headings within each group. He must consider:

- which order of presentation will be most interesting,
- which will be most logical, and
- which will be simplest to understand.

The result will become his final writing plan or report outline. Figure 2-6 shows Dan's final writing plan. Depending on how he prefers to work, Dan can use a hard copy of his outline, or work directly onscreen from a word-processing file.

Let the final outline evolve from the subject matter...

Building condition:
- OK – needs strengthening (shaft area)
- Existing elev shaft too small
- Remove adjoining staircase
- Shaft size now 35 × 8 ft

Tenants' needs:
- Sent out questionnaire
- Identified 5 major requests
- Requests we must meet:
  - Cut waiting time: 32 sec (max)
  - Handle freight up to 7 ft 6 in. long
- Requests we should try to meet:
  - Express elev to top 4 floors
  - Deluxe models (for prestige)
  - Private elev (for executives)

Budget: must be within $950,000

Elevator manufacturers:
- Researched 3
- Only YoYo Co. offers discount
- Only YoYo Co. has Montrose office

... rather than force the subject matter into a prescribed pattern

Figure 2-6 Topic headings arranged into a writing outline.

A final comment about outlining: If you have already developed an outlining method that works well for you, or you are using outlining software successfully, then we suggest you continue as you have been doing. The outlining method suggested here is for people who are seeking a simpler, more creative way to develop outlines than the one they are currently using.

Writing the First Draft

As we sit at our desks, with the heading "Writing the First Draft" at the top of Ron's computer screen and "Focus the Letter" at the top of Lisa's (Lisa is working on Chapter 3), we find we are experiencing the same problem that every writer encounters from time to time: an inability to find the right words—*any* words—that can be strung together to make coherent sentences and paragraphs. The ideas are there, circling around inside our skulls, and the outlines are there, so we cannot excuse ourselves by saying we have not prepared adequately. What, then, is wrong?

The answer is simple. Ten minutes ago the telephone rang and Jack, a neighbor, announced he would shortly bring over a "Neighborhood Block Watch" plan for Ron to sign. Ron paused to switch on the coffee, for we know that Jack will expect a cup while we talk, and now we can hear the percolator grumbling away in the distance. We cannot concentrate when we know our continuity of thought is so soon to be broken.

Write where you won't be disturbed: no telephone, no pager, no cellphone

Continuity is the key to getting one's writing done. In our case, this means writing at fairly long sittings during which we *know* we will not be disturbed. We must be out of reach of the telephone, visiting friends, and even family, so we can write continuously. Only when we have reached a logical break in the writing, or have temporarily exhausted an easy flow of words, can we afford to stop and enjoy that cup of cappuccino!

It is no easier to find a quiet place to write in the business world. The average technical person who tries to write a report in a large office cannot simply ignore the surroundings. A conversation taking place in an adjacent cubicle will interfere with one's creative thought processes. And even a co-worker collecting money for the pool on that night's NHL game between the New York Rangers and the L.A. Kings will interrupt writing continuity.

Write whatever way works best *for you*

The problem of finding a quiet place to write can be hard to resolve, particularly now that most people type their reports on a computer, so cannot move away from their desks (unless they are fortunate enough to own a laptop). For technical students, who frequently have to work on a tiny writing space in a crowded classroom, or in a roomful of computer terminals, conditions are even worse. Outlining in the classroom, followed by typing at home or in the seclusion of a library cubicle, is a possible alternative.

Before you start writing you need to consider the page layout and make decisions like these:

- What font you will use, and whether it should be serif or sans serif. A serif type has tops and tails on its ascenders and descenders (this book is set in Sabon, which is a serif type). A sans serif type is much plainer (Helvetica is a typical sans serif type).
- Whether you will print the report in 10 or 12 point type (i.e. with 10 or 12 characters to the linear inch). Generally, 11 or 12 point is better for serif fonts, and 10 or 11 point is better for sans serif fonts.
- The number of lines you want on a page, and the width of your planned typing lines.
- The width of the margins you want on either side of the text and at the top and bottom of the page.
- Whether you want the right margin to be justified (straight) or ragged. Research shows that paragraphs set with a ragged right margin are easier to read than paragraphs set with a justified right margin.
- Where you want the page numbers to be positioned (top or bottom of the page, and either centered or to one side of the page); on most systems page numbers are printed automatically, but you can select where they are to appear.
- The line spacing you want (single or double), and how many blank lines you want between paragraphs (normally one or one-and-a-half).
- Whether the first line of each paragraph is to be indented or set "flush" with the left margin; and, if indented, how long the indentation is to be.
- For long words at the end of a line, whether you or the computer will decide where the word is to be hyphenated (you can also select no hyphenation).
- The levels of headings you will use, and how you will use different font sizes and boldface type to differentiate between them. (See page 326 and Figure 12-1 of Chapter 12 for guidelines.)

You have to set up page parameters only once; the first time

Most popular word-processing programs provide default settings for these options, but you should be aware of them and how to customize the page layout for your particular needs. Every program is different, so consult the documentation that comes with your word-processing software for instructions on how to change an option.

Dan Skinner is ready to start writing, but now he encounters another difficulty. Equipped with his outline and the keyboard in front of him, he finds that he does not know where to begin. Or he may tackle the task enthusiastically, determined to write a really effective introduction, only to find that nothing he writes really says what he wants to say.

Tips for combatting "writer's block"

We frequently advise technical people who encounter this "no start" block that the best place for them to start writing is at paragraph two, or even somewhere in the middle. For example, if Dan finds that a particular part of his project interests him more than other parts, he should write about that part first. His interest and familiarity with the subject will help him write those few first words, and keep him going once he has started. The most important thing is to *start* writing, to put any words at all down, even if they are not exactly the right words, and to let them lead naturally into the next group of ideas.

Write without stopping to revise; that comes later

This is where continuity becomes essential: don't interrupt the writing process to correct a minor point of construction, write perfect grammar, find exactly the right word, fiddle with page layout, or construct sentences and paragraphs of just the right length. That can be done later, during revision. The important thing is to keep building on that rough draft, so that when you stop for a break you know you have written something you can work up into a presentable document.

If, as he writes, Dan cannot find exactly the word he wants, he should jot down a similar word and type in a question mark enclosed in parentheses immediately after it, as a reminder to change the word when the first draft is finished. Similarly, if he is not sure how to spell a certain word, he should resist the temptation to turn to a dictionary, for that will disrupt the nat-ural flow of his writing. Again, he should draw attention to the word as a reminder that he must consult his dictionary later. See Figure 2-7.

Get on a roll...and keep rolling!

We cannot stress this too strongly: writers should not correct their work as they write. Writing and revising are two entirely separate functions, and they call for different approaches. They cannot be done simultaneously. Writing calls for creativity and total immersion in the subject so the words tumble out in a constant flow. Revision calls for lucidity and logic, which force a writer to reason and query the suitability of the words he or she has written. The first requires excluding every thought but the subject; the second demands an objectivity that challenges the material from the reader's point of view. Writers who try to correct their work as they write soon become frustrated, because creativity and objectivity are constantly fighting for control.

The length of each writing session will vary, depending on the writer's experience and the complexity of the topic. If a document is short, it should be written all at one sitting. If it is long, it should be divided into several medium-length sessions that suit the writer's staying power.

In Dan's case, at the end of each session he should glance back over his work, note the words he has circled or questioned, and make a few necessary changes (Figure 2-7 is a page from a typical first draft). He must not yet attempt to rewrite paragraphs and sentences for better emphasis. He must leave such major changes until later, when enough time has elapsed for him to read his work objectively. Only then can he review his work as a complete document and see the relationship among its parts. Only then can he be completely critical.

Taking a Break

When Dan has written the final paragraph of his report, he has to resist the temptation to start revising it immediately. He knows some sections are weak, he is not happy about some passages, and the desire to correct them is strong. But it's too soon. Certainly he can pass the draft through a spell-checker, make a safety copy, and print out the pages (we create a double-spaced draft, so we will have room to write in revisions when we are editing). But then he needs to staple them together and set them aside while he tackles a completely unrelated task.

Let time "distance" you from your writing

Reading without a suitable waiting period encourages writers to look at their work through rose-tinted spectacles. Sentences they would normally recognize as weak or too wordy appear to contain words of wisdom. Gross inaccuracies that under other circumstances they would pounce on go unnoticed. Paragraphs that might not be understood by a reader new to the subject, seem abundantly clear. Their familiarity with their work blinds them to its weaknesses. The remedy is to wait.

Tenants' Needs

To find out what the building's tenants most needed in elevator service, we asked each company to fill out a questionaire (sp?). From their answers we were able to identify 5 factors needing consideration:

1. A major problem seems to be the length of time a person must wait for an elevator. Every tenant said we must cut out lengthy waits. A survey was carried out to find out how long people had to wait (during rush hours). This averaged out at 70 sec, more than twice the 32 sec established by Johnson (Ref?), before people get impacient (sp?). From this we calculated we would need 3 or 4 passenger elevators.

2. At first it seemed we would be forced to include a full-size freight elevator in our plan. Two companies (which?) both carry large but light displays up to their floors, but both later agreed they could hinge them, and if they did this they would need only 7 ft 6 in. width (maximum). They also said they did not need a freight elevator all the

Use question marks as a search tool

Figure 2-7 Part of the author's first draft. Note that the author has not stopped to hunt up minor details. Several revisions were made between this first draft and the final product (see pages 173 to 176).

Reading with a Plan

Read all the way through without a pen in your hand

Dan Skinner's first reading should take him straight through the draft without stopping to make corrections, so he can gain an overall impression of his report. Subsequent readings should be slower and more critical, with Dan writing changes in as he goes along. As he reads he should check for clarity, correct tone and style, and technical and grammatical accuracy.

Checking for Clarity

Checking for clarity means searching for passages that are vague or ambiguous. If the following paragraph remained uncorrected, it would confuse and annoy a reader:

Confusing!

Muddled Paragraph

When the owners were contacted on April 15, the assistant manager, Mr. Pierson, informed the engineer that they were thinking of advertising Lot 36 for sale. He has however reiterated his inability to make a definite decision by requesting his company to confirm their intentions with regard to buying the land within two months, when his boss, Mr. Davidson, general manager of the company, will have come back from a business tour in Europe. This will be June 8.

The only facts you can be sure about are that the owners of the land were contacted on April 15 and the general manager will be returning on June 8. The important information about the possible sale of Lot 36 is confusing. The writer was probably trying to say something like this:

Clear!

Revised Paragraph

The engineer spoke to the owners on April 15 to inquire if Lot 36 was for sale. He was informed by Mr. Pierson, the assistant manager, that the company was thinking of selling the lot, but that no decision would be made until after June 8, when the general manager returns from a business tour in Europe. Mr. Pierson suggested that the engineer submit a formal request to purchase the land by that date.

The more complex the topic, the more important it is to write clear paragraphs. Although the paragraph below is quite technical, it would be generally understood even by nontechnical readers:

Technical, but still clear

Clear Paragraph

A sound survey confirmed that the high noise level was caused mainly by the radar equipment blower motors, with a lesser contribution from the air-conditioning equipment. Tests showed that with the radar equipment shut down the ambient noise level at the microphone positions dropped by 10 dB, whereas with the air-conditioning equipment shut down the noise level dropped by 2.5 dB. General clatter and impact noise caused by the movement of furniture and personnel also contributed to the noisy working conditions, but could not be measured other than as sudden sporadic peaks of 2 to 5 dB.

This writer has made sure that

- the topic is clearly stated in the first sentence (the topic sentence),
- the topic is developed adequately by the remaining sentences, and
- no sentence contains information that does not substantiate the topic.

If any paragraph meets these basic requirements, its writer can feel reasonably sure the message has been conveyed clearly.

Writers who know their subject thoroughly may find it difficult to identify paragraphs that contain ambiguities. A passage that is clear to them may be meaningless or offer alternative interpretations to a reader unfamiliar with the subject. For example:

> Our examination indicates that the receiver requires both repair and recalibration, whereas the transmitter needs recalibration only, and the modulator requires the same.

Muddled writing

This sentence plants a question in the reader's mind: Does the modulator require both repair and recalibration, or only recalibration? The technician who wrote it knows, because he has been working on the equipment, but readers will never know unless they write, phone, or email the technician. The technician could have clarified the message by rearranging the information:

> Our examination indicates that the receiver requires both repair and recalibration, whereas the transmitter and modulator need only recalibration.

Clear writing

Sometimes ambiguities are so well buried they are surprisingly difficult to identify, as in this excerpt from a chief draftsperson's report to a department head:

> The drafting section will need three Nabuchi Model 700 CAD computers. The current price is $3175 and the supplier has indicated his quotation is "firm" for three months. We should therefore budget accordingly.

The department head took the message at face value and inserted $3175 for CAD computers into the budget. But two months later the company received an invoice for $9525. Unable by then to return two of the three computers, the department head had to overshoot his budget by $6350. This financial mismanagement was caused by the chief draftsperson, who had omitted to insert the word "each" immediately after "$3175."

Although many ambiguities can be sorted out by simple deduction, a reader should not have to interpret a writer's intentions. It's the writer's job to make reading a document as easy and stress-free as possible, by eliminating confusing statements and alternative meanings.

Checking for Correct Tone and Style

How do you know when your writing has the right tone? One of the most difficult aspects of technical writing is establishing a tone that is correct for the reader, suitable for the subject, and comfortable for you, the writer. If you know your subject well and have thoroughly researched your audience, you will most likely write confidently and will often automatically establish the correct tone. But if you try to set a tone that does not feel natural, or if you are a little uncertain about the subject and the reader, your reader will sense unsureness in your writing. And no matter how skillfully you edit your work, that hesitancy will show up in the final sentences and paragraphs.

Finding the Best Writing Level

Keep coming back to your readers: plant yourself in their shoes

If Dan Skinner is writing on a specific aspect of a very technical topic, and knows that his reader is an engineer with a thorough grounding in the subject, he can use technical terms and abbreviations. Conversely, if he is writing on the same topic for a nontechnical reader who has little or no knowledge of the subject, Dan may have to write a simplified narrative rather than state specific details, explain technical terms, and generally write more informatively.

For example, when engineer Rita Corrigan wrote the following in a modification report, she knew her readers would be electronics technicians at radar-equipped airfields:

> We modified the MTI by installing a K-59 double-decade circuit. This brightened moving targets by 12% and reduced ground clutter by 23%.

But when Rita reported on the same subject to the airport manager, she wrote this:

> We modified the radar set's Moving Target Indicator by installing a special circuit known as the K-59. This increased the brightness of responses from aircraft and decreased returns from fixed objects on the ground.

Adjust the level of writing to suit the reader

For the airport manager Rita included more description and eliminated technical details that might not be meaningful. In their place she made a general statement that aircraft responses were "increased" and ground returns "decreased." She also knew that the airport manager would be familiar with terms such as *Moving Target Indicator*, *responses*, and *returns*.

Now suppose that Rita also had to write to the local Chamber of Commerce to describe improvements to the airport's air traffic control system. This time her readers would be entirely nontechnical, so she would have to avoid using *any* technical terms:

> We have modified the airfield radar system to improve its performance, which has helped us to differentiate more clearly between low-flying aircraft and high objects on the ground.

Keeping to the Subject

Having established that he is writing at the correct level, Dan Skinner must now check that he has kept to the subject. He must take each paragraph and ask: Is this truly relevant? Is it direct? And is it to the point?

If Dan prepared his outline using the method described earlier, and followed it closely as he wrote his report, he can be reasonably sure that most of his writing is relevant. To check that his subject development follows his planned theme, he should identify the topic sentences of key paragraphs and check them against the headings in his outline. If the topic sentences follow the outline, he has kept to the main theme; if they tend to diverge from the outline, or if he has difficulty identifying them, he should read the paragraphs carefully to see whether they need to be rewritten or even eliminated. (For more information about topic sentences, see Chapter 12.)

Technical writing should always be as direct and specific as possible. Technical writers should convey just enough information for their readers to understand the subject thoroughly. Technical writing, unlike literary writing, has no room for details that are not essential to the main theme. This is readily apparent in the following descriptions of the same equipment.

Literary Description The new cabinet has a rough-textured dove gray finish that reflects the sun's rays in varying hues. Contrary to most instruments of this type, its controls are grouped artistically in one corner, where the deep black of the knobs provides an interesting contrast with the soft gray and white background. A cover plate, hardly noticeable to the layperson's inexperienced eye, conceals a cluster of unsightly adjustment screws that would otherwise mar the overall appearance of the cabinet and would nullify the esthetic appeal of its surprisingly effective design.

Technical Description The gray cabinet is functional, with the operator's controls grouped at the top right-hand corner where they can be grasped easily with one hand. Subsidiary controls and adjustment screws used by the maintenance crews are grouped at the bottom left-hand corner, where they are hidden by a hinged cover plate.

Technical writing is *functional* writing

A technical description concentrates on details that are important to the reader (it tells *where* the controls are and why they have been so placed), and so maintains an efficient, businesslike tone.

Using Simple Words

Don't use a 90-cent word when an equally suitable 25-cent word exists

A writer who uses unnecessary superlatives sets an unnaturally pompous tone. The engineer who writes that a design "contains ultrasophisticated circuitry" seems to be justifying the importance and complexity of his or her work instead of just saying that the design has a very complex circuit. The supervisor who recommends that technician Johannes Schmitt be "given an increase in remuneration" may be understood by the company controller but will only be considered pompous by Johannes. If the supervisor had written that Johannes should be "given a raise," both would have understood him. Unnecessary use of big words, when smaller, more generally recognized and equally effective synonyms are available, clouds technical writing and destroys the smooth flow that such writing demands.

Removing "Fat"

During the reading stage Dan should be critical of sentences and paragraphs that seem to contain too many words. He should check that he has not inserted words of low information content; that is, phrases and expressions that add little or no information. Their removal, or replacement by simpler, more descriptive words, can tighten up a sentence and add to its clarity. Low-information-content words and phrases are often hard to identify because the sentences in which they appear seem to be satisfactory. Consider this sentence:

> For your information, we have tested your spectrum analyzer and are of the opinion that it needs calibration.

The expressions "for your information" and "are of the opinion that" are words of low information content. The first can be deleted, and the second replaced by "consider," so that the sentence now reads:

> We have tested your spectrum analyzer and consider it needs calibration.

The same applies to this sentence:

> If you require further information, please feel free to telephone Mr. Thompson at 489-9039.

Weed out unnecessary expressions

The phrase "if you require," although not wrong, could be replaced by the single word "for"; but "please feel free to" is archaic and should be eliminated. The result:

> For further information please telephone Mr. Thompson at 489-9039.

See Tables 12-2 and 12-3 in Chapter 12, which contain lists of low-information-content words and wordy expressions.

Inadvertent repetition of information can also contribute to excessive length. For example, Dan may write:

> We tested the modem to check its compatibility with the server. After completing the modem tests we transmitted messages at low, medium, and high baud rates. The results of the transmission tests showed...

If he deletes the repeated words in sentence 2 ("After completing the modem tests") and sentence 3 ("...of the transmission tests..."), the result is a much tighter paragraph.

> We tested the modem to check its compatibility with the server, and then transmitted messages at low, medium, and high baud rates. The results showed...

Checking for Accuracy

Maintain top-level quality control

Nothing annoys readers more than to discover that they have been given inaccurate information (particularly if they have been using the information before they discover the error). Readers of Dan's report assume that he knows his facts and has checked that they have been correctly transcribed into the report. Discovering even a single technical error in his report can undermine Dan's credibility in their minds.

There is no way to prevent some errors from occurring when copying quantities and details from one document to another. Therefore, Dan must carefully check that he copies all facts, figures, equations, quantities, and extracts from other documents correctly.

Checking for accuracy also means ensuring that grammar, punctuation, and spelling have not been overlooked. Dan must check spelling with care, because his familiarity with the subject may blind him to obvious errors. (How many of us have inadvertently written "their" when we intended to write "there"? And "too" when we meant "two"?)

Be wary when using spell-check programs

Dan has to recognize that spell-check programs are not 100% reliable. He may use a word—particularly a technical word—that is not in the spell-checker's memory, or he may type in a word inaccurately and inadvertently form another word that the spell-checker recognizes. For example, if he typed in "departure" when he meant to type in "department," the spell-checker would not recognize it as an error. (Neither would it flag "their" and "too" as errors.) A spell-check program can not comprehend the context of the words; it simply examines each word and compares it to its master list. If it finds a "match," it takes no action; if it does not find a match, it highlights the word and sometimes also emits an audible warning.

Revising Your Own Words

Proofread on hard copy

We recommend that Dan print out his report and read and revise it on paper rather than on his computer screen. Our experience, and also that of many report writers, is that you catch many more typographical errors that way. We suggest that Dan mark up a hard copy of his report and then, as a separate step, transfer the changes to the online document.

As he reads, Dan should continually ask himself five questions:

1. **Can my readers understand me?**
   Will the person I am writing for be able to read my report all the way through without getting lost?
   What about other readers who might also see my report? Will they understand it?
2. **Is the focus right?**
   Is my report reader-oriented?
   Are the important points clearly visible?

Have I summarized the main points in an opening statement that the reader will see right away?

3. **Is my information correct?**
   Is it accurate?
   Is it complete?
   Is all of it necessary?

Make yourself a checklist, then *use* it!

4. **Is my language good?**
   Is it clear, definite, and unambiguous?
   Are there any grammatical, punctuation, or spelling errors?
   Does every paragraph have a topic sentence (preferably at the start of the paragraph)?
   Have I used any big, "overblown" words where simpler words would do a better job?
   Are there any low-information-content words and phrases?

5. **Have I kept my report as short as possible while still meeting my readers' needs and covering the topic adequately?**

By now Dan's draft should be in good shape and any further reading and revising will be final polishing. The amount will depend on the importance of the report. If his report is for limited or in-company distribution, a standard-quality job will normally suffice. But if the report is to be distributed outside the company, or submitted to an important client, Dan will spend as much time as necessary to ensure that it conveys a good image of both him and his employer.

Dan Skinner will now be able to issue his report with confidence, knowing that he has fashioned a good product. The approach described here will not have made report writing a simple task for him, but it will have helped him through the difficult conceptual stages, and helped him to read and revise more efficiently. When he writes his next report, he will be less likely to put it off until it is so late that he has to do a rush job.

ASSIGNMENTS

Exercise 2.1
Describe why the "pyramid" method of writing will help you become a better presenter of information.

Exercise 2.2
Which do you feel is the better way for you to develop an outline for a report: the organized method or the "random" method? Explain why.

Be comfortable with your writing method

Exercise 2.3
(a) What are the seven stages advocated for planning a report?
(b) Which is the most important stage? Explain why.
(c) Must the stages be followed exactly in the sequence listed?

Exercise 2.4
If several people are likely to read a report, how would you identify which one is your primary reader?

Exercise 2.5
Is it better to write a report without stopping to "clean up" the construction along the way, or to write a page at a time and edit that page before going on to the next? Explain why.

Exercise 2.6
What two factors will help you write more confidently, and probably help you set the right tone?

Keep referring back to the reader

Exercise 2.7
From the list of five main questions that you, as a writer, should ask yourself during the revision stage (see the boldface questions on pages 21–22), which do you think is the most important? Explain why.

Chapter 3
Letters, Memos, and Emails

How to Write Business Letters That Get Results www.bly.com/Pages/documents/File136.doc Well-known copywriter Robert W. Bly provides valuable advice about writing correspondence. "Failure to get to the point, technical jargon, pompous language, misreading the reader—these are the poor stylistic habits that cause others to ignore the letters we send."

Readers want to know *right away* what you most need to tell them

When you write a personal letter to a friend or relative, you probably don't worry whether your letter is too long or contains too much information. You assume your reader will be pleased to hear from you, so you launch into a general discourse, inserting comments and items of general news without concerning yourself very much about organization.

But when you write a business letter you have to be disciplined. Your readers are busy people who want only the details that concern them. Information they do not need irks them. For these people your letters must be focused, well planned, brief, and clear.

Using the Pyramid

Anna King, technical editor at H. L. Winman and Associates, teaches the technical staff at the firm what she refers to as the Pyramid Method of Writing. She finds this technique valuable because it helps the staff visualize their documents. Figure 3-1 shows what the basic pyramid looks like. You can see that only so much information can fit into the top part of the pyramid and is followed by the supporting details and facts. The pyramid helps you focus your information so your readers will know right away why you are writing to them.

Identifying the Main Message

If you write your letters pyramid-style, you will automatically focus the reader's attention on your main message. Before you place your fingers on the keyboard, fix clearly in your mind *why you are writing* and *what you most want your reader to know*. Then focus on this information by placing it right up front, where it will be seen immediately.

If you begin a letter with background information rather than the main point, your reader will wonder why you are writing until he or she has read well into the letter. Don McKelvey's letter to Jim Connaught is a typical example of an unfocused letter.

Dear Mr. Connaught:

I refer to our purchase order No. 21438 dated April 26, 2004, for a Vancourt microcopier model 3000, which was installed on May 14. During tests following its installation your technician discovered that some components had been damaged in transit. He ordered replacements and in a letter dated May 20 informed me that they would be shipped to us on May 27 and that he would return here to install them shortly thereafter.

It is now June 10, and I have neither received the parts nor heard from your technician. I would like to know when the replacement parts will be installed and when we can expect to use the microcopier.

Sincerely,

Don McKelvey

Jim had to read more than 70 words before he discovered what Don wanted him to do. If Don had written pyramid-style, starting with a main message, Jim would have known immediately why he was reading the letter:

Readers don't want to plough through paragraphs of background information before they encounter your main message

Dear Mr. Connaught:

We are still unable to use the Vancourt 3000 microcopier we purchased from you on April 26, 2004. Please inform me when I can expect it to be in service.

And placing the main message up front would have helped Don write a shorter explanation that would have been simpler to follow:

The microcopier was ordered on P.O. 21438 and installed on May 14. During tests, your technician discovered that some components had been damaged in transit. He ordered replacements, then in a letter dated May 20 informed me that they would be shipped to us on May 27, and that he would return here to install them. To date, I have neither received the parts nor heard from your technician.

Sincerely,

Don McKelvey

Unfortunately, knowing you should open every letter with a main message is not enough. You also need to know how to find exactly the right words to put at the top of the pyramid. And that is where many technical people have trouble.

Getting Started

To overcome this block, try using another technique recommended by Anna King. She suggests that when you start a letter, first write these six words:

This proven technique will never fail you!

I want to tell you that...

And then finish the sentence with what you *most* want to tell your reader. For example:

Dear Ms. Reynaud:

I want to tell you that...the environmental data you submitted to us on October 8 will have to be substantiated if it is to be included with the Labrador study.

Then, when your sentence is complete, *delete* the first six words (the *I want to tell you that...* expression). What you have left will be a focused opening statement:

Dear Ms. Reynaud:

The environmental data you submitted to us on October 8 will have to be substantiated if it is to be included with the Labrador study.

Often you can use an opening statement formed in this way just as it stands when you remove the six "hidden" words. At other times, however, you may feel the opening statement seems a bit abrupt. If so, you can soften it by inserting a few additional words. For example, in the letter to Ms. Reynaud, you might want to add the expression "I regret that...":

Dear Ms. Reynaud:

I regret that the environmental data you submitted to us on October 8 will have to be substantiated if it is to be included with the Labrador study.

Figure 3-1 depicts this convenient way of starting a letter and concurrently creating a main message. It also shows that in business letters the main message is more often referred to as the **Summary Statement**.

The writer's pyramid helps draw attention to the most important information

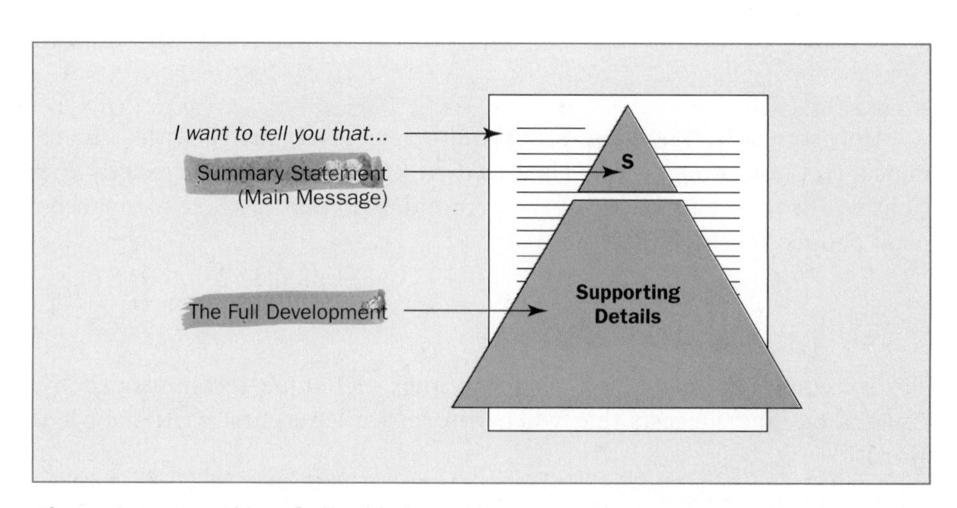

Figure 3-1 Creating a letter's summary statement.

Here are three more examples of properly formed Summary Statements:

Dear Colonel Watson:

We will complete the XRS modification on June 14, eight days earlier than scheduled.

This technique is similar to the newspaper-style of writing

Dear Ms. Mohammed:

Your excellent paper "Export Engineering" arrived just in time to be included in the program for the Pacific Rim Conference.

Dear Mr. Voorman:

Seven defective castings were found in shipment No. 308.

(You can check that *I want to tell you that…* was used to form these three opening sentences by mentally inserting the six hidden words at the start of each sentence.)

Avoiding False Starts

If you do not use the six hidden words to start a letter, you may inadvertently open with an awkwardly constructed sentence that seems to be going nowhere. For example:

Dear Mr. Corvenne:

In answer to your enquiry of December 7 concerning erroneous read-outs you are experiencing with your Mark 17 Analyzer, and our subsequent telephone conversation of December 18, during which we tried to pinpoint the fault, we have conducted an examination into your problem.

A "dragged out" start

Anna King refers to a long, rambling opening like this as "spinning your wheels," because such a sentence does not come to grips with the topic early enough. She has prepared a list of expressions (see Figure 3-2) that can easily cause you to write complicated, unfocused openings. In their place she recommends starting with the *I want to tell you that…* expression, which will help you focus your reader's attention on the main message. If the letter referring to the Mark 17 Analyzer had started this way, it would have been much more direct:

Dear Mr. Corvenne:

(I want to tell you that…) The problem with your Mark 17 Analyzer seems to be in the extrapolator circuit. Following your enquiry of December 7 and your subsequent description of erroneous read-outs, we examined… (etc.).

A direct start

Planning the Letter

Once you have identified and written the main message, your next step is to select, sort, and arrange the remaining information you want to convey

Try inserting *I want to tell you that...* in front of these openings: it doesn't work!

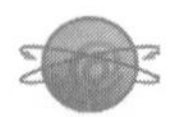

Strategies for Writing Persuasive Letters
www.washburn.edu/services/zzcwwctr/persuasive_menu.html
This step-by-step guide covers the purpose of the persuasive letter, prewriting questions for the writer, writing strategies, and revision tips.

When You Write A Letter...

Never start with a word that ends in "ing":

Referring...
Replying...

Never start with a phrase that ends with the preposition "to":

With reference to...
In answer to...
Pursuant to...
Due to...

Never start with a redundant expression:

I am writing...
For your information...
This is to inform you...
The purpose of this letter is...
We have received your letter...
Enclosed please find...
Attached herewith...

IN OTHER WORDS...

Don't Spin Your Wheels!

Figure 3-2 Anna King's suggestion to H. L. Winman engineers.

to your reader. This information should amplify the message you have already presented in the Summary Statement and provide evidence of its validity. For example, when Paul Shumeier wrote the following Summary Statement, he realized he would be presenting his reader with costly news:

Dear Mr. Larsen:

Tests of the environmental monitoring station at Wickens Peak show that 60% of the instruments need to be repaired and recalibrated at a cost of $7265.

He also realized that Mr. Larsen would expect the remainder of the letter to tell him why the repairs were necessary, exactly what needed to be done, and how Paul had derived the total cost. To provide this information, Paul first had to identify which questions would be foremost in Mr. Larsen's mind after he had read the Summary Statement. This meant asking himself six questions, all based on *Who?*, *Where?*, *When?*, *Why?*, *What?* and *How?*:

> *Who* (was involved)?
> *Where* (did this happen)?
> *When* (did this happen)?
> *Why* (are the repairs necessary)?
> *What* (repairs are needed)?
> *How* (were the costs calculated)?

Six questions: six answers. The answers provide the facts, form the body of a letter

To insert the answers to these questions into his letter, Paul now had to open up the lower part of the writer's pyramid. This becomes the **Full Development** (or *supporting details*) shown in Figure 3-1.

Opening Up the Pyramid

To help Paul—and you—organize a letter's Full Development, the lower part of the pyramid is divided into three compartments known as the **Background**, **Facts**, and **Outcome** (see Figure 3-3).

The **Background** covers *what* has happened previously, *who* was involved, *where* and *when* the event occurred or the facts were gathered and, sometimes, for whom the work was done. Paul wrote:

A well-developed *Background* section leads into direct, uncomplicated details

> Our electronics technicians examined the Wickens Peak Monitoring station on May 16 and 17, in response to your May 10 request to Patrick Friesen.

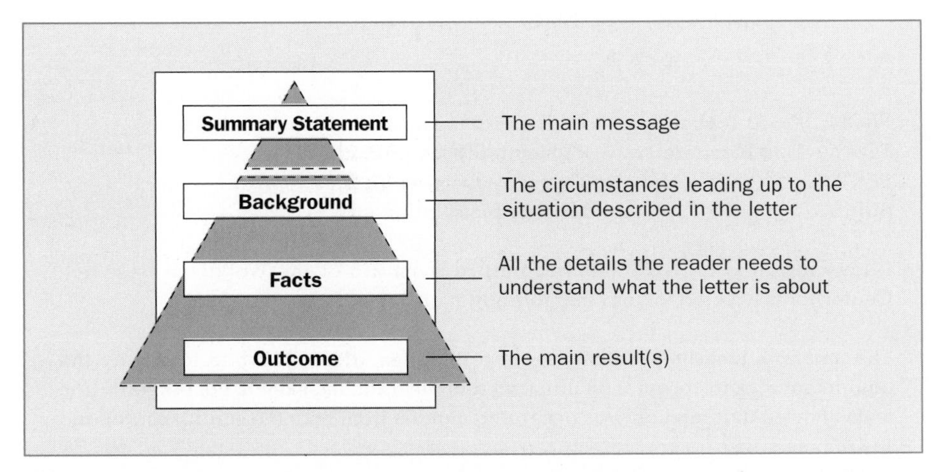

Figure 3-3 Basic writing plan for an informative business letter, interoffice memo, or email.

The **Facts** amplify the main message. They provide specific details the reader needs to fully understand the situation or to be convinced of the need to take further action. Here, Paul wrote:

Use attachments to simplify a letter

> Most of the damage was caused by a tree northwest of the site that fell onto the station during a storm on April 23 and damaged parts of the roof and north and west walls. Instruments along these walls were impact-damaged and then soaked by rain. Other instruments in the station were affected by moisture.
>
> Major repairs and recalibration are required for the 16 instruments listed in attachment 1, which describes the damage and estimated repair cost for each instrument. This work will be done at our Shepperton repair depot for a total cost of $4485. Minor repairs, which can be performed on site, are necessary for the 27 instruments listed in attachment 2. These on-site repairs will cost $2780.

These two paragraphs clearly answer the *Why?, What?,* and *How?* questions. Note that, rather than clutter the middle of his letter with a long list, Paul placed the details in two attachments and summarized only the main points in the body of his letter. (The attachments are not shown here.)

The **Outcome** describes the result or any effect the facts have had or will have. If the letter is purely informative and the reader is not expected to take any action, the Outcome simply sums up the main result. Paul would have written:

I have obtained Ms. Korton's approval to perform the repairs and a crew was sent in on May 23. They should complete their work by May 31.

Sincerely,

Paul Shumeier

Persuasive Communications: Using You-Attitude and Reader Benefit
www.washburn.edu/services/zzcwwctr/you-attitude.txt
Receivers of communications are usually more concerned about themselves than about the writer or the company that person represents. This article describes how to use the "you-attitude" and show reader benefit in your persuasive communication.

MACRO
ENGINEERING INC.
600 Deepdale Drive, Phoenix, AZ 85007

FROM: Kevin Toshak <K.Toshak@macro.com>
TO: Tina Mactiere <T.Mactiere@macro.com>
SENT: Thursday, October 22, 2003
SUBJECT: Monitor Installation at WRC

I have installed a TL-680 monitor unit in room 215 at the Wollaston Research Center, as instructed in your memorandum of October 15.

The unit was installed without major difficulties, although I had to modify the equipment rack to accept it as illustrated in the attached sketch. Post-installation tests showed that the unit was accepting signals from both the control center and the remote site.

Figure 3-4 An informative email.

But if the reader is expected to take some action, or approve somebody else taking action (often, the letter writer), then the Outcome becomes a *request for action*. Because Paul wanted an answer, he wrote:

If these repair costs are acceptable, please telephone, fax, or email your approval so I can send in our repair crew.

Sincerely,

Paul Shumeier

Write an *Action Statement* if you want your reader to act or react

These three parts can help you arrange the Full Development of any letter, memo, or email into a logical, coherent structure. Before starting, however, you have to decide whether you are writing to inform or persuade.

Writing to Inform

Normally, letters and memos that purely inform, with no response or action required from the reader, can be organized around the basic Summary Statement-Background-Facts-Outcome writing plan shown in Figure 3-3. Kevin Toshak's email to Tina Mactiere, in Figure 3-4, falls into this category.

Another example is a confirmation letter, in which the writer confirms previously made arrangements. In the following Macro Engineering Inc. memorandum, general manager Wayne Robertson ensures that he and chief buyer Christine Lamont both understand the arrangements that will evolve from a decision made at a company meeting:

Christine:

Statement Summary I am confirming that you will represent both Macro Engineering Inc. and H. L. Winman and Associates at the Materials Handling conference in Houston on May 15 and 16, 2004, as agreed at

Background the Planning Meeting on March 23. At the conference you will

Facts

- take part in a panel discussion on packaging electronic equipment from 10:00 to 11:15 a.m. on May 15, and
- host a wine-and-cheese reception for delegates from 5:00 to 7:00 p.m. on May 16.

Janet Kominsky is making your travel and hotel reservations, and the catering arrangements for the reception. Anna King will provide brochures from Cleveland, and my secretary will make up packages for you to distribute.

Outcome I'll brief you on other details before you leave.

Wayne

An informative letter *tells* the reader what has been done or what has to be done...

...it doesn't expect the reader to respond

Although the basic writing plan for letters has four compartments (see Figure 3-3), you do not have to write exactly four paragraphs. As both Wayne's and Kevin's memos show, you may combine two compartments into a single paragraph, or let one compartment be represented by several paragraphs. When you first use the Pyramid you'll find yourself writing separate paragraphs for each compartment, but as you become more comfortable with this method you'll understand that keeping the compartments in the correct order is the most important concept.

Writing to Persuade

A persuasive letter *sells* the reader to take some form of action

In a persuasive letter you expect your reader either to respond to your letter or to take some form of action. Consequently, the writing plan's Outcome compartment is renamed **Action**, as shown in Figure 3-5, to remind you to end a persuasive letter with an action statement. A request and a complaint are typical examples of persuasive letters, and so is the informal proposal described in Chapter 7.

Making a Request

Many writers hesitate to open with a request

Many technical people claim that placing the message at the start of a letter is not a problem until they either have to ask for something or to give the reader bad news. They then tend to lead gently up to the request or unhappy information.

Bill Kostash is no exception. He is service manager for Mechanical Maintenance Systems Inc., and he has to write to customers to ask if they will accept a change in the preventive maintenance contracts his company has with them. He starts by writing to Ms. Bea Nguyen, the contracts administrator for Multiple Industries in St. Cloud, Minnesota:

Except for the *Outcome/Action* compartment, tell and sell writing plans are similar

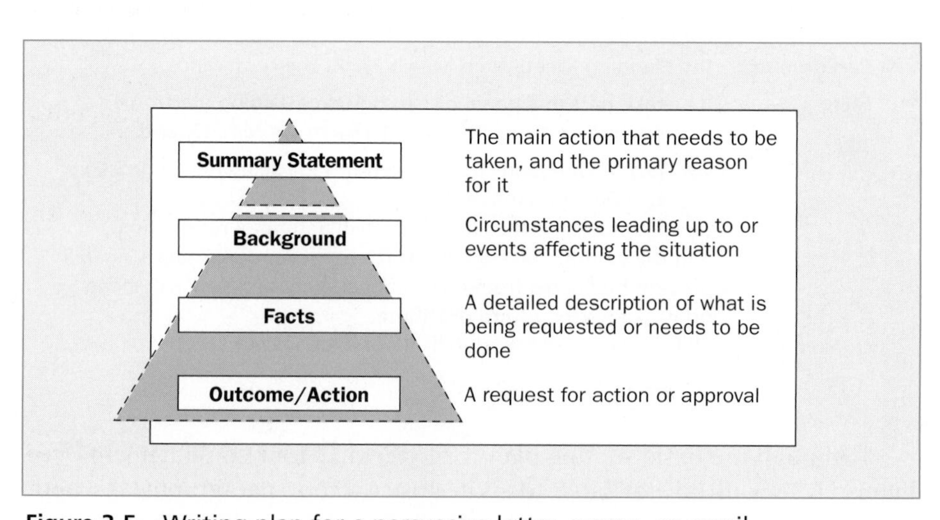

Figure 3-5 Writing plan for a persuasive letter, memo, or email.

June 18, 2003

Dear Ms. Nguyen:

I am writing with reference to our contract with you for the preventive maintenance services we provide on your RotoMat extruders and shapers. Under the terms of the current contract (No. RE208) dated January 2, 2003, we are required to perform monthly inspection and maintenance "...on the 15th day of each month or, if the 15th falls on a weekend or holiday, on the first working day thereafter."

(Bill is off to a bad start. Instead of opening with a Summary Statement he has inserted all the background details first, so Bea Nguyen does not yet know why he has written to her. He has also opened with one of the expressions Anna King lists as an awkward start in Figure 3-2. Let's see how he continues.)

An unfocused, meandering request letter

Our problem is that almost all of our clients ask that we perform their maintenance service between the 5th and 25th of each month, to avoid their end-of-month peak accounting periods. This in turn created difficulties for us, in that our service technicians experience a peak workload for 20 days and then have virtually no work for 10 days.

(Bea Nguyen still does not know why he is writing.)

Consequently, to even out our workload, I am requesting your approval to shift our inspection date from the 15th to the 29th of each month. If you agree to my request, I will send our technician in to service your machine on June 29—a second time this month—rather than create a six-week period between the June and July inspections. Could you let me know by June 25 if this change of date is acceptable?

Sincerely,

William J. Kostash

(*Now* Bea knows why Bill has written to her—but she had to read a long way to find out. And she probably had to reread his letter to fully understand the details.)

If Bill had used the writing plan in Figure 3-5 to shape his letter, his request would have been much more effective. The revised letter is shown in Figure 3-6, which

1. contains his **Summary Statement** (he states his request and what the effect will be),
2. contains the **Background** (the contract details),
3. contains the **Facts** (it describes the problem), and
4. contains the **Action** statement, in which he mentions *two* actions: what he wants Bea to do (to call him) and what he will do (schedule a second visit).

Writing with a plan creates a coherent request

Mechanical Maintenance Systems Inc.
2120 Cordoba Avenue
St. Paul, Minnesota 55307

June 18, 2004

Ms. Bea Nguyen
Contracts Administrator
Multiple Industries Inc.—Manufacturing Division
18 Commodore Bay
St. Cloud, MN 54018

Dear Ms. Nguyen:

A focused, definite, direct request

1 I am requesting your approval to change the date of our monthly preventive maintenance visits to service your RotoMat extruders and shapers to the 29th of each month. This will help spread my technicians' workload more evenly and so provide you with better service.

2 Our contract with you is No. RE208 dated January 2, 2004, and it requires that we perform a monthly inspection and maintenance on the 15th day of each month. 3 Unfortunately, almost all of our clients ask that we perform their maintenance service between the 5th and the 25th. This creates a problem for us in that our service technicians experience a peak workload for 20 days and then have very little work for 10 days.

4 Could you let me know by June 25 if you can accept the change? Then I will send a technician to your plant on June 29 for a second visit this month, rather than create a six-week space between the June and July inspections.

Sincerely,

William J. Kostash

William J. Kostash
Service Manager

Figure 3-6 A request letter written pyramid-style.

Registering a Complaint

The approach is the same if you have to write a letter of complaint or ask for an adjustment. The third compartment is relabeled as shown in Figure 3-7.

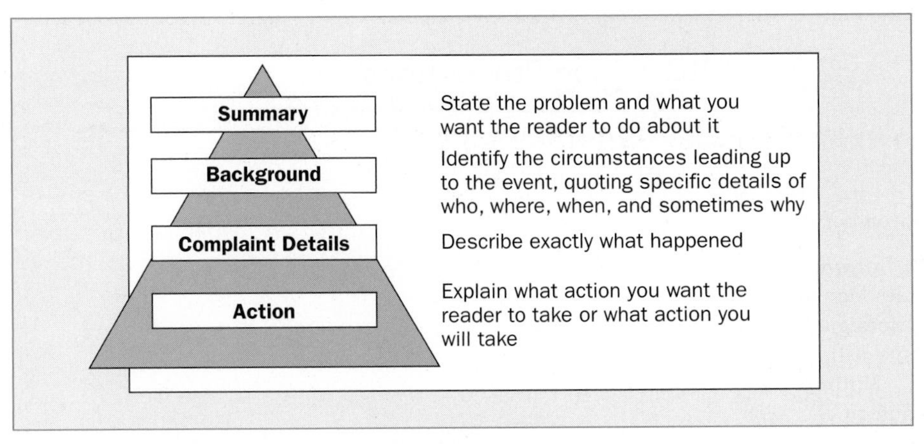

Figure 3-7 Writing plan for a complaint letter.

The parts of the complaint letter shown in Figure 3-7 are identified in Suzanne's letter in Figure 3-8 on page 36, with circled numbers. The corresponding comments are listed here:

1. In the **Summary**, it is often better to generalize what action is needed and then later, in the Action compartment, state exactly what has to be done.

2. If there are only a few **Background** facts, you may combine them with either the Summary Statement or the Complaint Details rather than place them in a very short paragraph by themselves.

3. In the **Complaint Details**, describe in chronological order what happened so the reader will understand the reason for your complaint or request for adjustment.

4. The **Action Statement** must be strong and confident and specifically identify what action you want the reader to take, or in some cases, what action you will take.

Responding to a Complaint

You may need to answer a complaint someone has written. It is easier when you agree with the complaint and can perform the requested action. This will be a much shorter response than if you don't agree with the complaint, because you don't have to go into as much detail about why you are agreeing. However, if you disagree with the complaint your response is more difficult to write and you have to be sure to provide a detailed description of why you cannot act as requested. Figure 3-9 shows the pyramid for responding both positively and negatively to complaints.

RGI Video Productions
316 St. Mary's Road
Brighton, NY 14639

November 12, 2003

Mr. Bruyere
Sales Manager
Professional Image Business Equipment
Suite 100
1675 Mattingly Drive
Brighton, NY 14639

Dear Mr. Bruyere:

❶ The Nabuchi 700 portable computer you recently sold me had a defective lithium-ion battery that had to be replaced while I was in Europe. Consequently I am requesting reimbursement of the expenses I incurred to replace the battery.

Set the scene

❷ I bought the computer and a Nabuchi 701PC international power converter from your Willows Mall store on September 4, 2003. (See attached sales invoice No. 14206A.)

The computer worked satisfactorily for the first six weeks, but during that time I had no occasion to use it solely on battery power.

...offer the details,...

❸ On October 25 I left for Europe, first giving the batteries an 18-hour charge as recommended in the operating instructions. While using the computer in flight, after only 35 minutes the low-battery lamp lit up and the screen warned of imminent failure. I recharged the batteries the following day, in Rheims, France, but achieved less than 25 minutes of operating time before the batteries again became fully discharged.

As the Nabuchi line is neither sold nor serviced in France, I had to buy and install a replacement lithium-ion battery (a Mercurio Z7S), which has since worked fine. I have enclosed the defective battery, plus a copy of the sales receipt for the replacement battery I purchased from Lestrange Limitée, Rheims.

...and end with a firm Action Statement

❹ Please send me a check for $244.30, which at the current rate of exchange is the US equivalent of the 1190 francs shown on the sales receipt.

Sincerely,

Suzanne Dumont, P.E.
enc 3

Figure 3-8 A complaint letter written pyramid-style.

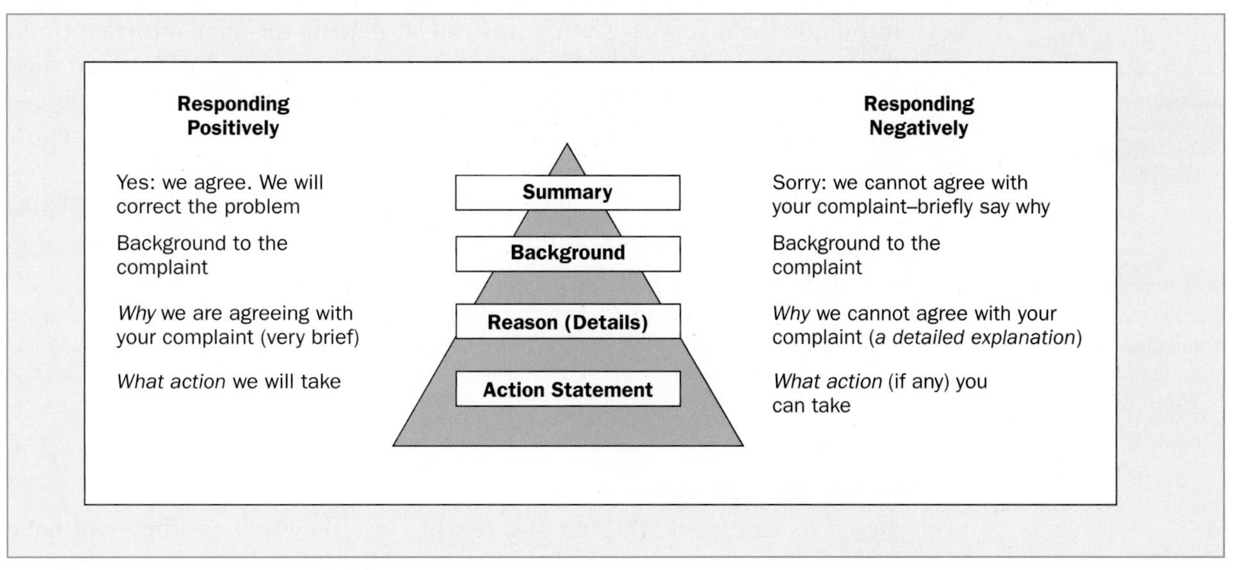

Figure 3-9 Writing plan for responding to a complaint.

Creating a Confident Image

Readers react positively to letters, memos, and emails in which the writer conveys an image of a confident person who knows the subject well and has a firm idea of what he or she plans to do, or expects the reader to do. Such an image is conveyed by both the quality of the writing and the physical appearance of the piece of correspondence.

Be Brief

For technical business correspondence, brevity means writing short letters, short paragraphs, short sentences, and short words.

Short Letters

A business reader will tend to react readily to a short letter, viewing its writer as an efficient provider of information. In contrast, the same reader may view a long letter as "heavy going" (even before reading it) and tend to put it aside to deal with it later. A short letter introduces its topic quickly, discusses it in sufficient depth, and then closes with a concluding statement, its length dictated solely by the amount of information that needs to be conveyed.

The key word here is "short"

We know of a company in which the managing director has ruled that no letter or memo may exceed one page. This is an effective way to encourage staff to be brief, and it works well for many people. But for letter writers who have more to say than they can squeeze onto a single page, that limitation can prove inhibiting. For them, we suggest borrowing a technique from report writing. Instead of placing all their information in the letter, they should change the letter into a semiformal report and then summarize the highlights—particularly the purpose and the outcome—into a one-page letter placed at the front of the report (so that the report becomes *an attachment* to the letter, as depicted in Figure 3-10).

If you use this device, refer to the attachment in the body of the letter *and* insert a main conclusion drawn from it, as has been done here:

A short cover letter is like an executive summary (see page 147)

> During the second week we measured sound levels at various locations in the production area of the plant, at night, during the day, and on weekends. These readings (see attachment) show that a maximum of 55 dB was recorded on weekdays, and 49 dB on weekends. In both cases these peaks were recorded between 5 and 6 p.m.

Short Paragraphs

Novelists can write long paragraphs because they assume they will have their readers' attention, and their readers have the time and patience to make their way through leisurely descriptions. But in business and industry, readers are working against the clock and so want bite-size paragraphs of easy-to-digest information.

Think of a paragraph as a miniature pyramid

Let the first sentence of each paragraph introduce just one idea, then make sure that subsequent sentences in that paragraph develop the idea adequately and do not introduce any other ideas. In technical business writing the first sentence of each paragraph should be a "topic sentence," so that the reader immediately knows your main idea. Consequently, a busy reader who skims your document by reading just the first sentence of

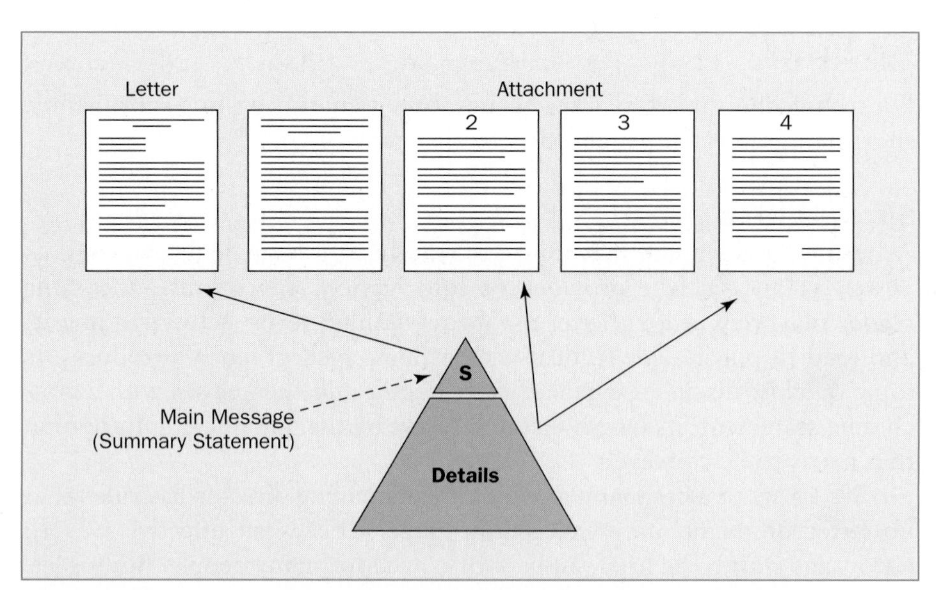

Figure 3-10 A short letter with attachments is an adaptation of the pyramid method of writing. An example can be seen in Figure 5-4 (page 108).

every paragraph will still gain a good understanding of the main points. Let the first sentence of each paragraph summarize the paragraph's contents, and the remaining sentences support the first sentence by providing additional information:

> *We have tested your 15 Vancourt 801 CD-ROM drives and find that 11 require repair and recalibration.* Only minor repairs will be necessary for 6 of these drives, which will be returned to you next week. Of the 5 remaining drives, 3 require major repairs which will take approximately 20 days, and 2 are so badly damaged that repairs will cost $180 each.

If an idea you are developing results in an overly long paragraph, try dividing the information into a short introductory paragraph and a series of subparagraphs, as has been done here:

> My inspection of the monitoring station at Freedom Lake Narrows revealed three areas requiring attention, two immediately and one within three months:
>
> 1. The water stage manometer is recording erratic readouts of water levels. A replacement monitor needs to be flown in immediately so that the existing unit can be returned to a repair depot for service.
> 2. The tubing to the bubble orifice is worn in several places and must be replaced (90 feet of ½ inch tubing will be required). This work should be done concurrently with the monitor replacement.
> 3. The shack's asphalt roof is wearing and will need resurfacing before freeze-up.

Paragraphs that are longer than eight or nine printed lines are too long

We are not suggesting that your letters should contain a series of small, evenly sized paragraphs. These would appear dull and stereotyped. Instead, paragraphs should vary from quite short to medium-long to give the reader variety. How you can adjust paragraph and sentence length to suit both reader and topic, and also to place emphasis correctly, is covered in Chapter 12.

Short Sentences

If you write short, uncomplicated sentences, you help your readers quickly grasp and understand each thought. Sometimes expert literary writers can successfully build sentences that develop more than one thought, but such sentences are confusing and out of place in the business world. Compare these two examples of the same information:

Convoluted sentences create the impression that their writer is confused

Complicated	There has been intermittent trouble with the vacuum pumps, although the flow valves and meters seem to be recording normal output, and the 5 inch pipe to the storage tanks has twice become clogged, causing backup in the system.
Clear	There has been intermittent trouble with the vacuum pumps, and twice the 5 inch pipe to the storage tank has become clogged and caused backup in the system. The flow valves and meters, however, seem to be recording normal output.

The first example is confusing because it jumps back and forth between what the trouble is and what is working normally. The second example is clear because it uses two sentences to express the two different thoughts.

Short Words

Short words are especially important for readers whose first language is not English

Some engineers and engineering technicians feel that the technical environment in which they work and the complex topics they have to write about, demand that they use long, complex words in their correspondence. Similarly, some people feel that long words build credibility and respect for their position; the opposite is true. They write "an error of considerable magnitude was perpetrated," rather than simply "we made a big mistake." In so doing, they make a reader's job unnecessarily difficult.

Because the engineering and scientific worlds encompass many long and complex terms that have to be used in their original form, make your correspondence more readable by surrounding such technical terms with simple words. Be aware, too, that in today's global society, many of your readers may read and write English as a second language. Long words that are not in the average person's vocabulary may cause confusion and misunderstanding. Chapter 12 has more information about writing for an international audience.

Be Clear

A clear letter conveys information simply and effectively, so that the reader readily understands its message. Writing clearly demands ingenuity and attention to detail. As a writer you must consider not only how you write your letters, but also how you present them.

Create a Good Visual Impression

Clarity depends on appearance as well as clarity of expression

The appearance of a letter tells much about the writer and the company he or she represents. If a letter is sloppily arranged or contains strikeovers, visible erasures, or spelling errors, then its readers imagine a careless individual working in a disorganized office. But if readers are presented with a neat letter, placed in the middle of the page and printed by a quality printer, then they imagine a well-organized individual working for a forward-thinking company noted for the quality of its service. Most people prefer to deal with the latter company and will read its correspondence first.

Develop the Subject Carefully

The key to effective subject development is to present the material logically, progressing gradually from a clear, understood point to one that is more complex. This means developing and consolidating each idea so the reader can fully understand it before presenting the next idea. The sections on paragraph unity and coherence in Chapter 12 (see pages 332 and 333) provide examples of coherent paragraphs. Using the pyramid technique will also help you structure your information in a logical sequence.

Be Definite

Know clearly what you want to say *before* you start writing

People who think better with their fingers on a keyboard or a pen in their hand sometimes make decisions as they write, producing indecisive letters that are irritating to read. These writers seem to examine and discard points without really grappling with the problem. By the time they have finished a letter, they have decided what they want to say, but it has been at the reader's expense. We call this a "brain dump."

Decision making does not come easily to many people. Those of us who hesitate before making a decision, who evaluate its implications from all possible angles and weigh its pros and cons, may allow our indecisiveness to creep into our writing. We hedge a little, explain too much, or try to say how or why we reached a decision before we tell our reader what the decision is. This is particularly true when we have to tell readers something unfavorable or contrary to their expectations.

As before, the key is to use the pyramid:

1. Decide exactly what you want to say (i.e. develop your main message), and then
2. Place the main message right up front (use *I want to tell you that…* to get started).

If you also write primarily in the active voice, you will sound even more decisive. Active verbs are strong, passive verbs are weak. For example:

Write directly from person to person, and name the "doer"

These passive expressions	***Should be replaced with***
it was our considered opinion	we considered
it is recommended that	I recommend
an investigation was made	we investigated
the outage was caused by a defective transmitter	a defective transmitter caused the outage

For hints on how to use the active voice, see "Emphasis" in Chapter 12 (pages 339 to 341).

Close on a Strong Note

You may feel you should always end a letter with a polite closing remark, such as: *I look forward to hearing from you at your earliest convenience*, or *Thanking you in advance for your kind cooperation*. In contemporary business correspondence—and particularly in technical correspondence—such closing statements are not only outdated but also weaken your impact on the reader. Today, you should close with a strong, definite statement.

The Outcome part of the letter provides a natural, positive close, as illustrated by the final sentences in the letter to Ms. Nguyen (Figure 3-6). You should resist the temptation to add a polite but uninformative and ineffective closing remark. Simply sign off with "Regards" or "Sincerely."

Adopting a Pleasant Tone

There is no quick and easy method to make your letters sound sincere, nor is there a checklist to tell you when you have imparted the right tone. Both qualities are extensions of your own personality that cannot be taught. They can only be shaped and sharpened through self-knowledge and which of your attributes you most need to develop.

To achieve the right tone, your correspondence should be simple and dignified, but still friendly. Approach your readers on a person-to-person basis, following the five suggestions below.

Know Your Reader

Reminder: Know who you are writing to!

If you have not identified your reader properly, you may have difficulty setting the correct tone. You need to know your reader's level of technical knowledge and whether he or she is familiar with the topic you are describing. Without this focus you may seem condescending to a knowledgeable reader if you explain too much and use overly simple words when the reader clearly expects to read technical terms. Conversely, you may overwhelm a reader who has only limited technical knowledge, if you confront him or her with heavy technical details.

Ideally, you should select just the right terminology to hold the reader's interest and perhaps offer a mild challenge. By letting readers feel they are grasping some of the complexities of a subject (often by using analogies within their range of knowledge), you can present technical information to nontechnical readers without confusing or upsetting them.

Sometimes you will know the person you are writing to, and then you will probably find it much easier to adopt a pleasant tone. Be careful, though, not to make your correspondence *too* chatty or informal. In business and technical writing you should consistently sound professional. You can never tell when your letter may be passed on to someone else!

Be Sincere

At one time it was considered good manners not to permit one's personality to creep into business correspondence. Today, business letters are much less formal and, as a result, much more effective.

Sincerity is the gift of making your readers feel that you are personally interested in them and their problems. You convey this by the words you use and the way you use them. A reader would be unlikely to believe you if you came straight out and said, "I am genuinely interested in your project." The secret is to be so involved and interested in the subject that you automatically convey the ring of enthusiasm that would appear in your voice if you were talking about it.

***Care* about both your topic and your reader**

Be Human

Too many letters lack humanity. They are written from one company to another, without any indication that there is a human being at the sending end and another at the receiving end. The letters might just as well be sent from computer to computer.

Do not be afraid to use the personal pronouns, "I," "you," "he," "she," "we," and "they." Let your reader believe you are personally involved by using "I" or "we," and that you know he or she is there by using "you." Contrary to what many of us were told in school, letters may be started in the first person. If you know the reader personally, or you have corresponded with each other before, or if your topic is informal, let a personal flavor appear in your letters by using "I" and the reader's first name:

Let your presence be apparent

Dear Ben:

I read your report with interest and agree with all but one of your conclusions.

If you do not know your reader personally and are writing formally as a representative of your company, then use the first person plural and the person's last name:

Dear Mr. Wicks:

We read your report with interest and agree with all but one of your conclusions.

Avoid Words That Antagonize

In writing, you only have one chance to explain your point. If your reader interprets your words differently from the way you had intended, you don't have the opportunity to rephrase them. You also don't have the variety of body language, voice inflection, or facial expressions that you do in face-to-face communication.

Be wary: you may unknowingly upset or antagonize your reader

If you use words that imply that the reader is wrong, has not tried to understand, or has failed to make him- or herself understood, you immediately place the reader on the defensive. For example, when field technician Des Tanski omitted sending motel receipts with his expense account, Andy Rittman (his supervisor) had to write to Des and ask for them. Andy wrote:

> You have failed to include motel receipts with your expense account.

This antagonized Des, because the words *you have failed* seemed to imply that he is something of a failure! Andy should have written:

> Please send motel receipts to support your expense account.

Other expressions that may annoy readers or put them on the defensive are listed in Table 3-1 below with a suggested way to make them more positive.

When a reader has to be corrected, the words you use should clear the air, not electrify it. Tell readers gently if they are wrong, and demonstrate why; reiterate your point of view in clear terms, to clarify any possibility of misunderstanding; or ask for further explanation of an ambiguous statement, refraining from pointing out that his or her writing is vague.

Table 3-1 Expressions that may prove abrasive.

	Sentences containing abrasive words	Much more positive sentences
1. Words that make a reader feel inadequate or guilty:	When completing your application *you neglected* to include your tax number.	Your tax number needs to be included on your application.
	Clearly, *you have not understood the* implications.	Let me explain the implications in more detail.
	We could not accept your bid because *you failed* to submit a complete price proposal.	We could not accept your bid because it did not include a complete price proposal.
	You ought to know that staff working after 11 p.m. have to be sent home by taxi.	It's company policy that staff working after 11 p.m. be sent home by taxi.
	When rejecting the request *you overlooked* human rights legislation.	Before rejecting the request you needed to consider human rights legislation.
2. Words that provoke and so create resistance in a reader:	*I am sure you will agree* that our decision is correct.	Please note that our decision is correct.
	We must insist that you return the form by November 30.	Please return the form by November 30.
	To ensure prompt payment *we demand* that you file your invoice within three days of job completion.	To ensure prompt payment please file your invoice within three days of job completion.
	You must bring the application to room 117.	Please bring the application to room 117.
	In your letter *you claim* that the food processor was incorrectly priced.	In your letter you state that the food processor was incorrectly priced.
3. Words that imply the writer is "talking down" to the reader:	We *have to assume* that you understand the problem.	We assume that you understand the problem.
	Undoubtedly you will be present at the hearings.	We request that you attend the hearings.
	We *simply do not understand* how you misinterpreted our instructions.	Apparently you misinterpreted our instructions.
	You *must understand* that we cannot reopen the file.	I regret we cannot reopen the file.
	If you are applying for reassessment, then *I must request* that you attend a preliminary hearing on October 5.	If you are applying for reassessment, then I request that you attend a preliminary hearing on October 5.

When your goal is to achieve some sort of action or response from the reader, using words that may antagonize will hinder communication. You can still be clear and direct without using these words, and you will find that you are more likely to get the result you expect.

Know When to Stop

When a letter is short, you may feel it looks too bare and be tempted to add an extra sentence or two to give it greater weight. If you do, you may inadvertently weaken the point you are trying to make. This is particularly true of short letters in which you have to apologize, to criticize, to say "thank you," or to pay a compliment (i.e. to "pat the reader on the back").

In all of these cases the key is to be brief: Know clearly what you want to say, say it, and then close the letter *without repeating what you have already said*. The following writer clearly did not know when to stop:

Simple words are much more meaningful than flowery expressions

Dear Mr. Farjeon:

I want to say how very much we appreciated the kind help you provided in overcoming a transducer problem we experienced last month. We have always received excellent service from your organization in the past, so it was only natural that we should turn to you again in our hour of need. The assistance you provided in helping us to identify an improved transducer for phasing in our standby generator was overwhelming, and we would like to extend our heartfelt thanks to all concerned for their help.

Sincerely,

Paul Marchant

Paul's letter would have been much more believable if he had simply said "thank you":

Dear Mr. Farjeon:

Thank you for your prompt assistance last month in identifying an improved transducer for phasing in our standby generator. Your help was very much appreciated.

Regards,

Paul Marchant

If a writer says too much when saying thank you or apologizing, the reader begins to doubt the writer's sincerity. You cannot set a realistic tone if you overstate a sentiment or overwhelm your reader with the intensity of your feelings.

Using a Businesslike Format

There are many opinions about what comprises the "correct" format for business correspondence. Most popular word-processing packages include templates for writing business letters, memos, faxes, and proposals. Some are good and easy to use; others are less practicable. The examples illustrated here are those most frequently used by contemporary technical organizations.

Letter Styles

Most business letters in North America are written full-block style

There are two letter formats: the full block and the modified block (see Figures 3-11 and 3-12). Full block is more widely used and is the format Anna King has adopted for H. L. Winman and Associates' correspondence (see Figure 5-5 on page 110 and Figure 6-5 on page 154). Anna is also aware that letter styles are continually changing. Some companies now write dates in European style (e.g. day-month-year: "27 January 2004"), omit all punctuation from names and addresses (e.g. "Ms. Jayne K Tooke"), and use interoffice memos and email for informal correspondence.

The following comments apply to both the full block format (Figure 3-11) and the modified block format (Figure 3-12):

1. The Post Office now requests one space between the city and the state, and two spaces between the the state and the zip code. The state is always printed as two capital letters (e.g. "NY" for New York) and the zip code is on the same line as the state.

2. Today's trend toward informality encourages writers to use first names in the salutation of the full block format: "Dear Jack." However, in the more traditional modified block format the last name is often retained in the salutation: "Dear Mr. Sleigh."

3. Subject lines should be informative (not just "Production Plan" or "Spectrum Analyzer"); they may be preceded by Subject:, Ref:, or Re:. They should be set in boldface type and not underlined. In the modified block format the subject line is centered.

The differences between the two formats are:

A. In the full block format every line starts at the left margin. In the modified block format the first word of each paragraph is indented.

B. In the modified block format, the date is set off to the right if a file number or reference is used. If there is no file number, the date starts at the centerline.

C. In the modified block format, the signature block starts at the centreline. Notice that in this example, the writer has signed "for" his manager.

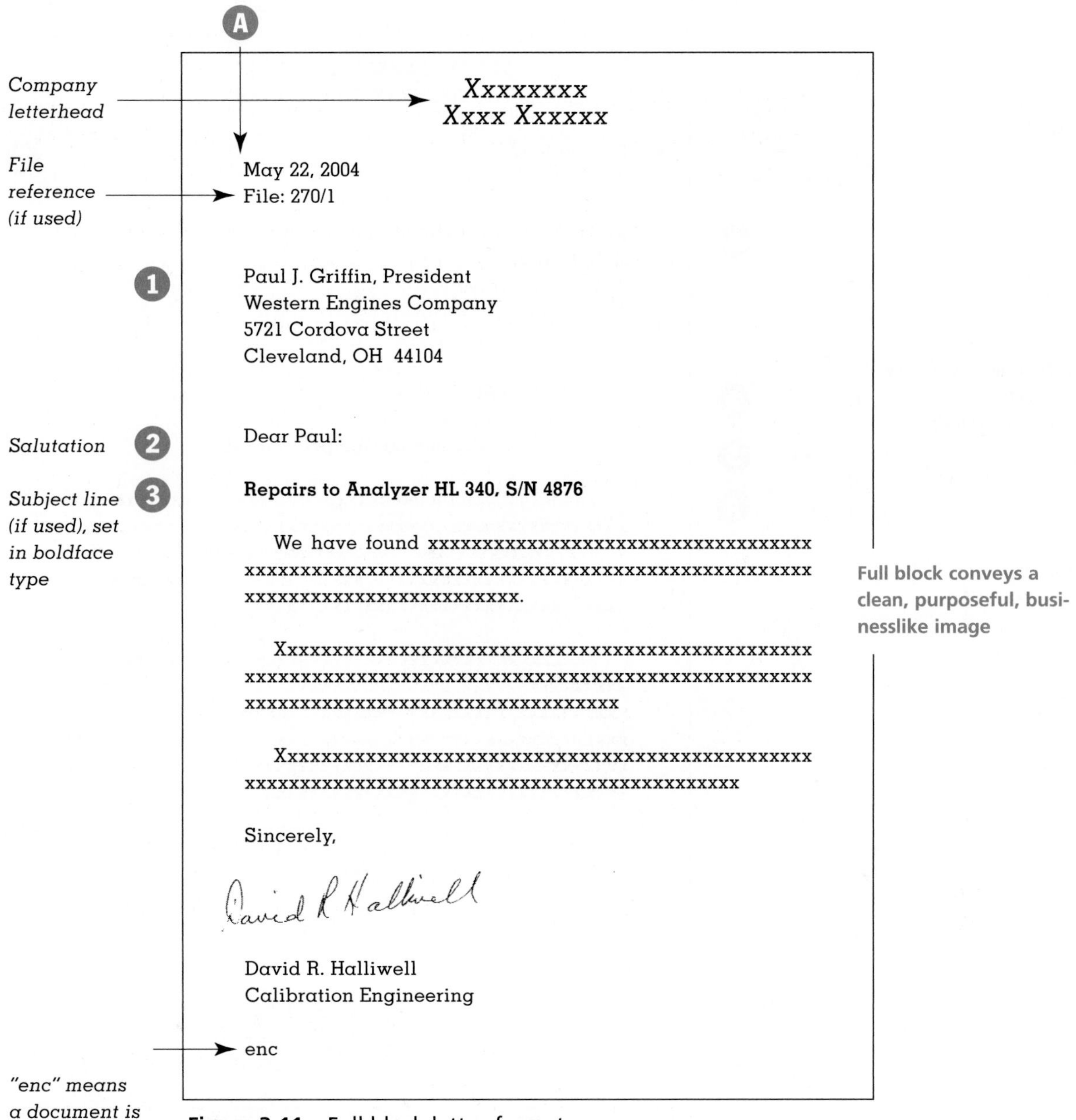
Xxxxxxxx
Xxxx Xxxxxx

May 22, 2004
File: 270/1

Paul J. Griffin, President
Western Engines Company
5721 Cordova Street
Cleveland, OH 44104

Dear Paul:

Repairs to Analyzer HL 340, S/N 4876

We have found xxxxxxxxxxxxxxxxxxxxxxxxxxxxxxxxxxx xxxxxxxxxxxxxxxxxxxxxxxxxxxxxxxxxxxxxxxxxxxxxxxxxxxx xxxxxxxxxxxxxxxxxxxxxxxxx.

Xxxxxxxxxxxxxxxxxxxxxxxxxxxxxxxxxxxxxxxxxxxxxxxxx xxxxxxxxxxxxxxxxxxxxxxxxxxxxxxxxxxxxxxxxxxxxxxxxxxxx xxxxxxxxxxxxxxxxxxxxxxxxxxxxxxxxxx

Xxxxxxxxxxxxxxxxxxxxxxxxxxxxxxxxxxxxxxxxxxxxxxxxx xxxxxxxxxxxxxxxxxxxxxxxxxxxxxxxxxxxxxxxxxxxxx

Sincerely,

David R. Halliwell

David R. Halliwell
Calibration Engineering

enc

Figure 3-11 Full block letter format.

Modified block conveys a more relaxed, approachable image

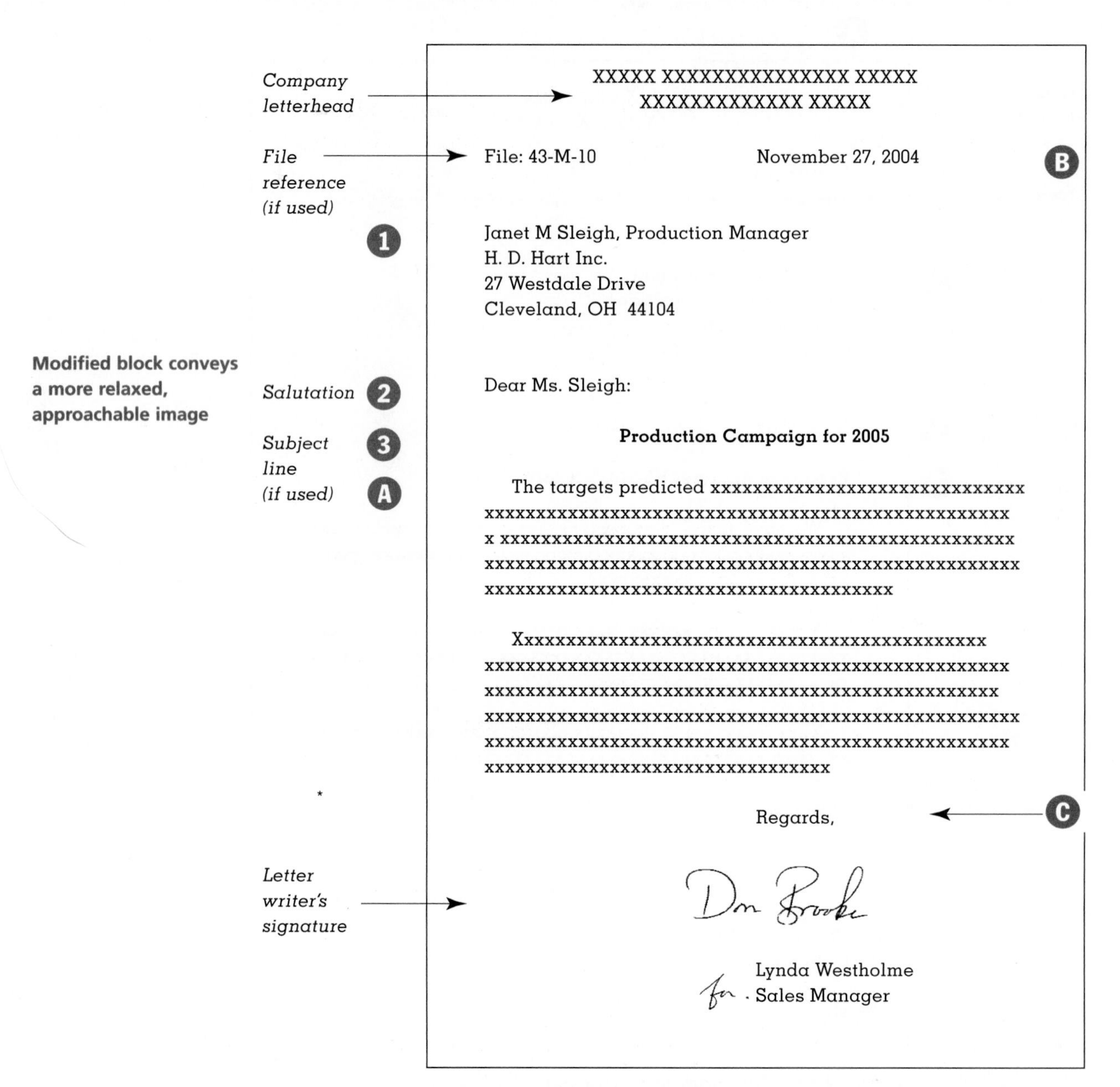

XXXXX XXXXXXXXXXXXXXX XXXXX
XXXXXXXXXXXXX XXXXX

File: 43-M-10 November 27, 2004

Janet M Sleigh, Production Manager
H. D. Hart Inc.
27 Westdale Drive
Cleveland, OH 44104

Dear Ms. Sleigh:

Production Campaign for 2005

The targets predicted xxxxxxxxxxxxxxxxxxxxxxxxxxxxxxx xxxxxxxxxxxxxxxxxxxxxxxxxxxxxxxxxxxxxxxxxxxxxx x xxxxxxxxxxxxxxxxxxxxxxxxxxxxxxxxxxxxxxxxxxxxx xxxxxxxxxxxxxxxxxxxxxxxxxxxxxxxxxxxxxxxxxxxxxx xxxxxxxxxxxxxxxxxxxxxxxxxxxxxxxxxxxx

Xxxxxxxxxxxxxxxxxxxxxxxxxxxxxxxxxxxxxxxxxxx xxxxxxxxxxxxxxxxxxxxxxxxxxxxxxxxxxxxxxxxxxxxxx xxxxxxxxxxxxxxxxxxxxxxxxxxxxxxxxxxxxxxxxxxxxx xxxxxxxxxxxxxxxxxxxxxxxxxxxxxxxxxxxxxxxxxxxxxxx xxxxxxxxxxxxxxxxxxxxxxxxxxxxxxxxxxxxxxxxxxxxxx xxxxxxxxxxxxxxxxxxxxxxxxxxxxxx

Regards,

Don Brooke

for Lynda Westholme
Sales Manager

Figure 3-12 Modified block letter format.

Interoffice Memo

The memo is an internal document normally written on a prepared form similar to that shown in Figure 3-13. Formats vary according to the preference of individual companies, although the basic information at the head of the form is generally similar. Examples of memos appear throughout Chapters 3, 4, and 5. The following comments refer to the memo in Figure 3-13.

The simplest of reporting mediums, the memo is slowly being replaced by email

1. The informality of an interoffice memorandum means titles of individuals (such as Office Manager and Senior Project Engineer) may be omitted.

2. No salutation or identification is necessary. The writer can jump straight into the subject.

3. Paragraphs and sentences are developed properly. The informality of the memo is not an invitation to omit words so that sentences seem like extracts from telegrams.

4. The subject line should offer the reader some information; a subject entry such as "Paychecks" would be insufficient.

H. L. WINMAN AND ASSOCIATES

INTER-OFFICE MEMORANDUM

(1) To: A. Rittman | From: R. Davis
Date: December 3, 2004 | Subject: Early Mailing of Paychecks for Field Personnel (4)

(2) (3) I need to know who you will have on field assignments during the week before Christmas so that I can mail their paychecks early. Please provide me with a list of names, plus their anticipated mailing addresses, by December 8.

Checks will be mailed on Tuesday, December 14. I suggest you inform your field staff of the proposed early mailing.

RD (5)

Figure 3-13 Interoffice memorandum.

5 The writer's initials are sufficient to finish the memorandum (although some organizations repeat the name in type beneath the initials). Some people prefer to write their initials beside their name on the "From" line, instead of signing at the foot of the memo.

Fax Cover Sheet

A fax cover sheet may carry a message in addition to being a transmittal document

Any document sent by facsimile machine is normally preceded by a single-page fax cover sheet that identifies both addressee and sender, and their contact information (see Figure 3-14). The cover sheet usually has a space for the sender to write a short explanatory note. A sender who has only a short message to send may write the message directly onto the fax cover sheet and then transmit just the single page.

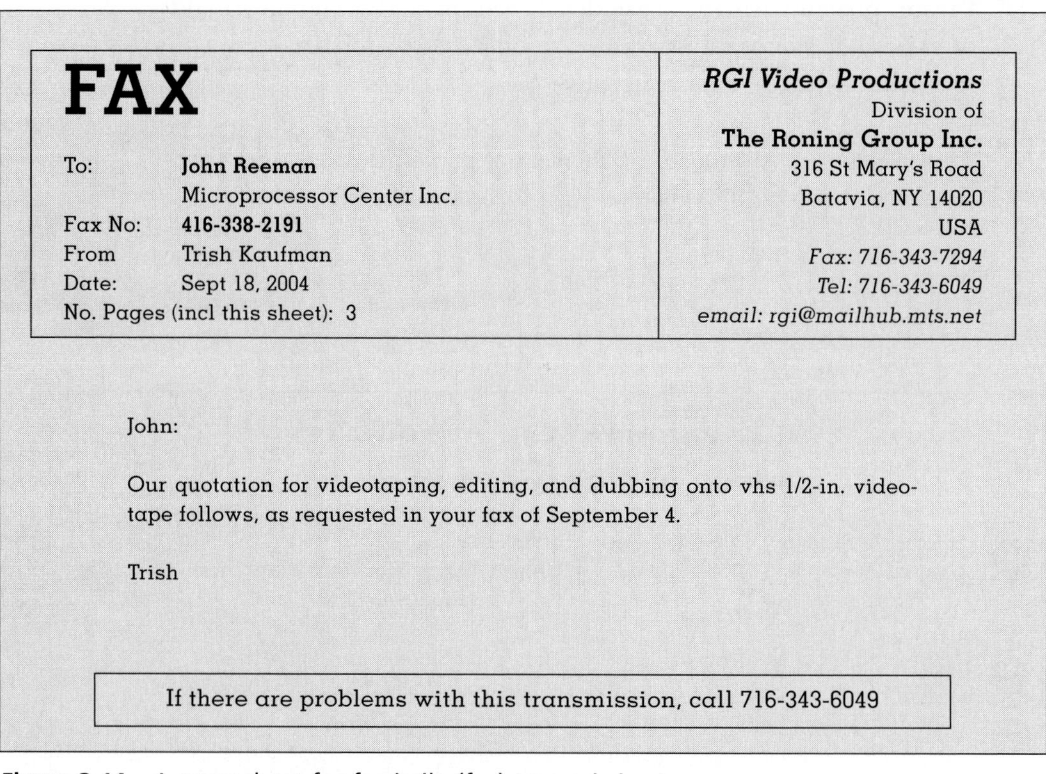

FAX

To: **John Reeman**
Microprocessor Center Inc.
Fax No: **416-338-2191**
From Trish Kaufman
Date: Sept 18, 2004
No. Pages (incl this sheet): 3

RGI Video Productions
Division of
The Roning Group Inc.
316 St Mary's Road
Batavia, NY 14020
USA
Fax: 716-343-7294
Tel: 716-343-6049
email: rgi@mailhub.mts.net

John:

Our quotation for videotaping, editing, and dubbing onto vhs 1/2-in. videotape follows, as requested in your fax of September 4.

Trish

If there are problems with this transmission, call 716-343-6049

Figure 3-14 A cover sheet for facsimile (fax) transmissions.

Writing Electronic Mail

Electronic mail (email) has become one of the fastest and most widely used means of communication. The criterion for writing email remains the same as for all other forms of written communication: keep the message brief but clear. Never allow an overzealous desire for brevity to cloud your message, because it can cost more to question an obscure communication than it would have cost to write a slightly longer but clearer message in the first place. You want to avoid having your reader make a telephone call or email you to find out exactly what you were trying to say.

The fast new way to communicate...

When Mike Toller in Columbus, Ohio, opened up a shipment of parts from Carlson Distributors, he found the order was incomplete and contained some items he had not ordered. He made a note of the deficiencies, sat at his keyboard, and typed in this message:

> To: Carlson Distributors, St. Louis:
>
> Your inv 216875 Oct 19, our P.O. W1634. Short-shipped
> 10 toolsets MKV, 4 801 sockets plus 2 doz mod 280A lathe bits unordered.
> Advise.
>
> M. Toller,
> Crown Manufacturing, Columbus

...needs just as much care and attention

A Beginner's Guide to Effective Email
www.webfoot.com/advice/email.top.html
This useful guide includes an introduction to email and a discussion about why it differs from ordinary correspondence.

In St. Louis, Chantal Goulet puzzled over the message on her computer screen, and then typed this brief reply:

> To: M. Toller, Crown Manufacturing, Columbus
> From: C. Goulet, Carlson Distributors, St. Louis
>
> The message you sent regarding our invoice 216875 and your P.O. W1634
> was difficult to understand. Please explain your concerns.
>
> Chantal Goulet

Mike was surprised: he thought his message was crystal clear. So he again sat at his keyboard and wrote:

What Mike Toller should have written the first time

> To: Chantal Goulet, Carlson Distributors, St. Louis
> From: Mike Toller, Crown Manufacturing, Columbus
>
> My message was quite clear: You short-shipped us 10 toolsets type MKV and 4 No. 801 sockets. You also shipped 2 dozen model 280A lathe bits we did not order. Please ship the missing items and advise how you want the bits returned.
>
> Mike Toller

Chantal replied in six words:

> Mike:
>
> Thanks. I will investigate the situation.
>
> Chantal

Keep the message short but keep it clear

If Mike had been more explicit when he wrote his original message, both he and Chantal would have saved time.

Email Netiquette

The word "Netiquette" means "the etiquette of writing electronic mail on the net (the Internet)." There are no established guidelines for Netiquette, but we can give you some suggestions that will help you be an efficient email communicator.

First, email does *not* give you a license to

- write snippets of disconnected information,
- write incorrectly constructed sentences,
- forget about using proper punctuation,
- ignore misspelled words, or
- be abrupt or impolite.

Neither, however, is it a forum for telling long stories, anecdotes, or jokes.

Writing Effective Email
www.uwec.edu/jerzdg/orr/handouts/TW/e-mail.htm
This document offers 10 tips to help you write effective professional emails. If you want to be taken seriously by professionals, you should know email etiquette.

Adopt the Right Tone

Too often, we hear people say "It's only email. It's supposed to be casual and quick." That's true. With email you can be less formal in tone but you still need to be professional. You still need to address the recipient and you still need to "sign" your name to the message. Even if you have a signature file attached to every message you should still type your name at the end. This helps humanize this very technical mode of communication.

Use a Specific Subject Line

The subject line for your email message must serve as a mini Summary Statement. It is the first indication to the reader of what your message is about. People receive so much email these days that our Subject line must be specific so we can help them identify the content and importance of our information.

Mark Hoylston, an engineering technician responsible for installing a new network at a client site, was writing to his supervisor to explain there was going to be a delay in the project because they had discovered some additional work that needed to be done first. His subject line on the email was simply

Subject: Progress

and his supervisor decided to read it later, when she had more time. To her, the word progress indicated the project was moving along nicely, but Mark was really writing to discuss the delay instead. If he had used

Subject: Delay in Project Progress

as the subject line, his message would have received the attention it deserved.

Write "Pyramid Style"

You can use the pyramid method for writing email messages, just as you do for ordinary letters:

1. Start with what you most want your reader to know and, if appropriate, what action you want the reader to take.
2. Follow with any background information the reader may need to understand the reason for your message, and provide details about any point that may need further explanation.

Check that each message contains *only* the information your reader will need to respond or to act—and no more. That is, take care to separate the essential *need to know* information from the less important *nice to know* details. Your email will still have four compartments (Summary, Background, Details, Action/Outcome) but they may be much shorter than they are in a letter format. For example, the Summary and Background may be in the same sentence.

If you need to include extensive details, use the email message as the Summary and then put the details in a file attachment so your message uses the structure shown in Figure 3-10. Your readers will appreciate this since they are not forced to read the entire document; they read the highlights in the email message and turn to the details in the attachment if and when they need them.

Proofread with Care

Reread what you have typed, even for a one-sentence reply

Proofread email *very* carefully: the informality of the medium and the speed with which you can create and answer messages can invite carelessness. It may sound contradictory to suggest that you print your email messages and edit them on paper before you send them, but we recommend you do so if a message is long or if its contents are particularly important. This is especially true if you are replying to a message immediately after you have read it.

Be Prudent

If you are annoyed or irritated by a message you receive, *wait* before replying. Let your irritation cool down. Email is ideal for transmitting facts; it's the wrong medium for sending emotionally charged messages.

Remember that email is not a good medium for conveying confidential information, and it is particularly not a medium for making uncomplimentary remarks about other people. Never put anything in an email that you would feel uncomfortable saying to someone in person. Email messages can too easily be forwarded or copied to other readers, and then you have no control over who else may see what you have written. Be just as professional as you are when writing regular letters and memos.

Similarly, be just as sensitive when deciding to copy a message to another person. Be sure that the original sender would want his or her message distributed to a wider audience.

Email Guidelines

Here are some suggestions that will help you write more effective email messages:

- Remember that busy readers want messages to be concise yet complete. Feed their needs.
- If you are writing to multiple readers, consider sending *two* messages rather than a single all-embracing message. Write
    1. a short summary, which you send to readers who are interested only in the main event and the result, and
    2. a detailed message, which you send to readers who need all the details.

Avoid Overloading the System

Limit how many readers receive your message

Be selective when replying to a multiple-reader message. It may be tempting to simply click the "Reply All" button rather than take the time to address your reply only to those readers who need it, but if you do your reply will go to everyone. And if other people reply in the same way, the system—and everyone else's In Basket—will quickly become overloaded.

When accessing email, unless you are using a high speed Internet connection (T1 line, LAN, or cable modem), consider downloading it immediately into your In Basket so that you remain online only briefly.

Then read and answer your mail offline (i.e. when you are not connected to the service). But avoid letting messages accumulate for too long in your In Basket. If you want to keep a message or may need to refer to it later, store it in an electronic "project folder" in the "filing cabinet" (or an electronic receptacle of a similar name, depending on the service you are using).

Avoid routinely printing copies of messages you want to keep: creating extra paper defeats the aim of email!

Help Identify the Originator

Make sure the originator's name is evident

When replying to a message, particularly if your reply is going to multiple addressees, quote a line or two from the original message to help put your reply in context. Identify the excerpt by placing a ">" sign before each line, like this:

Dan Reitsma wrote on May 12:

> The Society's constitution was last updated in
> 1984 and needs amending.

I agree, but first we need to check how much editing was done by Karen Ellsberg before she retired in 1997.

This reduces the frustration your reader will experience from having to scroll down through all the attached messages (often called the "history" or the "train.")

Write your name at the foot of every message you create, even though your name appears in the "To-From" list at the top or in a signature file. If a recipient decides to forward the message to other people, frequently only the text will be forwarded and recipients will not be able to identify the originator.

Avoid Complex Formatting

Use only simple formatting if you are sending messages outside your email system. Bold, italic, and color formatting may not convert correctly in transmission or may not be available in the recipient's system.

Write short paragraphs with line lengths of no more than 60 characters, and separate each paragraph with a blank line. Avoid creating columns and indenting subparagraphs, because most likely what you see on screen will not be what your readers see. For example, your screen may look like this:

Most email systems do not transmit tables and charts well

Facilities are located as follows:

Facility	Location	Distance
Master Control	Chicago, IL	28.6 mi south of transmitter
Remote Site 1	Des Moines, IA	Downtown
Remote Site 2	Nashville, TN	2.5 mi north of university

But your readers may see something like this:

Facilities are located as follows:

Facility Location Distance

Master Control Chicago, IL 28.6 mi south of transmitter

Remote Site 1 Des Moines, IA Downtown

Remote Site 2 Nashville, TN 2.5 mi north of university

If you need to format columns, consider creating the message as a word-processing file and sending it as an attachment to an email message.

Indicate Emphasis with Care

If you want to emphasize a word but are not sure if bold or italic type will convert, insert an asterisk on both sides of the word or expression:

> This service is available *only* to first-time software buyers.

Don't Shout!

Use upper- and lower-case letters, just as this sentence has been written (not like the one below).

> PARAGRAPHS COMPOSED OF ALL CAPITAL LETTERS ARE HARD TO READ. YOU CANNOT EASILY IDENTIFY WHICH ARE THE KEY WORDS.

This may be perceived as if you are shouting or that you are angry. The opposite is also true:

> paragraphs composed of all lower-case letters are hard to read. you cannot easily identify which are the key words.

Finally, avoid inserting "cute" graphics or humorous remarks into your email. They make you appear unprofessional.

ASSIGNMENTS

Most of the letter and memo writing projects below include all the details you need to do the assignment. You are encouraged, however, to introduce additional factors if you feel they will increase the depth or scope of your letter.

Project 3.1: Request for Free Parts

You are an engineering technologist employed by H. L. Winman and Associates, and you are engaged in a lake-level measurement program in northern New York State, working under Government Environmental Studies contract WM-23357.

At a critical moment in the program your Hektik Model 370 Water Stage Manometer breaks down. You take it apart and identify that it needs a replacement spring and drive assembly.

You're frustrated with repeated failures.

This is the third time this has occurred in the past six months, and each time the thread on the drive shaft has been stripped. You previously purchased replacement spring and drive assemblies from the manufacturer's US office, Hektik America Inc., 21 Lincoln St., Chester, Mass., on May 23 and June 17, at a cost of $218.50 each. (The Model 370 Water Stage Manometer is manufactured by Hektik Industries GmbH in Dusseldorf, Germany.)

Yesterday you faxed purchase order No. 26019 to the Chester office, requesting immediate shipment of a replacement spring and drive assembly.

Today you decide to write to Hektik America Inc., to complain about the repeated failures (you may attribute the cause to any condition you wish, if you feel you need to point out the cause), and to request that the current replacement part be supplied free of charge.

Write a letter to customer service.

Project 3.2: Revising a Letter

At 4:15 p.m. Norm Behouly comes to you with a problem. "I'm going on vacation tomorrow," he announces, "and I'll be away for three weeks. The trouble is, I've typed two letters into the computer, and now the system has gone down and I can't get them out!"

Norm asks you to print and mail them for him when you come in tomorrow morning. He gives you two file names: SURVEY.TXT and FENCE.TXT. "You'll have to sign them for me," he adds, "and I would appreciate it if you would take the time to read them first, just in case there's a typo I have missed."

Now it is 9:15 a.m. on the following morning and the computer system is again operational. You bring Norm's two letters up and immediately see that they need much more than just a cursory check for typographical errors.

Revise or correct each letter. Insert a full address for each recipient, including the name of your city and a hypothetical postal code.

Part 1: File SURVEY.TXT

Simplify a letter that has too many words

Dear Mr. Antony:

In response to your letter of June 7, 2004, and our meeting at your residence at 960 Bidwell Street on June 14, when you showed me the plan of your Lot (Lot 271-06) and the position of the fence bordering the Lot to the south, at 964 Bidwell Street, which is Lot 271-07. You claimed there is a discrepancy between the city site plan and the physical position of the fence, and asked me to do a survey of your Lot so as to establish the correct position.

Your Lot was surveyed by me and an assistant on June 21 and while there I hammered in two markers to delineate the southeast and southwest corners. (No markers were placed on the north side because the position of that fence is not in question.) Your neighbors to the south—Mr. and Mrs. Beamish—will not be happy when they find out that the fence between Lots 06 and 07 encroaches on your property. You will note from the positions of the markers that the east end of the fence is 14 inches inside your territory, but is angled toward the south so that at the west end, where it stops at the garage, it is correctly positioned.

It is assumed that you recognize that the south fence is yours, and the fence to the north is the responsibility of your neighbor to the north. Consequently you have the right to move the fence if you wish or to leave the fence where it now stands until repairs are necessary and then rebuild it in its correct position. As obviously you are aware, the fence is in good condition.

As per your request, I am writing to your neighbors today to inform them of the discrepancy and attaching our invoice.

Yours sincerely,

Part 2: File FEzNCE.TXT

Dear Mr. and Ms. Beamish:

As I am sure you must have been aware, a survey of Lot 271-06 was done recently, on June 14, to determine the exact borders of the Lot at 960 Bidwell Street, to your north. While the survey was being done, markers were positioned at the southeast and southwest corners of the Lot, to establish the exact dividing line between your Lot and that of Mr. and Ms. Antony at 960 Bidwell Street. No markers were placed at the northeast and northwest corners of the Antony's Lot.

Unfortunately the fence is incorrectly positioned between your Lot (No. 271-07) and Lot 271-06. At the southeast corner of Lot 271-06 the fence is 14 inches too far to the north and so encroaches onto your neighbors' Lot. (Actually, the fence slants toward your property as it progresses westward and at the garage end is properly positioned.)

I can only assume that you are unaware of this discrepancy, so at Mr. Antony's request I am writing to you so that you will know of the circumstances should Mr. Antony choose to reposition his fence. I am equally sure that you and the Antonys can come to an amicable agreement. Please feel free to contact me at your convenience if you need more information concerning this matter.

I remain, yours truly,

Make this letter more direct and easier to understand

Part 3: File GARAGE.DFT

Norm calls you from the airport: "I forgot to tell you," he says. "There's a third file—GARAGE.DFT. It's some notes about the garages on the Antonys' and the Beamishes' Lots, and I think the owners should know about them. Could you write to each of them for me? It shouldn't wait until I return."

From the notes in file GARAGE.DFT you gather that:

Create a letter from notes

1. The two garages are parallel to each other and the space between the adjacent walls is only 17.5 inches.
2. There is a pile of lumber stacked between the garages to a height of 47 inches.
3. City by-law 216, subparagraph 2(c) stipulates that garages must be a minimum of 24 inches apart.
4. City by-law 216, subparagraph 2(h) requires that passageways between garages must be accessible, for fire safety reasons.

You feel the homeowners could ignore the separation discrepancy for the moment, but should do something about the stacked lumber (the city inspectors may never notice the too-narrow distance between the garages, but almost certainly they will eventually notice that access between the garages is blocked and this may lead them to measure the separation distance). Write a letter to Mr. and Ms. Antony informing them of the problem. Tell them you are sending an identical letter to the Beamishes next door.

Project 3.3: Correcting a Billing Error

Today you receive a credit card statement from WorldCard, covering last month's purchases. There are eight debit entries, three personal and five

Nearly everyone has had to correct a billing error

for expenses incurred during a business trip you made to Wapiti Paper Mill between the 10th and the 14th. (You are an engineering technician employed by the local branch of H. L. Winman and Associates, and you went to the mill to investigate and rectify a problem in the process control system.) The five business expenses are:

Item	Date	Vendor/Location	Control No.	$
3	10	St. James Motel, Burntwood Lake	0134652	73.90
4	11	Burntwood Auto Service	0147162	305.60
5	12	Wapiti Autos	0203916	38.66
6	14	Wapiti Inn	0205771	256.50
7	14	Burntwood Auto Service	0211606	31.58

Item 4 puzzles you. You know you purchased gasoline three times and stayed one night on the road in a motel and three nights at another motel near the mill. But you could not have bought $305 of gasoline (your car's tank would not hold that much!).

Fortunately, you always keep a travel log and in it you recorded these entries:

11th	–	19.46 gal	@	$1.57/gal
12th	–	25.27 gal	@	$1.53/gal
14th	–	20.11 gal	@	$1.57/gal

You do not have the credit card vouchers because you attached them to the expense account you handed in to branch manager Vern Rogers on the 19th, and he has sent them on to head office in Cleveland. But from your records you can work out what the error is and can guess that it occurred during data entry at WorldCard's data center in New York.

Write to the manager of customer accounts at the credit card company, inform him or her of the error, and ask for an adjustment. WorldCard's address is: Suite 2160 – 24 Harley Avenue, New York, NY 10026.

Project 3.4: Letter of Thanks

Say thank you: elegantly but briefly

Last night you attended a talk delivered by Ms. Tina Mactiere to the local chapter of the Inter-State Engineering Association (ISEA). Today you have to write a letter of thanks to Ms. Mactiere, expressing your and the ISEA chapter's appreciation. (You are the chapter's technical program coordinator, and you arranged for Ms. Mactiere to give the talk.) Some details you may need are:

1. You are employed by Hogan Consultants Inc. at 212 Broad Avenue of your city, where your company president, Gavin Hogan, encourages his technical staff to participate in ISEA activities.
2. Tina Mactiere is president and chief executive officer of Macro Engineering Inc.

3. Her talk was given in the Prairie Room of the Chelmsford Hotel. The event was the Annual General Meeting (AGM) of the local ISEA Chapter. The program included a formal dinner at 6:30 p.m., Tina Mactiere's address at 8:15 p.m., and the AGM at 9:15 p.m. The affair concluded at 10:15 p.m.
4. Tina's talk was titled "Look After the P's and Q's." Her main thrust was that technical people are so concerned with keeping abreast of new technology that they omit other essential aspects of their professional development. She cited, for example, the need for scientists, engineers, and technicians to attend courses or seminars in supervisory management, interpersonal relations, and oral and written communication—topics she referred to as "people skills."
5. Tina proved to be a dynamic speaker. She used slides and a humorous three-minute videotape that neatly underscored the points she was making.
6. There were numerous questions from the audience after her talk, and a strong round of applause.
7. Many people came up to you after the AGM and congratulated you on your choice of speaker and the appropriateness of her topic.
8. Seventy-six ISEA members attended the dinner and meeting.

Project 3.5: Mis-ticketed for Flights

As an independent consultant you find yourself traveling frequently to different client sites. Most of your work is done remotely, from your home office, but sometimes an important meeting or presentation requires that you see people in person. You realize the value of developing a relationship with your clients. Even with all of today's technology you find the best way is still face-to-face.

Since you just recently established your own company, Pro-Active Consultants Inc., you don't have the resources or luxury of having a secretary to make your travel arrangements so you have to do it yourself.

The first step—booking the flights—was easy...

When you called Jet Express Airlines (you called them directly because you thought you might get a better price than if you used a travel agent) you spoke with a friendly representative named Joyce. You explained to her that you want to fly to St. Louis, Missouri, on Sunday, June 7, because you have a business meeting at the new site June 8 to 10, and then on Thursday, June 11, you want to fly from St. Louis to Nashville, Tenn., to visit a friend, returning to your city on Sunday, June 14.

"Wow," Joyce said. "Have I got a deal for you. I can get you to where you want to go for a total of $790.00. That's a great price considering it's not a straight, round-trip ticket but what we in the airline industry call an open-jaw ticket."

You said you needed to confirm your plans with the site manager and talk to your friend in Nashville, to make sure she is going to be available, before you give the agent your credit card number and pay for the flights.

"No problem," Joyce said. "I can hold these flights for 24 hours. Just call back before midnight tomorrow."

After a series of answering machine messages back and forth, you finally got in touch with the site manager and your friend: the dates and times you discussed with Joyce at JE Airlines were fine. When you called the airline to provide your credit card details and secure the flights, you were connected with a different representative named Jonathan.

"I'm sorry, but I can't find your reservations," he replied. "Are you sure you phoned back within 24 hours?"

"Yes, I'm sure," you said. "This is all I've spent my time on in the past 24 hours!" At this point you were getting a little annoyed. Every phone call seemed to eat away 30 to 40 minutes of your time.

...then frustration set in, one telephone call at a time

"Oh, wait a minute. There it is. It appears your reservations have been canceled," Jonathan said. "I don't know why, but they have gone." Luckily you wrote down the exact dates and flights that Joyce quoted. Here's what she had found:

Sunday, June 7	LV your city	1:00 p.m.	Flt. 832
	AR Chicago, IL	3:00 p.m.	
	LV Chicago, IL	4:26 p.m.	Flt. 808
	AR St. Louis, MO	6:18 p.m.	
Thursday, June 11	LV St. Louis, MO	6:30 p.m.	Flt. 2430
	AR Nashville, TN	8:30 p.m.	
Sunday, June 14	LV Nashville, TN	3:41 p.m.	Flt. 81
	AR Chicago, IL	5:13 p.m.	
	LV Chicago, IL	7:02 p.m.	Flt. 2160
	AR your city	9:17 p.m.	

Jonathan was patient and, although he couldn't get you the great deal Joyce did, he was able to get you on the exact same flights for only $38.00 more.

"Fine," you said, "I'll give you my credit card details to guarantee these flights. I'll put them on my company VISA card number 4321 1238 7898 5000, expiration date 9/06.

"OK," said Jonathan. "I'll email the confirmation to you today."

Two hours later when you open the email you are shocked. "Unbelievable!" you shout out loud. "JE hasn't included the June 11 leg from St. Louis to Nashville!"

So, you make *another* phone call to the airline (another hour of your time) and speak with a representative called Ashley, who isn't as friendly as the first two representatives. She explains that your only option is to purchase a one-way ticket from St. Louis to Nashville for $134.50.

"But that's $172.50 more than my original quote!"

"Well," says Ashley with a tone of sarcasm, "You could always take a bus from St. Louis to Nashville, couldn't you?"

With little choice you agree to purchase the additional ticket but you are not very pleased or impressed. So you decide to write to the airline and express your dissatisfaction, and ask Ashley for a name and address to write to. Here's the information she gives you:

Now it's time to write for an adjustment

Donavan Johnson
Director of Consumer Affairs
Jet Express Airlines
6001 Airport Highway
Raleigh, NC 27134

Write the letter. Ask for compensation for the trouble you have experienced and the expenses you have incurred.

Project 3.6: Request to Attend a Course

Assume that today is the second Monday of the *current* month, and that for the past four weeks you have been on a field assignment in San Antonio, Texas, where you have been conducting an extensive hardware and software installation program for Inter-State Telephones (IST). You have been assisted by two technicians (Ted McCourt and Carolyn Freedman), and you are now three days ahead of schedule. The task is to be completed by the 12th of *next* month. Today you receive a brochure from the University of Texas in Austin advertising a one-week course. Details are:

Course title:	Managing in a Technological Environment
Course dates:	Monday the 5th to Thursday 8th inclusive (of next month)
Type of course:	Maximum immersion: 9 a.m. to 5 p.m. daily, plus 7 to 10 p.m. Wednesday evening; approximately 20 hours of home assignments
Cost:	$495; includes materials, books, and lunches, plus a guest speaker from industry at each lunch
Registration:	No later than noon on Tuesday the 23rd; telephone registrations accepted
No. of Participants:	16

A course worth attending

You are impressed by the technical standard of the course described in the brochure and wish to attend. (Because of previous assignments, you missed a similar extension department evening course offered at your local college last winter. Your company sponsored four engineers to attend that course, for which the fee was $165 each.)

Ask for approval to go

Write an interoffice memo that you will fax to your department head, Denise Coltrane. In it, you should:

- describe the course (convince her it is a good one),
- ask if you can attend,
- ask if the company will pay the tuition fee plus travel and lodging,
- ask to be spared from the IST task for one week (be convincing), and
- ask for a quick reply (because time is short).

Assume that Denise can give technical and financial approval for you to attend. Also assume that you have a rental car for the IST project which you can use to drive to Austin, and that the hotel in Austin will cost $95 per night.

Project 3.7: A Faulty Home Entertainment Center

A music source you like...

Assume that recently you returned from a holiday in Waverly, where you visited your friends Martin and Joan Lamont. Martin gave Joan a PAM 98 Home Entertainment Center last Christmas, and you were impressed by its tone, appearance, and features. Martin told you privately that he bought it from Craven's Discount Center at 1837 Kelly Street in Waverly, and offered to go with you if you were interested in buying one.

You were, but you were disappointed to discover that Craven's had sold all of its PAM 98 entertainment centers, that no more were on order, and that no other stores in Waverly carried them. However, Harry Craven, the shop owner, suggested he had a demonstration model he could sell you at 5% off the regular discount price. You tested it, and it seemed to operate satisfactorily. Martin suggested a 15% price reduction would be more realistic for a demonstrator, but Mr. Craven wouldn't budge. He added, however, that he would have his technician give it a good check over if you left it with him for 48 hours. You agreed, and two days later you picked it up, paid $460.25, and received Craven's invoice No. C5603 stamped "Paid in Full." The following day you flew home.

...proves to have a problem

But when you plugged in the PAM 98 at home, you found that the CD player did not work. You also discovered that there was no local service center for the PAM line, so you took the entertainment center to Modern TV and Radio at 28A Waltham Avenue. When you picked it up three days later, shop manager Jim Williams handed you a circuit board with several bent and twisted pins.

"There's your problem," Jim said. "Craven's in Waverly must have replaced this PCB: you can tell it's one of theirs because the name CRAVEN is stamped on it." He explained that whoever inserted the circuit board did not align the pins properly and bent them by forcing the board into its socket.

You paid $83.50 for the repair job on Modern TV and Radio's Invoice No. 1796, and took both the PAM 98 and the ruined circuit board with you.

Write to Harry Craven, tell him what has happened, and ask for a refund of $___ (you decide how much). Assume you attach copies of the two invoices to your letter.

(*Note:* The PAM 98 was made by VICOM in Korea. It contains an AM/FM stereo receiver, a CD player, a dual cassette tape deck, and two eight-inch speakers.)

You don't think you should have to pay for the repairs

Project 3.8: Acknowledging a College Award

Assume that you are in the second year of the course you are enrolled in, and that three weeks ago the head of the department came to you and announced that you have been selected to be this year's winner of the Inter-State Engineering Association (ISEA) scholarship for "proficiency in technical studies." Yesterday you attended an awards luncheon with other scholarship winners and representatives of the firms donating the scholarships. You sat next to Calvin Wycks, vice-president of the local chapter of ISEA, who presented the award to you.

Expressing personal thanks is not always easy

Today, you write to ISEA to thank the association for the award. Use these details:

- Address your letter to the president, Marjorie McIvor.
- ISEA's address is 710 Durham Drive of your city.
- The award is a check for $500 and a wall plaque inscribed with your name.

Chapter 4
Short Informal Reports

Short Informal Reports
www.uwec.edu/jerzdg/orr/handouts/TW/reports.htm
This document introduces two basic principles of technical communication — meeting the reader's needs and using the inverted pyramid.

When you hear that someone has just finished writing a technical report, you might imagine a nicely bound formal document. In some cases you would be correct, but most of the time you would be wrong. Far fewer formal reports are issued than informal reports, which reach their readers as letters, memos, and email. This chapter describes the short informal reports you are likely to write as a technologist, engineer or engineering technician.

Internal Versus External

When you are reporting information to an audience inside your company you will be writing a memo-form report you can fax, mail, or email to the recipients. The length will vary depending on how much the audience needs to know about the subject. The tone will vary too, depending on how well you know the readers and how often you interact with them. Often a memo is going to several people, so although the tone can be informal it should always remain professional and respectful. See Chapter 3 for details on what a memo looks like.

When you are reporting information to an audience outside your company, you will be writing a letter-form report. Some letter reports may be as informal as a memo report, particularly if they are conveying information between organizations whose members know each other well or have corresponded frequently. Others may be more formal, presented as business letters conveying technical information from one company to another. The formality varies according to its purpose, the type of reader, and the subject being discussed. Like the memo, you can send the letter by fax, mail, or email. The format should still be a structured business letter. See Chapter 3 for examples of business letter formats.

Although there are many types of informal reports, all are based on the writing plan outlined in Figure 4-1. Each report contains

1. a brief statement describing what the reader most needs to know,

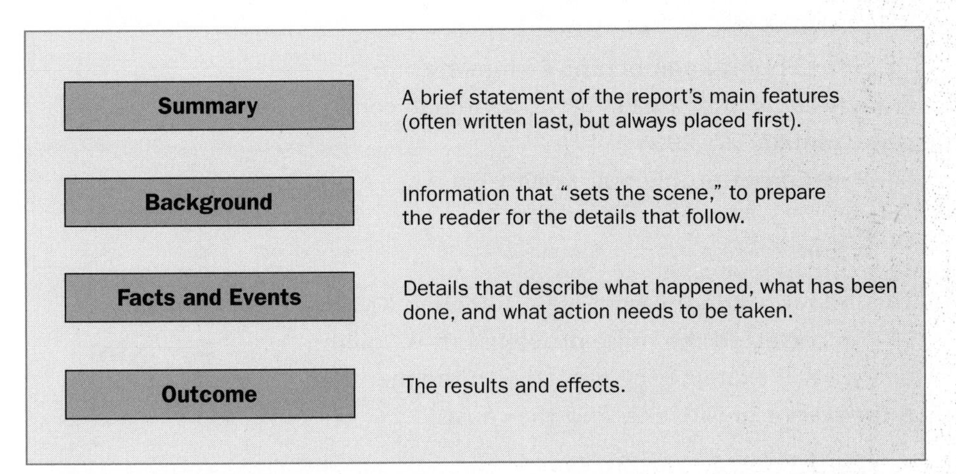

Figure 4-1 Basic writing plan for short reports. The plan is modified slightly to suit each situation.

The writing plan for basic reports is similar to the plan for business letters

2. a short introduction to the problem,
3. a discussion of the data, situation, or problem, and what has been done or could be done about it, and
4. a conclusion that sums up the results and possibly recommends what should be done next.

Keep in mind that these pyramids are just templates. You can use them as is or you can adapt them to your unique situations. For example, you may find that you don't need to discuss past activities in your progress report but need to focus only on present and future work. In that case, simply delete the Past Work compartment.

Writing Style

The reports described in this chapter are written in a direct, informative style that is crisp and to the point. The writers are usually describing events that have already occurred, so they write mostly in the past tense, which helps them to be consistent. They switch to the present or future tense only when they have to describe something that is presently occurring, outline what will happen in the future, or suggest what needs to be done. All three tenses occur in the report shown in Figure 4-2.

Dan Skinner has written the Background and most of the Facts paragraph mainly in the past tense because those sections deal with what has already been done. At the end of the Facts he has shifted into the present tense to report how the station manager feels now. For the Outcome he has jumped into the future tense to outline what he plans to do. This past-present-future arrangement is natural and logical; reader Don Gibbon will feel comfortable making the transitions from one tense to the next. Even Dan's Summary follows the same pattern.

To: Don Gibbon <dgibbon.ho@winman.com>
From: Dan Skinner <dskinner.fld@winman.com>
Date: January 24, 2004
~~Subject~~: Carpet problem at KMON-TV
RE:

Don

Summary of incident

The indoor/outdoor carpet we installed in KMON-TV has corrected the noise problem but is "pilling" badly. I will examine the carpet with the manufacturer's representative to find the cause.

Past tense — *Background*

The carpet was installed in the satellite studio control room during the night of January 8–9, to reduce the ambient noise level by 3.6 dB.

Mainly past tense — *Facts*

At the station manager's request, I returned to the control room today and checked the carpet's condition. After only two weeks it has tight little balls of carpet material adhering to its surface. I called the manufacturer's rep, who said that the condition is not unusual and does not mean that the carpet is wearing quickly. He suggested that it may be caused by improper cleaning techniques and probably can be easily corrected. However, our client is not pleased with the carpet's appearance.

Present tense

Future tense — *Outcome*

The manufacturer's rep and I will return to the control room between midnight and 2 a.m. on January 31 to study the carpet-cleaning techniques used by maintenance staff. I will email our findings to you later in the day.

Dan

Figure 4-2 A short report transmitted by electronic mail.

Incident Report

An incident report describes an event

Anytime you are involved in or witness an accident, whether equipment is damaged or people are injured, you need to write a report describing what you saw. An incident report informs management and others of what happened and is often kept on file.

Anna King is working late in the H. L. Winman office in Cleveland when the telephone rings. The caller is Bob Walton, a member of the electrical engineering staff, who is on a field trip to Tangwell. He tells Anna he has been involved in a traffic accident near Hadashville, his co-traveler has been injured, and some of his equipment has been damaged. He wants Jim Perchanski, his department head, to send out replacement equipment by air express.

Anna jots down notes while Bob talks. Because she will be out of the office the following day, she writes a report of the conversation and emails it to Jim Perchanski (see Figure 4-3), knowing he will access his email immediately after he arrives in the morning. She tells Jim what has happened to two members of his staff, where they are now, how soon they will be able to move on, and that one of them is injured. She also tells him that equipment is damaged and replacements are needed.

To: Jim Perchanski: perchnski.ho@winman.com
From: Anna King: aking.ho@winman.com
Date: September 16, 2003
Subject: Accident report and request for spare parts ❶

Jim:

Bob Walton and Pete Crandell have been involved in a highway accident, which will delay their inspection of the Sledgers Control project at Tangwell. They need replacement parts shipped to them tomorrow (Wednesday, September 17). ❷

Bob telephoned from Hadashville, at 7:35 p.m. to report the accident, which occurred at 5:15 p.m. some two miles west of Hadashville. Pete has been hospitalized with a fractured left knee and a suspected concussion. Bob was unhurt. The van and some of their equipment were damaged. ❸

Bob wants you to ship the following items by air express on a Remick Airlines Wednesday evening flight to Montrose, and to mark the shipment "HOLD FOR PICK UP BY R. WALTON SEP 18":

- 1 Spectrum analyzer, HK7741
- 1 Calibrator, Vancourt model 23R ❹
- 24 Glass vials, 300 mm long × 50 mm dia

He will rent a van and drive to Montrose to pick up the items Thursday morning. He will then drive on to Tangwell and expects to arrive there about 4:00 p.m. He has informed site RJ-17 at Tangwell of the delay. ❺

Bob is preparing an accident report for you. He is staying at Hunter's Motel in Hadashville (Tel: 614-453-6671). ❻

Anna

Address your recipient and "sign" your name in an email.

Figure 4-3 A third-person incident report.

Anna's message is an incident report, written pyramid style (see Figure 4-1), in which

- the **Summary Statement** is in the paragraph identified as (2),
- the **Background** is at the start of paragraph (3),
- the **Facts** are in the remainder of paragraph (3) and all of paragraph (4), and
- the **Outcome** is in paragraphs (5) and (6).

Because she will not be available to answer questions, Anna takes care to describe the situation clearly:

1 She knows that a subject line must be informative; it must tell what the message is about and stress its importance to the reader. If Anna had simply written "Transcript of Telephone Call from R. Walton," she would not have captured Jim Perchanski's attention nearly as sharply.

2 This brief summary gets right to the point by immediately telling Jim Perchanski in general terms what he most needs to know:

Why the message was written.
What happened.
What action has been taken.
What action he has to take.

3 In this paragraph Anna tells what she knows about the accident and its effects. That serves as background to the important facts that follow.

4 Anna knows that Jim Perchanski must act quickly to ship the replacement equipment, so she uses a list as an attention-getter: if Anna had described the items in a paragraph, they would not have been nearly as noticeable:

> He will need replacements for an HK7741 Spectrum Analyzer, a Vancourt 23R Calibrator, and 24 glass vials, each 300 mm long x 50 mm dia. He wants you to ship these items air express to the Remick Airlines terminal at Montrose, and to mark them...

Accuracy of information is essential in report writing

5 Instructions must be explicit, otherwise the equipment and Bob Walton may not meet in Montrose. Anna has identified specific days, and the date, to make sure that no misunderstandings occur. To state "tomorrow" or "the day after tomorrow" would be simple but might cause Jim Perchanski to assume a wrong date, since he will be reading the report one day later than it was written.

6 In this brief closing paragraph Anna indicates what further action is being taken and where Jim can contact Bob Walton if he needs more information.

H. L. WINMAN AND ASSOCIATES

INTER-OFFICE MEMORANDUM

To:	Jim Perchanski	From:	Bob Walton
Date:	September 17, 2003	Subject:	Report of Traffic Accident at Hadashville, Ohio

Pete Crandell and I were involved in a multiple-vehicle accident on September 16, which resulted in injuries to Pete, damage to our panel van and some equipment, and a two-day delay in our inspection of the Sledgers Control project. **A**

The accident occurred at 5:15 p.m. on Highway 1, about 2 miles west of Hadashville. We were traveling east in company panel van TLA 711, on our way to site RJ-17 at Tangwell. Pete was driving and we were approaching the intersection with Highway 459. **B**

Details of other people involved, and their vehicles, belong in the Background, not the Event

Other vehicles involved in the accident were:

- Toyota Tercel, license 881 FLM, driven by D. Varlick
- Ford truck, license TRB 851, driven by F. Zabetts
- Pontiac Grand Am, license 372 HEK, driven by K. Schmitt.

Positions of the vehicles and our panel van immediately before the accident are shown on the attached sketch.

As the Toyota attempted a right turn into Highway 459 it skidded into the Ford truck, which was standing at the intersection waiting to enter Highway 1. The impact caused the Toyota's rear end to swing into our lane, where Pete could not prevent our van from colliding with it. This in turn caused the van to slide broadside into the westbound lane, where the Pontiac approaching from the opposite direction collided with its left side. **C**

Pete was taken to Hadashville General Hospital with a broken left knee and a suspected concussion; he will be there for several days. The panel van was extensively damaged and was towed to Art's Autobody, 1330 Kirby Street, Hadashville. As some of our equipment was also damaged or shaken out of calibration, I telephoned Anna King on Tuesday evening and requested replacements (she has prepared a list for you). **D**

I have rented a replacement van from Budget, and have informed the duty engineer at site RJ-17 that my inspection of the Sledgers Control project will start on Friday, September 19, two days later than planned.

Bob

Figure 4-4 A first-person incident report.

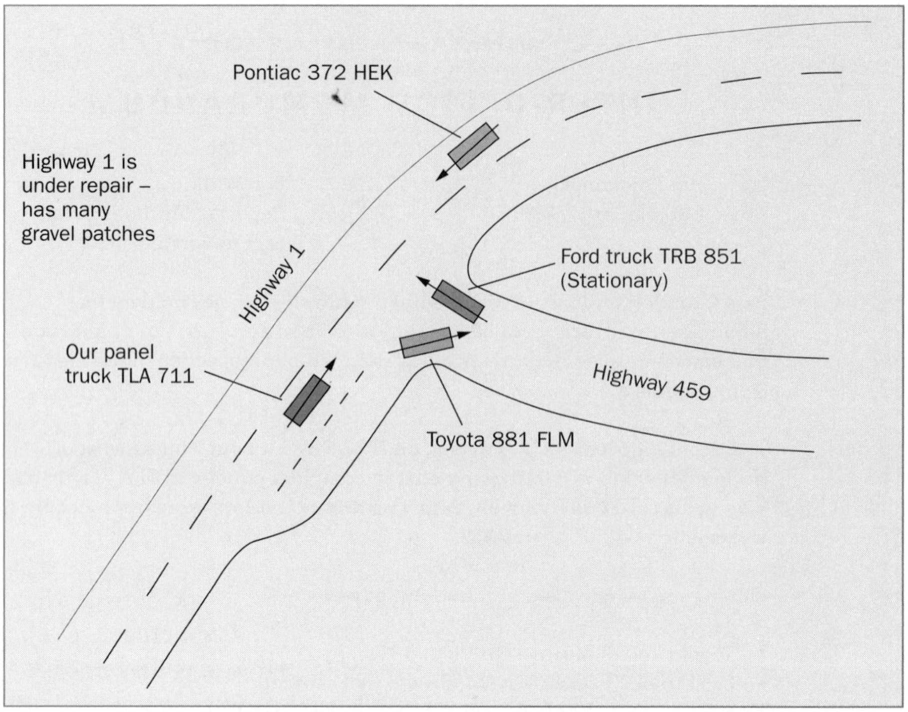

Figure 4-5 Attachment to Bob Walton's report (Figure 4-4).

A sketch helps a reader visualize the situation; many words would be needed for a written description of the same subject

When Jim walks in on Wednesday morning, he will know immediately what has happened and what action he has to take. He does not need to ask questions because he has been placed fully in the picture.

Where Anna King's report (Figure 4-3) expects reader action, Bob Walton's report (Figure 4-4) does not

Bob Walton also used the report writer's pyramid when he subsequently wrote to his supervisor from Hunter's Motel in Hadashville, to describe the accident and its effects. His report is in Figure 4-4 (page 71). Its focus and emphasis differ from those in Anna's earlier report, but it is still an incident report with the following parts:

A This is his **Summary**: it takes a main piece of information from each compartment that follows.

B This is the **Background**. By clearly describing the situation (*who? where? why? when?*) Bob helps Jim more easily understand what happened. Notice how he

A well-developed Background results in a succinct description of the Event

- establishes where they were, how they happened to be there, in what direction they were traveling, and who else was involved;
- itemizes vehicles, license numbers, and drivers' names in an easy-to-read list; and
- mentions that he is enclosing a sketch (Figure 4-5), so Jim can look at it *before* he reads on.

C Because his background information is complete, Bob's **Facts** can be concise. He simply provides a chronological description of what happened from the time the Toyota started to slide until all vehicles stopped moving.

D In the **Outcome** Bob describes the results of the accident (injuries, damage) and what he has done since (rented a van, requested replacement equipment). He closes on a strong note: he describes what is being done about the project, which was his reason for passing through Hadashville.

A test for an effective report: Does the reader have to ask questions? (No questions = a good report.)

Bob knows his role is to be an informative but objective (unbiased) reporter. No doubt he has an opinion of who is at fault, but to state it would have injected subjectivity into his report.

Trip Report

When you work offsite, you need to report your activities.

Whenever you are involved in an activity or perform work outside your normal working conditions (visiting a client, attending a conference, working remotely) you will be expected to keep your supervisor, manager, or co-workers informed of your activities. You will need to write a short report describing what you did. This is called a Trip Report.

You may have been absent only a few hours, inspecting cracks in a local water reservoir; you may have spent several days installing and testing a prototype pump at a power station in a nearby community; or you may have been far away for two months, overhauling communications equipment at a remote defense site.

You'll find that a trip report will have components of the other types of reports. For example, if you are sent to a remote site to conduct an inspection, your report will be a trip report/inspection report. Or if you have been involved in an accident, while onsite, your report will be a trip report/incident report.

Regardless of the length and complexity of your assignment, you will have to remember and transcribe many details into a logical, coherent, and factual report. Carry a pocket notebook to help you jot down daily occurrences. Without such a record to rely on, you may write a disorganized report that omits many details and emphasizes the wrong parts of the project. The simplest way to write a trip report is to answer the four questions shown in Figure 4-6, which, like all reports in this chapter, is a modification of the basic writing plan in Figure 4-1.

Short Trip Reports

Short trip reports do not need headings. A brief narrative following the Summary-Background-Facts-Outcome pattern carries the story:

Summary	A prototype automatic alarm has been installed at site RJ-17 for a one-month evaluation by the Roper Corporation.
Background	Dave Makepiece and I visited the site from January 15 to 17.
Facts	We completed the installation without difficulty, following installation instruction W27 throughout, and encountered no major problems. However, we omitted step 33, which called for connections to the remote control panel, because the panel has been permanently disconnected.
Outcome	The alarm will be removed by M. Tutanne on February 26, when he visits the site to discuss summer survey plans.

In practice, the very short Background can probably be combined with either the Summary or the Facts to form a single paragraph.

Longer Trip Reports

Long trip reports require headings to help their readers identify each compartment. Typical headings might be:

You also need to include any problems you experienced.

- **Summary.**
- **Assignment Details** (Background).
- **Work Accomplished** and **Problems Encountered** (Facts; best treated as two separate headings).
- **Suggested Follow-up** or **Follow-up Action Required** (Outcome Action).

Anna King's instructions to H. L. Winman and Associates' engineers (see Figure 4-7 on pages 75–76) tell them how to organize their longer trip reports, describes the information that would normally follow each heading, and includes excerpts and sample paragraphs.

Except for the Outcome section, trip reports should be written entirely in the past tense.

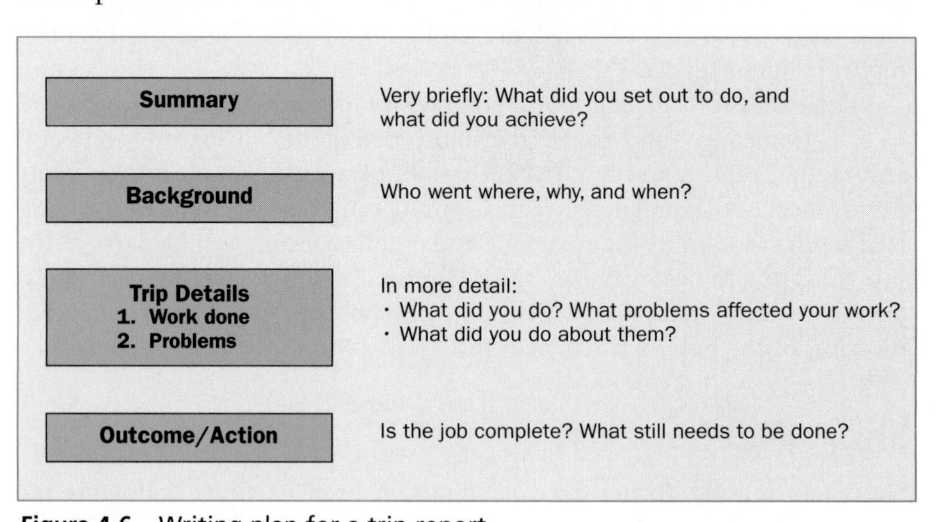

Figure 4-6 Writing plan for a trip report.

H. L. WINMAN AND ASSOCIATES

475 Reston Avenue, Cleveland, OH 44104

Guidelines for Writing Long Trip Reports

Use a standard format for long trip reports. These instructions suggest how you can organize your information under five main headings: ***Summary; Assignment Details; Work Accomplished; Problems Encountered;*** **and *Follow-up Action.*** You may omit the headings from very short reports.

This may look like a model report...

Summary

Make your summary a short opening statement that says what was and was not accomplished. Highlight any significant outcomes.

Assignment Details

State the purpose of the trip and include any other information the reader may want to know. If the information is lengthy, use subheadings such as

- Purpose of Trip
- Background
- Project No./Authority
- Personnel Involved
- Person(s) Contacted
- Date(s) of Field Trip

Work Accomplished

Describe the work you did. Present it in chronological order unless more than one project is involved, in which case describe each project separately. Keep it short: don't describe at great length routine work that ran smoothly. Whenever possible, refer to your work instruction or specification, and attach a copy to your report:

...but really it's an instruction

> The manual control was disconnected as described in steps 6 to 13 of modification instruction MI1403, enclosed as attachment 1.

Go into more detail only if you encountered difficulty, or if work was necessary beyond that anticipated by the job specification:

> At the request of the site maintenance staff, we installed a manual control in the power house as a temporary replacement for a defective GG20 control. I left the parts removed from the panel, together with instructions for returning the panel to its original configuration, with Frank Mason, the senior power house engineer.

If parts of the assignment could not be completed, identify them and explain why the work was not done:

1

Figure 4-7 Anna King's instructions for writing long trip reports.

We had to omit Test No. 46 because the RamSort equipment had been removed.

Problems Encountered

Describe problems in detail. Knowledge of problems you encountered and how you overcame them can be invaluable to the engineering or operating departments, which may be able to prevent similar problems elsewhere.

Avoid statements that do not tell the reader what the problem was or how it was overcome. For example:

> Considerable time was spent in trying to mount the miniature control panel. Only by fabricating extra parts were we able to complete step 17.

If this information is to be used by the engineering or operating department, it must be more specific:

> We spent three hours trying to mount the miniature control panel according to the instructions in step 17. Because the main frame had additional equipment mounted on it, which prevented us from using most of the parts supplied, we had to fabricate a small sheetmetal extension to the main frame and mount it with the miniature panel, as shown in attachment 2.

Follow-up Action

Tie up any loose ends here. If any work has not been completed, draw attention to it even though you may already have mentioned it under "Work Accomplished." Identify what needs to be done, if possible indicate how and when it should be done, and say whose responsibility it now becomes:

The Outcome looks forward, says "who will do what"

> The manual control mounted as a temporary replacement in the power house is to be removed when a new GG20 control panel is received on site. This will be done by Frank Mason, with whom I left instructions for doing the work.

In some cases you may direct follow-up action to someone else in your own or another department:

> The manual control is to be removed from the power house by R. Walton, who will visit the site on May 12.

If your report is very long, insert subheadings and use a paragraph numbering system to increase its readability.

Anna King
January 20, 2004

2

Progress or Status Reports

Progress or status reports keep management aware of what its project groups are doing. Even for a short-term project, management wants to hear how the project is progressing, especially if problems are affecting its schedule. Because delays can have a marked effect on costs, management needs to know about them early.

There are three major types of progress or status reports:

1. The **occasional progress report** is written to keep management informed about a project's progress. It usually isn't expected at a mutually agreed upon time. You decide when it is necessary to write one.
2. The **periodic progress report** is written as a regular update on project status. Some reporting periods may require a longer report than others. It depends on what is happening.
3. The **personal progress report** is usually written once a month and is used to keep management informed about your personal activities.

Progress Reports
www.io.com/~hcexres/tcm1603/acchtml/progrep.html
This document deals with the purpose, timing, format, and organization of progress reports.

Occasional Progress Report

Jack Binscarth, one of Macro Engineering Inc.'s technologists in Phoenix, has been assigned to Cantor Petroleums north of Lansing, Michigan, to analyze oil samples. The job is expected to take five weeks, but problems have developed that have prevented Jack from completing the work on time. To let his manager know what is happening, he writes the brief progress report in Figure 4-8, adapting the standard Summary-Background-Facts-Outcome arrangement into the five compartments shown in Figure 4-9 (see page 80). This type of progress report is similar to a field trip report.

Periodic Progress Report

If a project is to continue for several months, normally management will specify that progress reports be submitted at regular intervals.

A periodic progress report may be no more than a one-paragraph statement describing the progress of a simple design task, or it may be a multipage document covering many facets of a large construction project. (There are also form-type progress reports, which call for simple entries of quantities consumed, amount of concrete poured, and so on, with cryptic comments.) Regardless of its size, the report should answer the four main questions the reader is likely to ask:

Anticipate your reader's curiosity

1. Will your project be completed on schedule?
2. What progress have you made?

600 Deepdale Drive, Phoenix, AZ 85007

FROM:	Jack Binscarth (at Cantor Petroleums)	DATE:	October 14, 2003
TO:	Fred Stokes Chief Engineer, Head Office	SUBJECT:	Delay in Analysis of Oil Samples

Summary

My analysis of oil samples for Cantor Petroleums has been delayed by problems at the refinery. I now expect to complete the project on October 25, nine days later than planned.

Progress

The first problem occurred on September 23, when a strike of refinery personnel set the project back four working days. I had hoped to recover all of this lost time by working a partial overtime schedule, but failure of the refinery's spectrophotometer on October 13 again stopped my work. To date, I have analyzed 111 samples and have 21 more to do.

Progress reports follow a past-present-future arrangement

Situation Now

The spectrophotometer is being repaired by the manufacturer, who has promised to return it to the refinery on October 19. Today I informed the refinery manager of the delay, and he has agreed to an increase in the project price to offset the additional time. He will call you about this.

Future Plans

Providing there are no further delays I will analyze the remaining samples between October 20 and 24, and then submit my report to the client the following morning. This means I should be back in the office on October 26.

Jack

Figure 4-8 An occasional progress report. (Because the report is short, the Background compartment has been omitted.)

3. Have you had any problems?
4. What are your plans/expectations?

To answer these questions, a periodic progress report can use the standard Summary-Background-Facts-Outcome arrangement:

Summary	A brief overview of the project schedule, progress made, and plans (*answers the first question*).
Background	The situation at the start of the report period.
Facts	Progress made (*answers the second question*) and problems encountered (*answers the third question*).
Outcome	Plans/expectations for the next period (*answers the last question*).

Figure 4-10 on page 81 shows how survey crew chief Pat Fraser used these four compartments to write an effective progress report (the numbers below are keyed to parts of the report):

1. The **Summary** tells civil engineering coordinator Karen Woodford how closely the survey project is adhering to schedule, and predicts future progress. This is the information she wants to read first.

2. The **Background** section reminds Karen of the situation at the end of the previous reporting period and predicts what Pat expected to accomplish during this period. Background should always be stated briefly.

3. The **Facts** (or Discussion) section is broken into two parts:
   - Work done during the period (3a)
   - Problems affecting the project (3b)

The past-present-future structure is equally apparent here

Pat Fraser opens each paragraph of this compartment with a topic sentence (a summary statement) that states the main point of the paragraph in general terms:

- Dry, clear weather...enabled us to progress faster than anticipated.
- The electrical fault in the EDM equipment...recurred on May 23.
- I have had difficulty hiring reliable people to clear brush along the route.

Now the writing plan has extended beyond the four basic compartments

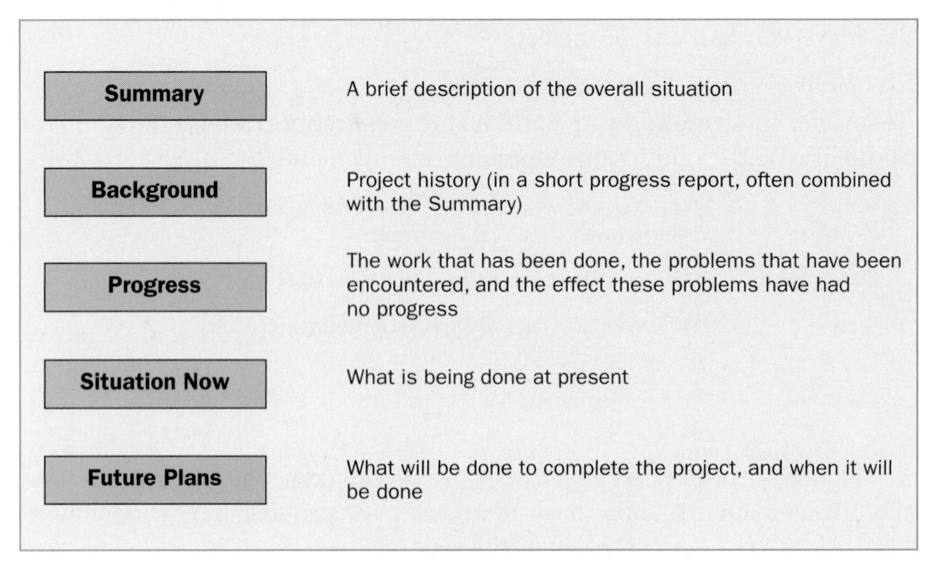

Figure 4-9 Writing plan for an occasional progress report.

Pat then describes in more detail what happened, using *facts* (exact dates and position numbers, for example) to support each topic sentence. To prevent the report from becoming too long, Pat attaches the survey results to it and simply refers to them in the narrative. (Because of their length, they have not been printed with Figure 4-10.)

4. In the **Outcome** paragraph Pat tells Karen what the crew expects to accomplish during the forthcoming period, and even suggests when they might eventually get back on schedule. This final statement clearly supports the opening paragraph, and brings the report to a logical close.

Other factors you should consider when writing periodic reports are:

Heading titles parallel the pyramid's parts

- If a progress report is long, use headings such as these to help readers *see* your organization:

  Adherence to Schedule (This is your Summary).
  Progress During Period (These are your Facts; state the Background information at the front of the Progress section.)
  Problems Encountered
  Projection for Next Period (This is the Outcome.)

- For lengthy progress or problems sections, start with a summarizing statement describing general progress, then write several subparagraphs each giving details of a particular aspect of the project.

H. L. WINMAN AND ASSOCIATES

INTER-OFFICE MEMORANDUM

To:	Karen Woodford, Coordinator Civil Engineering	Subject:	Progress Report No. 4— Allardyce Survey Report
From:	Pat Fraser, Survey Crew Chief	Date:	May 31, 2004

The Allardyce Route survey has progressed well during the May 16 to 31 period. The survey crew has regained two days, and now is only four days behind schedule. We expect to be back on schedule by June 30.

1

The Summary sums up key features from the report's body

Project plan AR-51 shows we should have surveyed positions 30 to 34 during this period. But, as stated in my May 15 report, we were six days behind schedule at the end of the previous period, having surveyed only as far as position 28. Consequently, we expected to survey only to position 32 by May 31.

2

Dry, clear weather from May 18 to 23 enabled us to progress faster than anticipated. We reached position 31 on May 23, carried out a terrain analysis for the Catherine Lake diversion scheme on May 24 and 25, resumed surveying on May 26, and reached position 32 at 09:00 a.m. on May 29, two days earlier than expected. At end of work on May 31, we were just 300 yards short of position 33. Survey results are attached.

3a

Two problems affected the project during this period:

1. The electrical fault in the EDM equipment, which delayed us several times early in the project, recurred on May 23. I had the unit repaired at Fort Wilson on May 24 and 25, while we conducted the terrain analysis, and it has since worked satisfactorily.
2. I have had difficulty hiring reliable local people to clear brush along the route. Most remain with us for only a few days and then quit, and I have had to waste time hiring replacements. This problem will continue until mid-June, when the college students we interviewed in March will join the crew.

3b

The "Present Work" compartment may be omitted from a progress report

We plan to advance to position 37 by June 15, which should place us only two days behind schedule. If we can maintain the same pace, I hope to make up the remaining two days during the June 16 to 30 period.

4

Pat Fraser

Figure 4-10 A periodic progress report.

For example:

4. Interior construction work has progressed rapidly but exterior work has been hampered by heavy rain.

   4.1 In the east wing, we erected all partitions, laid 80% of the floor tiles, and installed 20% of the light fixtures.

   4.2 In the west wing, we laid all remaining floor tiles, installed all light fixtures, bolted down 16 of the 24 benches and connected them to the water supply and drains.

   4.3 We started landscaping on September 16, but had to abandon the work from September 18 to 23 when heavy rains turned the soil into a quagmire. By the end of the month we had completed only the outer areas of the parking lot.

- Be as brief as possible when describing routine work. Quote specifics rather than generalizations, and place lengthy details in an attachment. If, for example, you are reporting an extensive analysis, in your progress section you might write:

  We analyzed 142 samples, 88 (62%) of which met specifications. Results of our analyses are shown in attachment 1.

  Attachment 1 would contain several pages of tabular data (numbers, quantities, measurements), which, if included as part of the report narrative, would inhibit reading continuity.

- Describe problems, difficulties, and unusual circumstances in depth. State clearly what the problem was, how it affected your project, what measures you took to overcome it, and whether the remedial measures were successful. For example:

Each problem description is shaped like a miniature pyramid

Topic Sentence	Juvenile vandalism has proved to be a petty but time-consuming problem. On September 3 (Labor Day) youths scaled the fence around the materials compound and stole about $300 worth of
Facts	building supplies. On September 16 they started up a front end loader, drove it into the excavation, then got it stuck in the mud and burned out the clutch. To prevent a recurrence, from
Outcome	September 18 I have doubled the night watch and have had the site policed by a patrol dog. There have been no further attempts at vandalism.

- Forewarn management of any situation that, although it may not yet affect your project, may become a future problem. Thus, management may be able to avert a costly work stoppage or equipment breakdown. Here is a typical situation:

Predict potential developments...

7.1 Unless the strike at Vulcan Steel Works ends shortly, it will soon curtail our construction program. Our present supply of reinforcing barmats will last until mid-October, after which we must find an alternative source of supply. I have researched other suppliers, but have been warned by union representatives that any attempt to obtain steel elsewhere may result in a walk-out at other plants.

Where should such an entry appear in your progress report? The best position would be at the end of the Facts (Problems) section, immediately before the Outcome.

- If your report is lengthy or comprehensive, number your paragraphs and subparagraphs (see the examples above). The paragraph numbers can help you refer to a specific part of a previous report, like this:

The possibility of a shortage of steel mentioned in para 7.1 of my September report was averted when the strike at Vulcan Steel Works ended on October 6.

...and then in a subsequent report describe the outcome

- Maintain continuity between reports. If you introduce a problem that has not been resolved in one report, then refer to it again in your next report, even though no change may have occurred or it was solved only a day later (see the example in the previous paragraph). You must never simply drop a problem because it no longer applies.
- If management expects you to include project cost information in your progress report, insert it in three places:

1. In the **Summary** (comment briefly on how closely you are adhering to projected costs).

2. In the **Progress** section (give more details of costs, and particularly cost implications of problems).

3. In the **Outcome** section (indicate future cost trends).
   Costs are usually closely linked with your adherence to schedule: the more you drop behind schedule, the more likely you'll have to report a cost overrun.

Personal Progress Report

A personal progress report serves two purposes:

(1) It keeps management informed of your monthly activities.

(2) It can be used to document your progress and help you manage your time.

Most organizations use performance evaluations as a tool to help employees grow in their careers and to help focus employee professional development. Managers or supervisors are asked to review each of their employees' performance and recognize their strengths and areas for improvement. As an employee, it is important that you are involved in this process. It is usually done once a year.

To help your manager understand what activities you have been involved in and what you have achieved, we recommend you write monthly personal progress reports. These reports will also help you with your time-management skills because you will have to plan your future activities. Set aside 30 minutes on the last day of each month and write a short report, following the writing plan in Figure 4-11. Eventually you will only need to focus on your Future Work compartment since the Future Work becomes the Present Work.

There is enough detail in Susan's report (Figure 4-12) for the manager to understand the situation but not too much detail to slow down the reading. If the manager needs more detail, he will ask Susan. However, it must have enough detail so both Susan and her manager can understand the points a year from now if they need to refer to it for her performance evaluation. She decided to use paragraphs rather than a list so there is more continuity when reading. Notice how each topic is a mini-pyramid with a topic sentence as the summary.

If you are writing regular project status reports you don't need to repeat specifics from earlier reports. Remember, this is about what *you* did, not about the project. Susan decided to break her report into topic areas so the manager understands her activities in each area. You can decide how you want to organize your report but be consistent each month.

Project Completion Report

It's mostly the Facts compartment that gets expanded and relabeled

A project completion report may be the only report evolving from a short project, or the last in a series of progress reports concerning a lengthy project. Thus the Summary-Background-Facts-Outcome arrangement shown in

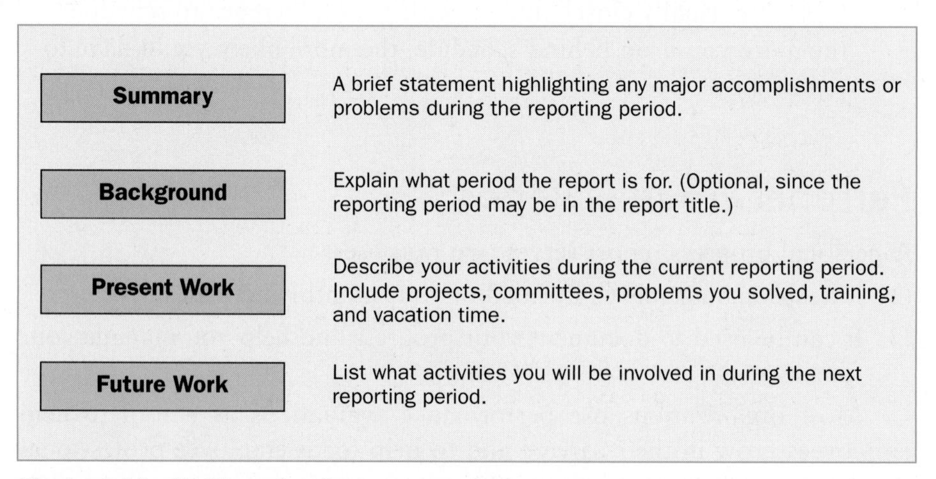

Figure 4-11 Writing plan for a personal progress report.

600 Deepdale Drive, Phoenix, AZ 85007

Monthly Progress Report for Susan Jenkins
November 1–30, 2004

Summary
Most of my efforts this month were spent diagnosing problems with the CI software. The problems we discovered will need further investigation next month. Although I spent time helping the new intern set up his workstation, it will pay off next month when he is able to help the JCL team begin focus groups for the MarTel project. I was also heavily involved in professional society activities.

Present Work
Centurion Insurance
I completed testing the CI conversion program on two browsers: Netscape and Internet Explorer. This took longer than I originally estimated because I discovered a problem with the program and had to determine if it was on our end or on the CI platform. Both browsers continue to freeze up immediately after initializing the program. I consulted with the lead technical specialist at CI and we agreed that the code must have not been passing the appropriate parameters. The project is still running two weeks behind schedule.

MarTel Corporation
I worked with the local MarTel account representative to determine who should be involved in the focus group. We determined the demographics and number of participants, the location, and the topics we need to explore.

Intern Orientation
I met several times with Dave Jankowski, the intern from City College. He joined the group this month and will be here for six months. I installed the required software, helped him access the server, and established his passwords. I went over our email guidelines with him so he understands how our team uses it.

Committee Work
I attended the E-learning Administrative Committee meeting in Lake Tahoe on November 13–14, representing the Southwestern region. We need to stay involved as a firm in what is happening at the national and international level so our computer engineering groups remain competitive.

As the Safety Council representative for our group, I attended a CPR refresher course and met with the other council members to revise our fire evacuation plans.

Future Work
Next month I plan to

- conduct a line item code review of the CI conversion program with the technical specialist to determine any bugs that may be causing the freeze problems,
- monitor the progress of the MarTel focus group,
- supervise the intern's activities,
- prepare a proposal to present a paper at the IEEE Computer Society annual conference,
- investigate adding client testimonials to our web site, and
- update all engineer resumes on the company server.

Figure 4-12 A personal progress report.

Figure 4-1 can be adhered to fairly closely, with the Facts compartment being separated into two compartments labeled **Project Highlights** and **Exceptions** (see Figure 4-13). The Exceptions section draws attention to deviations from the original project plan.

The project completion report written by Jack Binscarth at the end of his analysis of oil samples for Cantor Petroleums identifies the five writing compartments beside each part of the report. (See Figure 4-14; Jack's progress report for this project is in Figure 4-8.) Note particularly that in a short report like this it's acceptable to combine two, or sometimes more, writing compartments into a single paragraph. In Jack's project completion report, paragraph 1 contains both the **Summary** and the **Background**, and paragraph 2 contains both the **Project Highlights** and the **Exceptions**.

Inspection Report

An inspection report may be a type of trip report.

An inspection can range from a quick check of a small building to assess its suitability as a temporary storage center, to a full-scale examination of an airline's aircraft, avionics equipment, repair facilities, and maintenance methods. In both cases the inspectors will report their findings in an inspection report. The building inspector's report will be brief: it will state that the building either is or is not suitable, and give reasons why. The airline inspector's report will be lengthy: it will describe in detail the condition of every aspect of the airline's operations and list every deficiency (condition that must be corrected). In both cases the inspectors' reports can follow the Summary-Background-Facts-Outcome arrangement, as shown in Figure 4-15 on page 88.

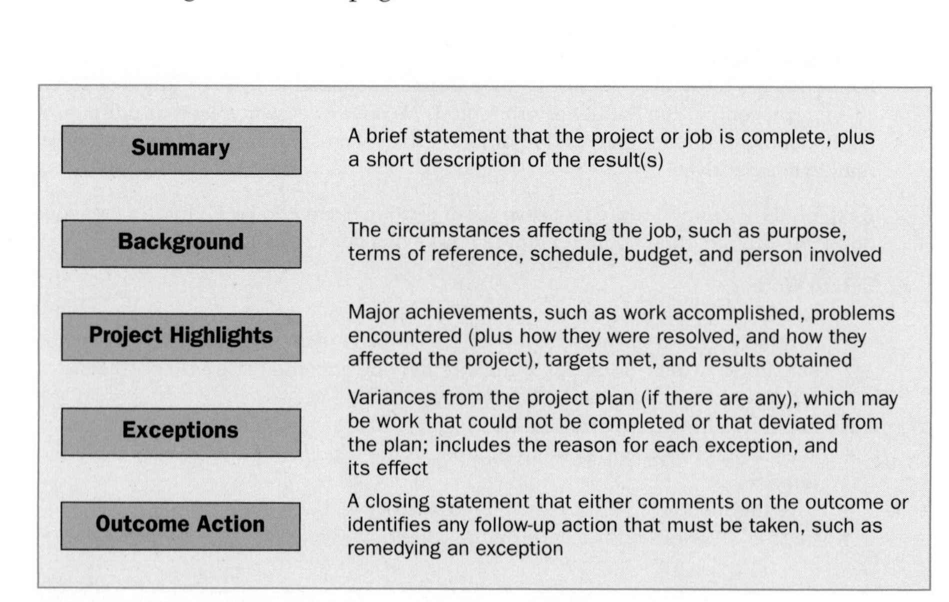

Figure 4-13 Writing plan for a project completion report.

600 Deepdale Drive, Phoenix, AZ 85007

FROM: Jack Binscarth DATE: October 25, 2003
TO: Fred Stokes SUBJECT: Finalizing Cantor Petroleums' Project

Summary Statement

I completed the analysis of oil samples for Cantor Petroleums on October 24, eight days later than planned. The work was done at the refinery, as requested in Cantor Petroleums' purchase order No. 376188 dated September 4, 2003, and was scheduled to start on September 11 and end on October 16. I was assigned to the project under work order No. 2716.

Background

Project Highlights

The work plan called for me to analyze 132 oil samples within the five-week period, but three problems caused me to overrun the schedule and complete four fewer analyses than specified. The delay was caused by a strike of refinery personnel and a faulty spectrophotometer that had to be sent out for repair and recalibration. The incomplete analyses were caused by four contaminated samples that could not be replaced in less than six weeks.

Exceptions

Outcome

Russ Dienstadt, the refinery manager, agreed to a cost overrun and has corresponded with you separately on this subject. He also agreed that it would be uneconomical for me to return to analyze replacements for the four contaminated samples. When I delivered the 128 analyses to him on October 24, he accepted the project as being complete.

Jack

Figure 4-14 A project completion report.

It's better to describe conditions and deficiencies in two separate compartments

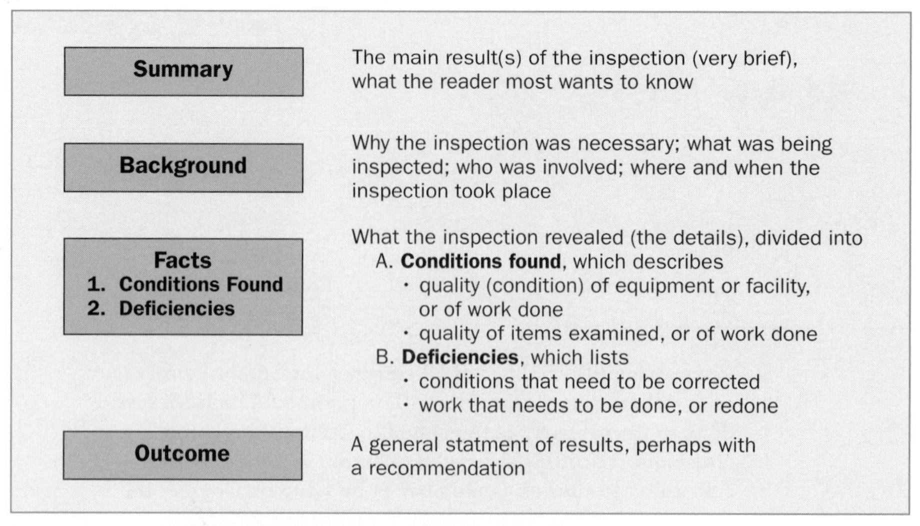

Figure 4-15 Writing plan for an inspection report.

Kevin Doherty's building inspection report in Figure 4-16 shows how these compartments helped him shape his report into a logical, easy-to-follow document. Note particularly how:

- His **Summary** (1) tells the Production Manager the one thing he most wants to know: can they use the building?
- The **Background** (2) describes who went where, why, and when.
- Kevin has opened the **Conditions** section (3) with a summarizing general statement, and then supported it with facts (3A). Because they are not in any particular order, he used bullets for the itemized list.
- He has presented the **Deficiencies** (3B) as a briefly stated list, which makes it easy to identify what has to be done, and has used active verbs to demonstrate that the actions *must* be performed. Because he may need to refer to these items later, he used numbers for this list.
- The recommendation in his **Outcome** (4) supports his summary.

For a short inspection report like this, Kevin was correct in presenting all the Conditions first and then listing all the Deficiencies. But such an arrangement could become cumbersome for a long report that covers many items. For example, if Fran Hartley followed this sequence for an inspection at Remick Airlines, the organization of the Facts section would look like this:

A plan for a short inspection report

A. **Conditions Found:**
   1. Electrical Shop
   2. Avionics Calibration Center
   3. Flammable Materials Storage
   (etc...)

B. Deficiencies:

1. Electrical Shop
2. Avionics Calibration Center
3. Flammable Materials Storage

 (etc...)

The more departments Fran inspects, the longer the report becomes and the further apart each department's Conditions and Deficiencies sections grow.

To overcome this difficulty, Fran should treat each department as a *separate* inspection and reorganize the report so that for each department the Deficiencies section immediately follows the Conditions section. The organization of the whole report would then become:

The plan for a longer, more detailed inspection report

Summary

Background

Facts:

1. Electrical Shop
   - A. Conditions Found
   - B. Deficiencies
2. Avionics Calibration Center
   - A. Conditions Found
   - B. Deficiencies
3. Flammable Materials Storage
   - A. Conditions Found
   - B. Deficiencies

 (etc...)

Outcome:

Conclusions

Recommendations

Laboratory Report

Lab reports are written frequently in colleges, less often in industry

There are two kinds of laboratory reports: those written in industry to document laboratory research or tests on materials or equipment, and those written in academic institutions to record laboratory tests performed by students. The former are generally known as "Test Reports" or "Laboratory Reports;" those written by students are simply called "Lab Reports."

Industrial laboratory reports can describe a wide range of topics, from testing a piece of metal to determine its tensile strength, through analyzing a sample of soil (a drill core) to identify its composition, to checking a microwave oven to assess whether it emits radiation. Academic lab reports can also describe many topics, but their purpose is different since they

MACRO
ENGINEERING INC.
600 Deepdale Drive, Phoenix, AZ 85007

FROM:	Kevin Doherty	DATE:	January 8, 2004
TO:	Hugh Smithson Production Manager	SUBJECT:	Inspection of Carter Building

1 The Carter Building at the corner of River Avenue and 39th Street will make a suitable storage and assembly center for the Dennison contract.

2 Christine Lamont and I inspected the Carter Building on January 6 to assess its suitability both for storage and as a work area for 20 assemblers for 15 months. We were accompanied by Ken Wiens of Wilshire Properties.

3 We found the interior of the building to be spacious and to have good facilities, but to be unsightly. Our inspection showed that:

3a
- There are 4200 ft^2 of usable floor space (see attached building plan, supplied by Mr. Wiens); we need 2400 ft^2 for the project.
- There are two offices, each 16 m^2, and a large unimpeded space ideal for partitioning into a storage area and four work stations.
- The building is structurally sound and dry, but it is very dirty and smells strongly (the previous tenant was a fertilizer distributor).
- There are numerous power outlets, newly installed with heavy-duty circuits, and the building has excellent overhead lighting.
- Several walls are damaged and many contain obnoxious graffiti.
- There is a new loading ramp on the north side of the building, suitable for semitrailers.
- Washroom facilities are adequate for up to 30 people, but one toilet and two washbasins are broken.

Before we rent the building, the rental agency will have to

3b
1. clean it thoroughly,
2. repair damaged walls, partitions, and toilet facilities, and
3. redecorate the interior.

Ken Wiens said his firm would be willing to do this.

4 I recommend we rent the Carter Building from Wilshire Properties, with the provision that the deficiencies listed above be corrected before we move in.

Figure 4-16 A short informal inspection report.

describe tests that are usually intended to help students learn something or prove a theory rather than produce a result for a client.

Laboratory reports generally conform to a standard pattern, although emphasis differs depending on the purpose of the report and how its results will be used. Readers of industrial laboratory or test reports are usually more interested in results ("Is the enclosed sample of steel safe to use for construction of microwave towers that will be exposed to temperatures as low as –40°C in a North Dakota winter?" a client may ask), than in how a test was carried out. Readers of academic lab reports are usually professors and instructors, who are more likely to be interested in thoroughly documented details, from which they can assess the student report writer's understanding of the subject and what the test proved.

A laboratory report comprises several readily identifiable compartments, each usually preceded by a heading. These compartments are described briefly below.

Engineering Lab Reports
www.engr.udayton.edu/Special/Writing/labrep/default.htm
The University of Dayton maintains this detailed site about writing engineering lab reports.

A generic writing plan for a lab report

Part	*Section Title*	*Contents*
Summary	**Summary**	A very brief statement of the purpose of the tests, the main findings, and what can be interpreted from them. (In short laboratory reports, the summary can be combined with the next compartment.)
Background	**Objective**	A more detailed description of why tests were performed, on whose authority they were conducted, and what they were expected to achieve or prove.
Facts	**Equipment Setup**	*There are four parts here:* A description of the test setup, plus a list of equipment and materials used. A drawing of the test hook-up may be inserted here. (If a series of tests is being performed, with a different equipment setup for each test, then a separate equipment description, materials list, and illustration should be inserted immediately before each test description.)

Part	*Section Title*	*Contents*
	Test Method	A detailed, step-by-step explanation of the tests. In industrial laboratory reports the depth of explanation depends on the reader's needs: if a reader is nontechnical and likely to be interested only in results, then the test description can be condensed. For lab reports written at a college or university, however, students are expected to provide a thorough description of their method.
	Test Results	Usually a brief statement of the test results or the findings evolving from the tests.
	Analysis (or Interpretation)	A detailed discussion of the results or findings, their implications, and what can be interpreted from them. (The analysis section is particularly important in academic lab reports.)
Outcome	**Conclusions**	A brief summing-up, which shows how the test results, findings, and analysis meet the objective(s) established at the start of the report.
Backup	**Attachments**	These are pages of supporting data such as test measurements derived during the tests, or documentation, such as specifications, procedures, instructions, and drawings, which would interrupt reading continuity if placed in the report narrative (i.e. in the Test Method section).

In practice, the writing plan is adapted to suit the industry and the circumstances

The compartments described here are those most likely to be used for either an industrial laboratory report or a college/university lab report. In practice, however, emphasis and labeling of the compartments will differ, depending on the requirements of the organization employing the report writer or, in an academic setting, the professor or instructor who will evaluate the report.

ASSIGNMENTS

Project 4.1: Checking an Insurance Claim

Assume that you are employed by the local branch of H. L. Winman and Associates. When you arrive at work this morning, branch manager Vern Rogers calls you into his office.

"We've had a call from Hugh Smithson in our Buffalo branch," he says. "Twelve cartons of special instruments they shipped yesterday were in a truck that rolled off the highway 8 miles east of Rochester, New York. The insurance company wants someone to look over the damage with one of their adjusters, to confirm how much can be repaired or salvaged."

In Rochester you meet Noella Redovich of Milltown Insurance Company. She takes you to a warehouse where the smashed crates tell their own story of the violence of the accident. Very few of the delicate instruments could have survived such an impact.

You examine the crates, which are a jumble of broken glass, tangled wire, and chipped and splintered instrument cases. As you check each container, the adjuster notes the numbers in her book: 10, 4, 12, 11, 6, 3, 1, 9, 8, and 2 are totally beyond repair and obviously have no salvage value. Crate number 5, surprisingly, is hardly marked: somehow it must have been cushioned. You examine its contents.

"This one seems okay," you say.

Noella adds up the totals: "Not very good for us," she says. "Ten out of eleven means a heavy claim."

"Twelve," you say. "There were twelve crates."

Noella checks her figures and you recheck and count the crates. There are 10 smashed ones and one good one. "One is missing," you say. "Number 7."

Noella suggests it might have been stolen before the accident occurred. "The police were on the scene immediately. There would not have been time after the accident."

When you return to your office, Vern Rogers asks you to write a report and fax or email it immediately to Hugh Smithson. Here is additional information you may need for your report:

- The shipping company was Merryhew Van Lines Albany.
- The waybill number was C2719.
- The 12 crates were being shipped to Melwood Test Labs, Syracuse.
- Milltown Insurance Company's local address is Room 14A, 22 Western Avenue, Rochester, New York.
- The crates are being held at C and J Storage Inc., 63 Crane Street, Rochester.

- You reported the missing crate to the police in Rochester at 2:25 p.m., immediately after completing your inspection.

Write the trip/inspection report.

Project 4.2: Accident at Cormorant Dam

You are an engineering technologist employed in the local branch of H. L. Winman and Associates. Currently you are supervising installation work at a remote construction project at Cormorant Dam.

The day before you left for the construction site, your branch manager (Vern Rogers) called you into his office. "I'd like you to meet Harry Vincent," he said, and introduced you to a tall, gray-haired man. "Harry is with the Environmental Protection Agency (EPA) and he wants you to take some air pollution readings while you're at Cormorant Dam."

Mr. Vincent opened a wooden box about 14 × 10 × 10 inches, with a leather shoulder strap attached to it. In the box, embedded in foam rubber, you could see a battery-powered instrument. "It's a Vancourt MK 7 Air Sampler," he explained, "and it's very delicate. Don't check it with your luggage when you fly to Cormorant Dam. Always carry it with you."

For the next hour Mr. Vincent demonstrated how to use the air sampler, and made you practice with it until he was confident you could take the twice-daily measurements he wanted.

Now it is 10 days later and you have just finished taking the late-afternoon air sample measurements. You are standing on a small platform halfway up some construction framework at Cormorant Dam, and are replacing the air sampler in its box.

A painful cause for writing an incident report

Suddenly there is a shout from above, followed immediately by two sharp blows, one on your hardhat and the other on your shoulder. You glimpse a 3-foot length of 4-inches square construction lumber tumble past you followed by the air sampler box, which has been knocked out of your hand. The box turns end over end until it crashes to the ground. When you retrieve it the box is misshapen and splintered and the air sampler inside it is twisted. Also, your arm is throbbing badly and you cannot grip anything. An examination at the medical center shows you have a dislocated shoulder, and now your arm is supported by a sling. (Fortunately, it is not your writing hand.)

Part 1

Write an incident report to Harry Vincent of the Environmental Protection Agency. Tell him

- what has happened,
- that you have shipped the damaged air sampler to him on Remick Airlines Flight 751, for him to pick up at your city's airport (you enclose the airline's receipt with your report), and
- that if he wants you to continue taking air pollution measurements, he will have to send you another air sampler.

Harry Vincent's title is Regional Inspector and his address is Environmental Protection Agency, Suite 306, 444 Waltham Avenue of your city.

Part 2

Write a memo-form incident report to Vern Rogers. You can mention that you were absent from the construction site for 24 hours, but that otherwise the incident has not affected your supervision work.

Project 4.3: Theft at Whiteshell Lake

You are the team leader of a four-person inspection crew en route to a remote site 508 miles from your office, where construction of a nuclear power generating station is in progress. You are traveling in a panel van and after 376 miles you and the crew agree to stop for the night. At 8:05 p.m. you pull into the Tow Path Inn, a small motel beside the road that skirts around Whiteshell Lake.

The following morning you are having breakfast in the motel's tiny dining room when Fran Pedersen, one of the crew, goes out to the van to fetch the road map. She returns almost immediately and gasps, "The van has been broken into!"

The four of you scramble out to the parking lot and can see right away that the window on the front passenger's door has been smashed.

"They were after the radio," Shawn Mahler observes, pointing to a gaping hole in the dash. "Check if anything else is missing," you suggest. Already you are expecting the worst, but to your surprise find that only two other items have been taken, one inconsequential, one important: about $6.00 from a tray in the dash (parking meter quarters), and a video camera and videotapes from a storage box in the rear of the van.

You try telephoning your office, but it is too early and no one answers. The motel has a fax machine, so you write a memo to your manager and send it by fax. In it you describe what has happened and ask for a replacement video camera to be sent to you. Here is some additional information you draw on to write your report:

Expensive equipment, for which you need replacements

- You are driving company panel van license number JCP 392; it is a Chevy.
- Your trip was authorized by Travel Order N-704, dated one week ago, and was signed by your manager.
- The power generating station is being constructed beside the Mooswa River, 18 miles north of the small town of Freehampton.
- The Tow Path Inn is 3.5 miles west of Clearwater Village, on highway A1136.
- The third member of your crew is Servi Dashi.
- The video camera is a Nabuchi TX350 "Portacam." You rented it

from Meadows Electronics at 2120 Grassmere Road of your city. Its serial number is 21784B.

- Your manager's name is M. B. Corrigan.
- The purpose of the video camera is to record construction progress visually. The videotapes will be edited and then shown at the Power Authority Directors' Meeting scheduled for the 15th of next month.
- You telephoned the police at Clearwater Village to report the break-in and theft. They ask you to drop in and make your report in person. You plan to do this at the start of your drive to the construction site (which will be *after* you have sent your fax).
- In your report you ask your manager to ship you a replacement video camera by Greyhound bus the day after tomorrow. One bus a day passes through Freehampton, but it stops only on request. (You will drive to Freehampton to meet the bus, and will telephone your manager tomorrow to check that the video camera *will* be on that particular bus.)
- You use today's date as the date of your report.

Part 1

A single incident can evolve into several reports or letters

Write the incident report to M. B. Corrigan. Prepare it as a memo with a fax cover sheet.

Part 2

Write a letter to Meadows Electronics, to explain the loss of their video camera. You may mention that M. B. Corrigan will be contacting them about insurance coverage.

Project 4.4: Effect of a Power Outage

H. L. Winman and Associates has been carrying out a series of extreme cold and heat tests on electronic and mechanical switches for Terrapin Control Systems of Palo Alto, California. The tests have been running for four months and will last another two months. The schedule is tight because of initial problems with measuring equipment, which delayed the start by nine days and used up any spare time the project had available.

Currently, you are testing the switches for continuous periods of from 8 to 14 hours. The tests have two parts:

1. For the first 6 hours each day you increase or decrease temperature in 2°C increments until a predetermined high or low temperature is reached. At each 2° increment you test the switches and record how they perform.
2. For the remaining 2 to 8 hours you bake or deep-freeze the switches at the preselected temperature. No monitoring is necessary during

this period (although the switches are tested at room temperature the following day).

To avoid having a technician stay throughout part 2, which on some evenings runs as late as 12:30 a.m., you have installed electrical timers in the circuits of the oven and freezer chamber. The timers are set to switch off at the end of the prescribed bake and deep-freeze periods.

This morning when you remove batches 64H and 66C from the oven and freezer chambers you notice that, instead of being close to room temperature, the oven is still hot and the freezer is still cold. You check the electrical timers, and both are "off." Then you notice that the electric clock on the lab wall reads only 3:39; your wristwatch reads 9:03—a different of 5 hours and 24 minutes. You telephone the local power company.

"Was there a power cut last night?" you ask.

"Yes, there was," the voice answers. "We had a transformer blowout at Penns Vale. It affected everyone in your area."

You ask when the power cut started and ended.

"The transformer blew out at 9:23 last night," the voice announces. "And we restored power to your area at 2:47 a.m."

You thank the voice, and consult your log for the previous day's tests:

An electrical fault renders a day's work useless...

- You started part 1 at 9:55 a.m.
- You started part 2 at 3:55 p.m., and set the timers to run for 8 hours (they were to switch off at 11:55 p.m.).

You consider what has happened:

- The continuous bake and deep-freeze periods were interrupted part way through.
- The oven temperature dropped, and the freezer temperature rose, for 5 hours and 24 minutes (but to what temperature?).
- The power was restored and the oven temperature again increased, and the freezer temperature decreased (but to what temperature?).
- The electric timers switched off at 5:19 a.m. (after their eight hours ***total*** **running time**).

You consider the implications of the power cut:

- The batches have had uncontrolled, nonstandard testing and will have to be discarded.
- Yesterday's tests will have to be run again (on two new batches). The cost:

Labor:	14 hours (7 hours per batch) = $336.
Materials:	Two complete batches at $92 each.
Time:	One day extra to be added to the program schedule.

...so you reassess the situation and report to management

Write an incident report to your project coordinator (J. H. Grayson). Describe what has happened and the implications, and suggest what might possibly be done to prevent a recurrence.

Project 4.5: Problem Connectors at Site 14

You are an independent consultant running a business you call Pro-Active Consultants from your home office. One of your clients is H. L. Winman and Associates. One week ago you received a telephone call from HLW's electrical engineering project coordinator Don Gibbon, who assigns you to conduct an investigation report. He told you that H. L. Winman and Associates is a management consultant to Interstate Power Company, and currently is supervising the installation of parallel HV DC power transmission lines and a microwave transmitting system along a corridor between Weekaskasing Lake and Flint Narrows. The microwave transmission towers are located approximately 40 miles apart, and are numbered consecutively from No. 1 at Weekaskasing Lake to No. 17 at Flint Narrows. Each tower site has a small residential community and a maintenance crew.

A field trip to investigate a problem

The maintenance crew supervisor at tower site No. 11 is Karen Wasalyshyn, and she telephoned Don Gibbon yesterday to say she had found nine faulty connectors type MT-27 at her site and had to replace them. Don wanted to know if the problem is purely local or is prevalent elsewhere, so he instructed you to fly to site No. 14, the nearest site to your office, to investigate whether there are any other faulty MT-27 connectors. "I'll email you a test procedure you can use to test the connectors," he added.

You flew to Site No. 14 two days ago and stayed there until this morning, when you flew back to your home city. The site maintenance crew supervisor was Don Sanderson, who asked you to send him a copy of your report. Here are the details you discovered:

1. There are 317 type MT-27 connectors on site, with 92 in stock and 225 installed along the lines and up the tower.
2. You tested 278 of the connectors.
3. You could not test the remaining 39 because they were along part of the transmission line that was powered-up throughout your visit.
4. You placed each connector under tension using test procedure TP-33.
5. 241 of the connectors were OK.
6. 37 of the connectors proved to be faulty.
7. You identified the fault as a hairline crack, which became visible when a faulty connector was placed under tension.

8. You also noticed that, although the connectors looked similar, there seemed to be two kinds of connectors on site. One batch of connectors had the letters GLA on the base. The other had the letters MVK on the base.
9. Of the 278 connectors you checked, 201 were stamped MVK, and 77 were stamped GLA.
10. All the faulty connectors had the letters GLA stamped on the base. There were no faulty connectors with the letters MVK on the base.
11. You figured that the letters must identify either different manufacturers or different batches made by the same manufacturer.
12. You recommended to Don Sanderson that he replace all installed GLA connectors with MVK connectors, and to place all the GLA connectors in a separate box marked NOT TO BE INSTALLED, until he receives instructions from Don Gibbon.

Part 1

You have returned to your home office. Write a trip report to Don Gibbon, as a letter attached to an email message. Write the email message too.

A trip report to describe your findings

Part 2

Don Gibbon telephones the next day. "Will you email the maintenance crew supervisor at each of the 17 sites?" he asks. "Tell them to test the connectors the same way you did, and to replace *all* GLA connectors—not just the faulty ones—with MVK connectors. They are to ship the GLA connectors to me: you know the Cleveland address. I'll email you a group address alias you can use. It will distribute your message to all 17 sites." Finally, he tells you to address a copy of the message to him.

Here are some email addresses you will need to complete this assignment:

- IPC Supervisors' group alias: ipc@winman.grp17.com
- Don Gibbon's address: dgibbon@winman.ho.com
- Your email address: pro_active@mbupline.net

Chapter 5
Longer Informal and Semiformal Reports

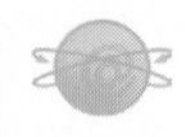

Longer Informal and Semiformal Reports
www.csee.umbc.edu/~sherman/Courses/documents/TR_how_to.html
This Web site gives some excellent advice on writing a technical report, covering how to write the thesis, the components of the technical report, organization as well as common mistakes to avoid.

The previous chapter discussed short reports that deal primarily with facts, in which the writer identifies the relevant details and presents them briefly and directly. This chapter describes longer reports that often deal with less tangible evidence, in which the writer analyzes a situation in depth before drawing a conclusion and, sometimes, making a recommendation. Longer reports may describe an investigation of a problem or unsatisfactory condition, an evaluation of alternatives to improve a situation, a study to determine the feasibility of taking certain action, or a proposal for making a change in methods or procedures. All are written in a fluid narrative style that is both persuasive and convincing; their writers have concepts or new ideas to present and they want their readers to understand their line of reasoning.

Investigation Report

The term "Investigation Report" covers any report in which you describe how you performed tests, examined data, or conducted an investigation using tangible evidence. You start with known data and then analyze and examine it so that the reader can see how the investigation was conducted and the final results reached. The report may be issued as a letter, as a memo, or as a semiformal report. It may travel to its reader by regular mail or as an attachment to an email message.

Although they are not always readily identifiable, a well-written investigation report contains standard parts that help shape the narrative and guide the reader to a full understanding of its topic. They are shown in Figure 5-1, which is an expanded version of the basic writing plan described at the start of Chapter 4. These parts are easy to recognize in long investigation reports, where headings act as signposts introducing each parcel of information. They are more difficult to identify in short reports that use a continuous narrative. In the three-page investigation report in Figure 5-2, the parts are identified by circled numbers:

1. This is the **Summary**.

2. The **Background** is only one sentence, which refers to the memo that instigated the investigation. Because the reader already knows the circumstances, the report writer can omit details.

3. The **Investigation Details** start here, with a very brief reference to the **Approach**.

4. These are the **Findings**.

5. This is the first of three **Ideas** for resolving the problem.

6. These are the **Criteria**: the requirements against which each idea will be measured.

7. The **Analysis** starts here. The table provides a convenient, easy-to-access summary of what each idea will achieve and cost.

8. In the Analysis, each idea is compared to the Criteria. (Note that the ideas are *not* compared one against another.)

Comparing each plan, idea, method, or product against the criteria helps a writer be objective

Summary	A brief statement of the situation or problem and what should be done about it.	
Background	**An introduction** to the situation or problem.	
Investigation Details	The **Facts**, comprising:	
	Approach	How the investigation was tackled.
	Findings	What the investigation revealed.
	Ideas	Different ways the situation can be improved or the problem resolved.
	Criteria	Factors that influence the analysis.
	Analysis	Evaluation of each idea.
Conclusions	The **Outcome**, or result of the investigation; a summing-up.	
Recommendation	A positive statement advocating action.*	
Attachments	Evidence: detailed facts, figures, and statistics that support the Discussion.*	

**Included only when appropriate.*

Figure 5-1 Writing plan for an investigation report.

An expanded writing plan is still based on the pyramid seen in earlier chapters

MEMORANDUM **KCMO-TV**

TO: Dennis Carlisle, Operations Manager
FROM: Phyllis van der Wyck, Engineering Department
DATE: October 21, 2002
SUBJECT: Investigation of High Ambient Sound Level, Satellite Studio Control Room

1

I have investigated the high ambient sound level reported in the control room of our satellite studio at 21 Union Road, and have traced it to the building's air-conditioning equipment. The sound level can be reduced to an acceptable level by replacing the blower motor and soundproofing the air-conditioning ducts and blowers. The cost will be $9800.

2

My investigation was authorized by your memo of August 28, 2002, in which you described the audio difficulties your production crews are experiencing when programming from the satellite studio.

3

Tests conducted with a sound level meter at various locations in the control room established that the average ambient sound level is 36.8 dB, with peaks of 38.7 dB near the west wall. This is approximately 7 to 9 dB higher than the sound levels measured in the control room for Studio 1 on Westover Road, where the average ambient sound level is 29.5 dB with peaks of 30.2 dB near the south wall.

This two-and-one-half page memo-report would benefit from having headings inserted at appropriate places

The unusually high sound level is caused by the air-conditioning equipment, which is in an annex adjacent to the west wall of the control room. Air-conditioner rumble and blower fan noise are carried easily into the control room because the short air ducts permit little noise dissipation between the equipment and the work area. The flat hardboard surface of the west wall also acts as a sounding board and bounces the noise back into the room.

Ideas are numbered consecutively for ease of reference

I have considered three methods we could use to reduce the ambient sound level:

1. Move the air-conditioning equipment to a storage room at the other end of the building, for an estimated 10–12 dB reduction in sound level. This would, however, require major structural alterations that will cost between $22,000 and $26,000.

Figure 5-2 A memo investigation report with primarily objective development.

Dennis Carlisle – page 2

2. Replace the existing blower fan assembly with a model TL-1 blower manufactured by the Quietaire Corporation of Detroit, and line the ducts with Agrafoam, a new soundproofing product developed by the automobile industry in Germany. Together, these methods would reduce the ambient sound level by about 6.5–7.5 dB. The cost will be $9800.
3. Cover the vinyl floor tiles with Monroe 200 indoor/outdoor carpet, a practice that has proved successful in air traffic control centers, and mount carpet on the control room's west wall, for a sound level reduction of about 4.0–4.5 dB. The cost will be $2400.

The remedy we select must

- reduce the ambient sound level by at least 7.3 dB, to provide conditions similar to those at the Westover Road control room,
- be implemented quickly (ideally by November 15, when the Christmas Pageant programs will be recorded), and
- cost no more than $10,000, if the modifications are to be completed within the 2002–03 budget year.

As the table shows, none of the three methods meets all of the above criteria, although Method 2 comes close to doing so.

	Required	***Method 1 Relocation***	***Method 2 Blower/Ducts***	***Method 3 Carpet***
Projected sound level reduction (min)	7.3 dB	10–12 dB	6.5–7.5 dB	4.0–4.5 dB
Time to implement (max)	3 weeks	12 weeks	3 weeks	1.5 weeks
Approximate cost (max)	$10,000	$22–26,000	$9800	$2400

A table simplifies a comparison, makes it easier to analyze

- Method 1—relocating the air-conditioning equipment—would reduce the sound level more than the required minimum but cannot be implemented quickly or within budget.
- Method 2—replacing the blower motor and lining the ducts—probably would reduce the sound level to an acceptable level, but only just. It could be implemented quickly and within budget.

- Method 3—covering the floor and one wall with carpet—would reduce the sound level by only one-third of the desired reduction. It could be implemented quickly and within budget.

9 The only method that comes close to meeting our immediate requirements is method 2. If we were to combine it with method 3, we could achieve a probable total sound reduction of 8.3–9.8 dB, which would meet the required reduction but would exceed the budget by $2200. (Note that, when combining methods, the total reduction will be *less* than the summation of the two individual sound level reductions.)

A major recommendation and a minor recommendation

10 Because method 2 comes close to the required minimum reduction in sound level, I recommend we replace the blower motor and line the ducts with Agrafoam for a total cost of $9800. However, because actual sound level reductions can differ from those projected, I suggest we retest the sound levels following installation. If a further reduction in sound level proves necessary, then I recommend we install Monroe 200 carpet on the floor and west wall in March 2003 at a total cost of $2400, using $200 from the 2002–03 budget year and $2200 from the 2003–04 budget year.

These modifications will provide the quieter working environment needed by your production crews.

P. Van der Wyck

11 Att: Specifications and cost estimates

(9) The **Outcome** starts by drawing **Conclusions**...

(10) ...and continues with a **Recommendation.**

(11) The **Attachments** contain drawings, specifications, and detailed cost estimates for each Idea.

Conducting a Comparative Analysis

Phyllis's report (Figure 5-2) includes a comparative analysis, in which she evaluates the three sound level reduction methods to identify which would be most suitable to implement. A comparative analysis can be written either objectively or subjectively:

A comparative analysis is also a justification

- In an *objective* comparative analysis you do not allow your opinions to intrude until the very end of the report, when you make your recommendation.
- In a *subjective* comparative analysis you allow your voice to be heard—your opinions to be apparent—much earlier in the report, usually when you analyze the alternatives.

Phyllis uses the objective method in her report, presenting only facts until the very end, when—at point (10)—she writes:

> Because method 2 comes close to the required minimum reduction in sound level, *I recommend* we replace the blower motor and... *I suggest* we retest the sound levels....

Always write recommendations in the active voice

The objective and subjective methods are shown side-by-side in Figure 5-3. The objective method is on the left side of the flow chart; the subjective method is on the right side.

There are three primary differences between the two methods:

1. In the objective method you describe each idea or product you are evaluating as a series of facts, without commenting on the product's quality or value. For example, when describing idea 1, Phyllis writes:

   > Move the air-conditioning equipment to a storage room at the other end of the building, for an estimated 10-12 dB reduction in sound level. This would, however, require major structural alterations that will cost between $22,000 and $26,000.

   Present only facts to maintain objectivity

   But in the subjective method you can insert comments and opinions about the idea or product. If Phyllis had been using the subjective method, she would have written this:

Both objective and subjective development methods provide limited opportunities to orchestrate information

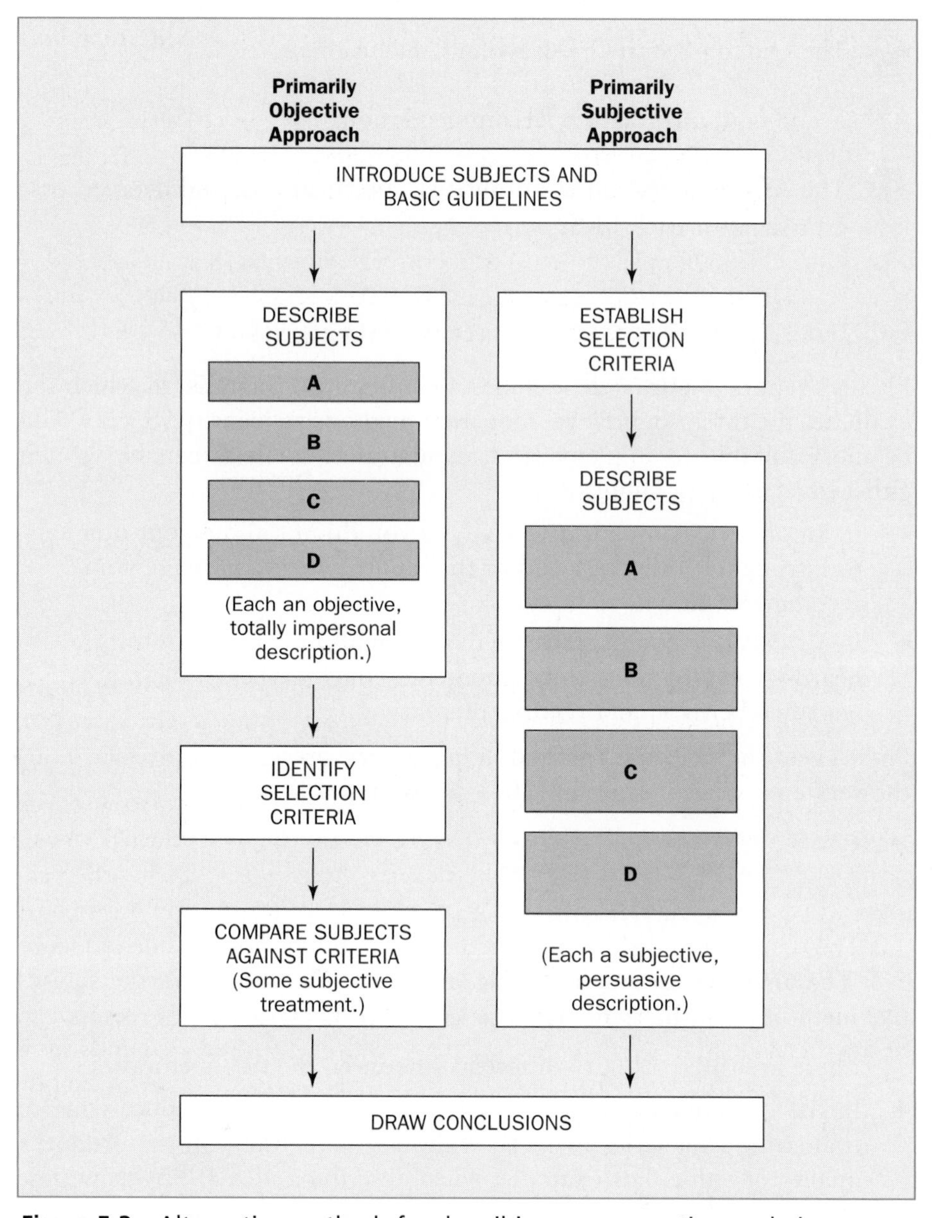

Figure 5-3 Alternative methods for describing a comparative analysis.

When you offer an opinion, you are immediately subjective

We could move the air-conditioning equipment to a storage room at the other end of the building for an estimated sound level reduction of 10 to 12 dB. This would achieve the greatest sound level reduction among the three methods—in fact it would be the only one to fully meet the requirements—but at $22,000 to $26,000 it would be the most costly.

2. When you establish your criteria, you place them in different positions:

   - When using the objective method, you place the criteria *after* you have described the alternative ideas or products. This helps ensure you do not start evaluating the ideas or products as you describe them.
   - In the subjective method you establish your criteria *before* you describe the alternative ideas. You do this precisely so you *can* evaluate the ideas and products as you describe them.

   It's also important to prove any criterion you establish that a reader might question. For example, if Phyllis had written only this statement:

   The remedy we select must...reduce the ambient sound level by at least 7.3 dB.

   her readers might have asked: "Why precisely 7.3 dB?" So she inserted a reason:

   Anticipate your readers' questions

   ... to provide conditions similar to those at the Westover Road control room.

3. Only the objective method has a clearly defined Evaluation compartment. In the subjective method, an evaluation occurs at the same time that the writer describes each idea.

If you ensure that when you evaluate each idea you compare it *only* against the criteria, you will have taken a significant step toward writing a clear comparative analysis. Comparing ideas against each other can result in a confusing analysis that is difficult to write and even more difficult to understand.

Which is better: to use the objective or the subjective method? Both are equally valid; however, for a beginning report writer, we recommend using the objective method. Using the subjective method demands more skill, because there is a danger that your comments will make you appear too opinionated and so offend the reader.

Opening with a Summary Page

Some companies preface their investigation reports with a standard title and summary page similar to the H. L. Winman and Associates' design illustrated in Figure 5-4. This page saves a reader the trouble of searching for the summary and the report's identification details. Subsequent pages (which have not been included with this example) contain the report narrative, starting with the Background (Introduction). Reports written in this way tend to adopt a slightly more formal tone than memos and letter reports and are less likely to be written in the first person. Sometimes they are called a form report, although the preferred name is semiformal investigation report.

H. L. WINMAN AND ASSOCIATES

475 Reston Avenue, Cleveland, OH 44104

Investigation Report

Authorization and other details are grouped together in an easy-to-find arrangement

Report No: 70/26 **File Ref:** 53-Civ-26
Date: March 20, 2003
Prepared for: City of Montrose, Ohio
Authority: City of Montrose letter Hwy/69/38, Nov 7, 2002

Report Prepared by: G. G. Waterston

Approved by: M. Warner

Subject or Title

Investigation of Stormwater Drainage Problem
Proposed Interchange at Intersection of
Highways 6 and 54

Summary of Investigation

The summary is a miniature pyramid: para 1 = background...

The proposed interchange to be constructed at the intersection of Highways 6 and 54, on the northern perimeter of Montrose, incorporates an underpass that will depress part of Highway 54 and some of its approach roads below the average surface level of the surrounding area. A special method for draining the stormwater from the depressed roads will have to be developed.

Two methods were investigated that could contend with the anticipated peak runoff. The standard method of direct pumping would be feasible but would demand installation of four heavy-duty pumps, plus enlargement of the ¾ mile drainage ditch between the interchange and Lake McKing. An alternative method of storage-pumping would allow the runoff to collect quickly in a deep storage pond that would be excavated beside the interchange; after each storm is over, the pond would be pumped slowly into the existing drainage ditch to Lake McKing.

...para 2 = investigation details, and para 3 = outcome

Although both methods would be equally effective, the storage-pumping method is recommended because it would be the most economical to construct. Construction cost of a storage-pumping stormwater drainage system would be $984,000, whereas that of a direct pumping system would be $1,116,000.

Figure 5-4 Title and summary page for a semiformal investigation report.

Evaluation Report/Feasibility Study

Evaluation reports are similar to investigation reports—both names are frequently used interchangeably. Evaluation reports often start with an idea or concept their authors want to develop, prove, or disprove. Their authors first establish guidelines to keep their report within prescribed bounds, and then research data, conduct tests, and analyze the results to determine the concept's viability. At the end of the evaluation they draw a conclusion that the concept either is or is not feasible, or perhaps is feasible in a modified form.

Two titles, but a similar function

The writing plan shown in Figure 5-1 can also be applied to an evaluation report. Morley Wozniak's evaluation of landfill sites in Figures 5-5 and 5-6 follows this plan.

Morley's report is significant in that it is preceded by a one-page cover letter, thus using the technique suggested on page 38 and illustrated in Figure 3-10. His cover letter is similar to the Executive Summary that often precedes a formal report (Executive Summaries are described in Chapter 6), since it *describes and comments on* key implications drawn from the report. Its addressee (Quillicom's town engineer) has the option of distributing it to the town councillors with the report or detaching it and replacing it with a cover letter of his own.

A cover letter can introduce a sensitive issue or confidential information

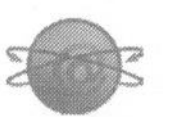

Recommendation and Feasibility Reports
www.io.com/~hcexres/tcm1603/acchtml/feas.html
Also from Online Technical Writing, this document describes feasibility reports in detail and includes several samples.

The parts of the writing plan shown in Figure 5-1 are identified in Morley's report by circled numbers beside the narrative; they are keyed to the additional comments provided here.

1. Although several factors affect site selection, in his **Summary** Morley focuses primarily on environmental impact because he believes it is of overriding importance.

2. The **Introduction** provides Background details leading up to the study assigned to H. L. Winman and Associates, and then to Morley.

3. The **Evaluation Details** start here, with a single paragraph in which Morley outlines his **Approach** (i.e. how he tackled the study). Note that he mentions the three components *in the same sequence* that he will describe them further on in the report.

4. These are Morley's **Findings**—the results of his research. He describes the findings in detail because his readers must fully understand the geology of the area if they are to accept the conclusions he will draw later in his report. He is totally objective here, reporting only facts without letting his opinions intrude.

H. L. WINMAN AND ASSOCIATES

475 Reston Avenue, Cleveland, OH 44104

May 23, 2003

City, state, and zip code, correctly placed all on one line

Robert D. Delorme, P.E.
Town Engineer
Municipal Offices
Quillicom MI 48716

Dear Mr. Delorme,

Our assessment of the three sites selected as potential landfills for the Town of Quillicom shows that each has a disadvantage or limitation. The most serious exists at Lot 18, Subdivision 5N, which is the site preferred by the Town Council. A distinct possibility exists that a landfill located here could contaminate the town's water supply.

The disadvantages of the two other sites affect only cost and convenience. Lot 47, Subdivision 6E, will be considerably more expensive to operate, while Lot 23, Subdivision 3S, will have a much lower capacity and so will have to be replaced much sooner than either of the other sites.

If the Town Council still prefers to use Lot 18, a drilling program must first be conducted to identify the soil and bedrock structure between the lot and Quillicom. Providing the boreholes show no evidence that contamination will occur, then the site would be a sound choice.

The enclosed report describes our study in detail. I will be glad to discuss it and its implications with you.

A cover letter accompanying an in-depth report or proposal may be signed by the author for the department manager

Regards,

Vincent Hrabi
Branch Manager
H. L. Winman and Associates
Lansing, Michigan
enc

Figure 5-5 The cover letter preceding an evaluation report. This letter is also an executive summary.

5 Morley presents the three possible landfill sites as his **Ideas** (even though they were originally presented to him by the client, the Town of Quillicom). In effect he is saying to his readers: "Now that I have described the geology of the land to you, here are three locations within the area for you to choose from." He is still totally objective.

6 Morley now identifies three general **Criteria** he will use to evaluate the sites. He does not identify specific criteria because they have not been defined.

7 In his **Analysis,** Morley must clearly establish the factors on which he will base his conclusions. Now he allows some subjectivity to appear in his writing (we can hear his voice behind his words). He is moving down the right side of the plan for a comparative analysis, as shown in Figure 5-3.

8 Morley's **Conclusions** identify the main features affecting each site. Note that he simply offers the alternatives without saying or even implying which is preferable. This part of the report, together with the Recommendations, is the Outcome (sometimes referred to as the *terminal summary*).

9 In the **Recommendations** Morley states specifically what he believes the Town Council must do. He must sound definite and convincing, so he starts with "We recommend..." rather than the passive "It is recommended that..."

10 The **Attachment** brings together all the site details in an easy-to-read form, and simultaneously provides readers with Evidence to support what Morley says about the landfill sites in the report narrative.

Morley's rationale for organizing and writing his report

Proposals
www.io.com/~hcexres/tcm1603/acchtml/props.html
This document is one chapter from the online textbook used in Austin Community College's online course, Online Technical Writing (www.io.com/~hcexres/tcm1603/acchtml/acctoc.html). It describes types of proposals, their organization and format, and the common sections in a proposal. Included are several sample proposals and a revision checklist.

H. L. WINMAN AND ASSOCIATES

475 Reston Avenue, Cleveland, OH 44104

Evaluation of the Proposed Landfill Sites for the Town of Quillicom, Michigan

Summary

Two of the three locations selected as potential landfill sites for the Town of Quillicom, both southeast of the town, are environmentally safe. There is insufficient data to determine whether the third site, to the north of Quillicom, poses an environmental risk. All three sites are financially viable although one, because of its greater distance from Quillicom, would be more costly to operate.

Introduction

The Background traces the history leading up to the present study

The Town of Quillicom in northern Michigan currently operates a landfill 2.2 miles southeast of the town. The landfill was constructed in 1958, and since 1974 has also served the mining community of Melody Lake, 1.7 miles to the southwest of the landfill. In a report dated February 27, 2003, Quillicom town engineer Robert Delorme identified that the existing landfill was nearing capacity and that a new landfill must be found and operational by April 30, 2005.

Previously, in 2000, the town had identified two sites as potential replacement landfills: Lot 18, Subdivision 5N, 2.1 miles north of Quillicom; and Lot 47, Subdivision 6E, 9.1 miles to the southeast. The costs to set up and operate both sites were determined, and Lot 18 proved to be more economical ($2000 more to purchase and develop, but $17,000 a year less to operate). It was favored by the Town Council. However, in a letter to the Council dated November 15, 2001, Mr. Delorme expressed his concern that leachate from the site could possibly contaminate the town's source of potable ground water, and recommended that the town first carry out an environmental study.

The town subsequently engaged H. L. Winman and Associates to examine the sites and determine both their financial viability and their environmental safety. In a letter dated March 15, 2003, Mr. Delorme commissioned us to carry out the study, and to include a third potential landfill site at Lot 23, Subdivision 3S, immediately adjacent to the existing landfill, in our assessment.

Figure 5-6 The evaluation report (6 pages).

Study Plan

❸ We divided our study into three components: (1) an examination of the area geology and its ability to constrain leachate movement; (2) an examination of the physical properties of the proposed landfill sites; and (3) an evaluation of the financial and environmental suitability of the sites.

Area Geology and Hydrogeology

❹ Bedrock at Quillicom and in the area of all three proposed landfill sites is chiefly granite and gneiss lying 15 to 30 meters below the surface. A layer of till varying in thickness from 10 to 20 meters covers the bedrock, and is itself covered by 1 to 15 meters of lacustrine silts and clays.

The whole area has experienced repeated glaciation, with the most recent occurring about 20,000 years ago with the advance of the Late Wisconsonian Ice Field. The advancing ice severely scarred this granite and gneiss. When the ice began to retreat 10,000 years later, till was deposited over the region. In addition, meltwater streams below the ice field deposited vast quantities of alluvial material, which today exist as eskers.

The melting ice also created Lake Agassiz and caused silts and clays to be deposited to a depth of up to 30 meters over the entire lake bed. (In the Quillicom area these lacustrine deposits range from 5 to 15 meters deep.) Then, as the lake drained and water levels receded, streams cut into the lacustrine and till deposits. These streams eventually dried up and their channels were filled with windblown silts and sands. Today the channels are known as buried stringers and, if they are water bearing, as stringer aquifers in the weathered bedrock and eskers.

Presenting technical details so they will be understood by all readers takes skill!

The Town of Quillicom obtains its potable water from a stringer aquifer on the surface of the weathered bedrock. Other stringer aquifers are known to exist in the area to the east and south of Quillicom, and likely also exist to the west and north.

The Michigan Water Resources Department provided us with logs obtained during drillings for ground water wells in the mid 1980's, all to the south and east of Quillicom. We have plotted the locations and types of materials on a topographic map, which shows that

- the bedrock in the area slopes downward, toward the south, from the Town of Quillicom, and
- a major 350-meter wide glacial esker starts half a mile southeast of Quillicom and continues for several miles southeast under Highway A806, to beyond the proposed landfill at Lot 47, Subdivision 6E.

As there are very few borehole records for the area north of Quillicom, we could not plot a similar map for that area.

2

The Proposed Landfill Sites

The attributes of the three proposed landfill sites are discussed briefly below and itemized in detail in the attachment. The anticipated life of each site is based on the 2002 population of Quillicom. Similarly, projected operating costs are based on 2002 prices.

5

Lot 23, Subdivision 3S

This narrow, 19.3 acre strip of land is immediately east of and adjacent to the existing landfill, 2.3 miles southeast of Quillicom on Highway A806. As it is the smallest site, it will cost only $9000 to purchase and develop. Its annual operating cost will be $47,000, the same as at the present landfill, and it will have an operational life of 12 to 14 years.

Lot 47, Subdivision 6E

Judicious use of white space makes technical details more readable...

The largest of the three proposed sites at 88.7 acres, but also the most distant, Lot 47 is a rectangular parcel of land 9.1 miles southeast of Quillicom on Highway A806. Its combined purchase and development price will be $20,000, and its annual operating cost will be $64,000. (The high operating cost is caused primarily by the much greater distance the garbage collection vehicles will have to travel.) It will have a lifespan of almost 60 years.

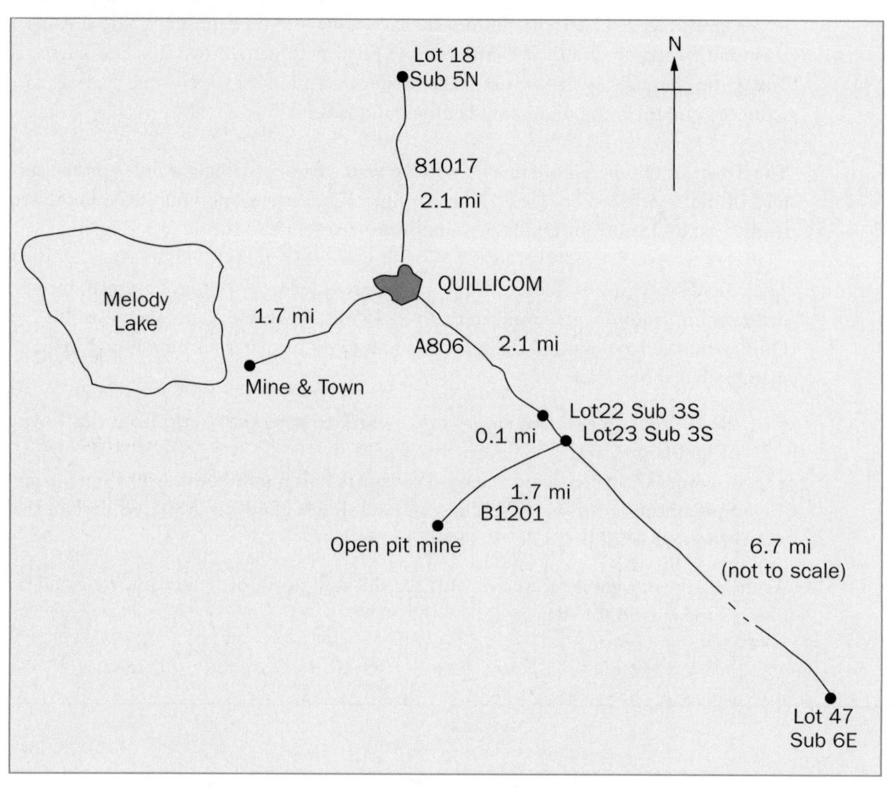

3

Lot 18, Subdivision 5N

A roughly square, 56.2 acre parcel of land, this lot is 2.1 miles directly north of Quillicom on Highway B1017. It will cost $22,000 to purchase and develop, and $47,000 a year to operate (the same as at present). At the current fill rate, it will last for 36 to 40 years.

Site Comparisons

We considered three factors when comparing the three proposed landfill sites: cost, environmental impact, and convenience.

Cost. We examined cost from two points of view: the immediate expense to purchase and develop the site, and the annual cost to operate it.

7

- Lot 23, adjacent to the existing landfill, offers the lowest purchase and development cost at $9000, compared with $20,000 and $22,000 for the two alternative sites.
- Lot 23 and Lot 18 (the site north of Quillicom) offer comparable operating costs at $47,000 per year, whereas Lot 47 (9 miles southeast of Quillicom) would have the highest annual operating cost of $64,000.

...as does the author's choice not to justify the right margin

However, if the purchase and development costs are spread over 10 years and added to the operating costs, Lots 18 and 23 show a more comparable cost structure:

Site:	Lot 18	Lot 23	Lot 47
Annual Cost:	$49,200	$47,900	$66,000

Environmental Risk. The primary environmental risk is the effect that leachate from the landfill could have on Quillicom's source of potable water. If a landfill lies on a glacial esker, leachate from the landfill will probably contaminate ground water aquifers in the esker. If these aquifers are connected hydraulically to stringer aquifers, the stringer aquifers also probably will become contaminated.

- Lots 23 and 47 (and the existing landfill) lie on a major esker south of Quillicom but offer no environmental risk because the slope of the bedrock in the area is to the south, away from the town. Consequently, even if leachate from the landfill contaminates the ground water, it will flow away from Quillicom and will not contaminate the town's water supply.
- Lot 18, however, is in the uncharted area north of Quillicom, where neither the presence of eskers nor the slope of the bedrock has been determined. Consequently it offers a potential risk that leachate from a landfill located here could contaminate the town's water supply. This will be particularly true if the slope of the bedrock south of Quillicom is the same north of the town, since then leachate will flow south, toward Quillicom.

This analysis sets the scene for the conclusions the report author will draw

Convenience. To establish convenience we considered the size of each landfill site (measured as the number of years it can be used before another site must be found) and its proximity to Quillicom.

4

- Lot 47 is the largest site, offering close to 60 years of use, but is four times farther from Quillicom than either of the two other sites.
- Lot 18, to the north, is next largest and can provide between 36 and 40 years of use. It is a comfortable 2.1 miles distant from Quillicom.
- Lot 23, although the same distance as Lot 18, has a life span of only 12 to 14 years.

Conclusions

The possibility of leachate contamination of the Town of Quillicom's water supply makes Lot 18, Subdivision 5N, a doubtful choice until sufficient drilling has been done to create a profile of the strata between the lot and Quillicom.

8

The remaining two sites are environmentally sound but have different advantages:

Remember: conclusions and recommendations must never introduce new information

- Lot 47, Subdivision 6E, provides the greatest space but will be costly to operate.
- Lot 23, Subdivision 3S, offers the lowest cost but will have only a limited life span.

Recommendations

We recommend that the Town of Quillicom purchases Lot 23, Subdivision 3S, and operates it as a temporary landfill from 2005–2015. We also recommend that the town concurrently conducts a drilling program to the north of Quillicom to determine whether Lot 18, Subdivision 5N, will be an environmentally sound site to use after the year 2015.

9

Morley Wozniak

Morley Wozniak, P.E.

5

Attachment

10

Comparison of Proposed Landfill Sites for the Town of Quillicom

Comparison Factor	*Lot 18 Sub 5N*	*Lot 23 Sub 3S*	*Lot 47 Sub 6E*
Distance from Quillicom (driving dist in miles)	2.1 N	2.2 SE	9.1 SE
Size (acres)	56.2	19.3	88.7
Life (years)	36–40	12–14	58–60
Environmental risk (the site's potential for contaminating the Quillicom water supply)	Unknown	None	None
Development costs:			
Purchase price ($)	10,000	2,000	5,000
Construction cost ($)	12,000	7,000	15,000
Ten-year cost ($/yr)	2,200	900	2,000
Operating costs ($/yr)	47,000	47,000	64,000
Combined development and operating costs:			
Year 1, w/o amortization ($)	69,000	56,000	84,000
Per year, w amortization ($)	49,200	47,900	66,000

6

An "open" table (no lines drawn around it) is cleaner for a simple presentation

We encourage using the first person in letter and report writing

Compare how Phyllis van der Wyck and Morley Wozniak each use the first person (see Phyllis's investigation report in Figure 5-2). Phyllis uses the informal "I" because she is writing a memo report to another member of the television station where she works. Morley uses the slightly more formal "we" because he knows his semiformal report will be distributed to the Quillicom town councillors. (Note, however, that he uses "I" in the personal cover letter to Robert Delorme that accompanies the report.)

Like an evaluation report, a feasibility study starts by introducing an idea or concept, and then develops and analyzes the idea to assess whether it is technically or economically feasible. The chief difference lies in the name and application of each document. An evaluation report is generally based on an idea that is originated and evaluated within the same company; hence, it is nearly always informal. A feasibility study is normally prepared at a slightly higher level: the management of company *A* asks company *B* to conduct a feasibility study for it, because company *A*'s staff is not experienced in a specific technical field. For example, if a national wholesaler engaged solely in the distribution of dry goods were to consider purchasing an executive jet, it would seek advice from a firm of management consultants. The consultants would examine the advantages and disadvantages, and publish their results in a feasibility study that they would issue as either a letter or a formal report. Often the differences between a feasibility study and an evaluation report are so slight that only personal preference dictates which label is used for a particular document.

ASSIGNMENTS

Project 5.1: Resolving a Landfill Problem

You are the assistant engineer for the Town of Quillicom in Michigan. Your boss is Robert D. Delorme, P.E., who is the town engineer.

Mr. Delorme calls you into his office and announces, "I have a project for you. The Town Council has finally decided to so something about the landfill problem, and they want it done in a hurry."

You know about the landfill problem. The existing landfill site—at Lot 22, Subdivision 3S—is nearly full and, recognizing that a new site will not be selected before the present site reaches its capacity, he has authorized dumping an additional layer of garbage on top of the compacted fill.

"Before you do anything, I want you to read this," Mr. Delorme continues, placing a report in your hands. "It's a study done by H. L. Winman and Associates a while ago, and it affects what you will be doing. Take it away and read it, and then come back to me for further instructions."

You read the report (you will find it in Figure 5-6, on pages 112 to 117) and then go back to see Mr. Delorme.

Start by researching and reading about what has gone before

"The town councillors have decided," he says, "that before they can make a decision they need to know how much it will cost to drill a dozen boreholes north of Quillicom and analyze the results. I want you to get some quotations that I can present to them. You should also be aware that the councillors very much prefer Lot 18, Subdivision 5N, rather than either of the other locations."

"Who will plot the results of the drilling?" you ask.

"Morley Wozniak, at H. L. Winman and Associates in Lansing, Michigan. He did the previous study and wrote the report I asked you to read." Mr. Delorme hands you a map (the same as that in Figure 5-6, page 114) and a list titled *Borehole Specifications for the Area North of Quillicom*, which contains the exact positions Morley has identified where the drilling must be carried out.

"And how many quotations should I get?" you ask.

"Two, as a minimum," Mr. Delorme suggests. "Three would be better."

The following week you call on the only two drilling companies you know of in the area, one in Quillicom and one in Marquette. They give you the following quotations:

Northwest Drillers, Inc. Quillicom, Michigan	$78,520 (tax incl)
M. J. Peabody Inc. Marquette, Michigan	$75,900 (tax extra)

You had almost given up hope that you would find a third company to give you a quotation, when Mr. Delorme telephones. "Go and see Bert Knowles," he instructs you. "He's the assistant to the superintendent at Melody Lake Mine, and he has a suggestion for an alternative landfill site."

An unusual but realistic option for a landfill site

Mr. Knowles comes right to the point: "We do both open-pit and underground mining. Our open-pit mine is nearly worked out and we will finish excavating there in less than two years. The problem is that it's unsightly, and the Environmental Protection Agency is leaning on Melody Lake Mines to do something about it. That's where you come in."

He explains that the Town of Quillicom can use the open-pit mine for a landfill, and the mine will lease it to the town for one dollar a year. "We have only two conditions. You must spread soil over the compacted garbage, and do it progressively as you go along so there will be no obnoxious smell for the people who live near the mine to contend with. And then you must seed it and plant trees."

You agree: it's something the Town would do anyway, before closing a landfill site.

Mr. Knowles drives you to the site and you stand on the lip of a shallow, roughly oval excavation varying from about 10 to 50 feet deep.

"How large is it?" you enquire.

"You'll have to talk to Inga Paullsen. She's the mine geologist."

When you visit Inga, she calculates the size of the excavation as 60.9 acres. "That's what it will be," she adds, "when the mining is complete. Why do you need to know?"

You describe the difficulty the Town Council is having in finding a landfill site, mention the three other sites, say the one north of the town could create an environmental problem, and explain you won't know until drilling has been completed there.

A surprising piece of information introduces a new aspect

"But drilling has already been done there," Inga exclaims. "When I was a junior at college I worked one summer with an exploration crew sinking boreholes north of Quillicom. We were looking for an alternative place to sink a mine shaft, but we found no ore deposits north of either Melody Lake or Quillicom. We drilled quite a few boreholes."

Inga tells you the mine does not have the records, only a report from the drilling company. She searches for it among the geology records, but cannot find it. "It's strange," she mutters, "It should be here. Someone must have removed it."

The drilling company Inga worked for was Mayquill Explorations, but she says it does not exist anymore. "When Ernie Mays retired he simply closed down the company. Maybe he still has the records. You could ask him. He still lives in Quillicom."

Ernie Mays is about 70 and he lives in a bungalow at 211 Westerhill Crescent.

"I quit eight years ago," he tells you. "I sold some of my accounts to Northwest Drillers—those that were still active—and kept the remainder."

He remembers drilling for Melody Lake Mines. "We sank about 20 boreholes, all north of Quillicom, but we didn't find anything."

You ask if he remembers whether the bedrock slopes, but he shakes his head. "Not really," he says. "Nothing definite."

But he adds that he does remember there was evidence of a large sand esker running roughly south-southwest toward Quillicom.

"Do you still have the records?"

"No," he says. "The mine has them. Mr Caldicott came to see me himself, about three years ago, and I gave them all to him."

A second piece of information introduces still another aspect!

Suddenly, everything falls into place. Frank Caldicott is not only general manager of Melody Lake Mines, but also a very influential Quillicom town councillor. And his youngest sister, Julie, is married to the town engineer—Robert Delorme, your boss.

Because you have so much new information to include, you decide to write a semiformal report of your findings. (You will have to decide whether you will include the information you now have about the previous drilling north of Quillicom, and the location of the records.)

Here is some additional data you may need to write your report:

1. You are concerned about groundwater contamination problems if the open-pit mine is used as a landfill, so call Morley Wozniak at H. L. Winman and Associates in Lansing. He tells you that it will not be a problem. "Both the lake and the mining community are north of the pit, and the bedrock slopes to the south."
2. You calculate that costs to develop the open-pit mine as a landfill will be only $3000, because you can use the buildings and approach roads that are already there.
3. The open-pit mine is 2.5 miles directly south of Quillicom, but 4.1 miles by road (2.4 miles southeast along highway A806, then 1.7 miles southwest along highway B1201).
4. The annual operating cost for using the open-pit mine as a landfill will be $49,500, which is $2500 more than the cost for operating the current landfill.
5. You obtain a third drilling estimate from Quattro Drilling and Exploration Company in Houghton, Michigan, which quotes $83,200, tax included.

Before starting to write your report you visit Thunder Bay on other business. On a hunch you visit the Land Titles Office and look up the surveys for the area north of Quillicom. Against Lot 18, Subdivision 5N, you find the owner listed as *Julie Sarah Caldicott, 207 Northern Drive, Quillicom, Michigan.*

A question of ethics: do you mention the "family" connection?

Now write your report.

Project 5.2: Identifying a Power Plant Problem

You are an independent consultant and operate a business known as Pro-Active Consultants Inc. from your home. Four days ago you received a telephone call from Paullette Machon, who is vice president, operations, of Baldur Agricultural Chemicals (BAC), a company with manufacturing plants across the country. She said she has a task for you and invited you to visit her at the BAC office at 1450 Disraeli Crescent (of the town or city where you live).

"I want you to drive over to our plant in Gordontown," she announced, "to look into a technical problem in the power house." (Gordontown is 43 miles from your city, has a population of 15,700, and its primary employer is the BAC plant.) "I'm concerned that power house costs are rising at Gordontown just at the moment when world fertilizer prices are dropping," Ms. Machon continues. "This is causing the company to be uncompetitive in both national and international markets."

Ms. Machon explains that BAC requires a lot of hot water and steam in its manufacturing operations. However, over the past two years fuel consumption at Gordontown has risen by 18%, numerous breakdowns have occurred that have interfered with production, and there has been a sharp rise in production costs. She has visited the power house repeatedly, but has never found anything that could be attributed to poor operation. In fact, the power house has always been immaculate.

Now Ms. Machon wants an independent consultant to take a look, talk to the people in the power house, and try to identify any production problems.

A technical problem affected by the personalities involved

She also hinted that the problem may not only be technical. "The present chief engineer at the BAC power house is Curt Hänness, and he is to retire in three months. BAC management has to decide whether to promote Harry Markham, the existing senior shift engineer, or to bring in a new chief engineer from outside the company. On paper, Markham is ideal for the job. He has worked in the power house for 15 years (he is now 36) and always under Hänness, so his knowledge of the plant and its operations cannot be challenged. Yet the rising costs indicate that all is not as it should be in the plant, and we want to be sure that the new chief engineer does not perpetuate the present conditions."

She said she would inform Hänness and Markham that she has engaged you to study the hot water and power generating system in their power house, and that they are to expect you.

You visit the BAC power plant in Gordtontown today. During your talks to plant staff and tours of the plant you make the following notes:

These are the "technical" details...

1. Housekeeping excellent—whole place shines (but is this only surface polish for impression of visitors?)
2. Maintenance logs are inadequately kept—need to be done more often. Need more detail. Equipment files not up to date and not properly filed.
3. Boiler cleaning badly neglected. Firm instructions re boiler cleaning need to be issued by head office.
4. Flow meters are of doubtful accuracy. May be overreading. Not serviced for three years. Manufacturer's service department should be contacted (these are Weston meters). Manufacturer needs to be called in to do a complete check and then recalibrate meters.
5. Overreading of meters could give false flow figures—make plant seem to produce more steam than is actually produced.
6. Good housekeeping obviously achieved by neglecting maintenance. Incorrectly placed emphasis probably caused by frequent visits from company president, who likes to bring in important visitors and impress them. Hänness likes reflected glory (so does Markham).
7. Shift engineers are responsible for maintenance of pumps and vacuum equipment. Not enough time given over to this. They seem to

prefer straight replacement of whole units on failure rather than preventive maintenance. Costly method! Obviously more breakdowns: they wait for a failure before taking action. A preventive maintenance plan is needed.

8. Markham seems O.K. Genial type; obviously knows his power house. Proud of it! But seems to resist change. Definitely resents suggestions. Does he lack all-round knowledge? Is he limited only to what goes on in his plant? Is he afraid of new ideas because he doesn't understand them? Young staff hinted at this: too loyal to say it outright, but I felt they were restive, hampered by his insistence that they use old techniques that are known to work but are slow. Nothing concrete was said—I just "felt" it.
9. Hänness has done a good job training Markham. Made him a carbon copy. Hänness doesn't do much now. Markham runs the show, and has for over a year. He *expects* to get the job when Hänness retires. It'll be a real blow to him if he doesn't! BAC might even lose a good company man.
10. Discussed microprocessor-controlled CORLAND 200 power panel with staff. Young engineers had read about it in "Plant Maintenance"—eager to have one installed (I described the one I'd seen at Pinewood Paper Mill). But Hänness and Markham knew nothing about it—didn't seem to be interested. Are they not keeping up-to-date with technical magazines?

...and these are the "personal" details

When you return to your office you write an evaluation report for Ms. Machon. You can either address both the technical problems and the personnel difficulties within the one report, or write two separate reports.

Chapter 6 Formal Reports

Technical Reports
www.io.com/~hcexres/tcm1603/acchtml/techreps.html
This document is one chapter from the online textbook used in Austin Community College's online course, Online Technical Writing (www.io.com/~hcexres/tcm1603/acchtml/acctoc.html). It describes types of technical reports and their general characteristics and audience, and provides a checklist that can be used when writing technical reports.

Formal reports require more careful preparation than the informal and semiformal reports described in previous chapters. Because they will be distributed outside the originating company, their writers must consider the impression the reports will convey of the entire company. Harvey Winman recognized long ago that a well-written, esthetically pleasing report can do much to convince prospective clients that H. L. Winman and Associates should handle their business, whereas a poorly written, badly presented report can cause clients to question the company's capability. Harvey also knows that the initial impression conveyed by a report can influence a reader's readiness to plow through its technical details.

The presentation aspect must convey the originating company's "image," suit the purpose of the report, and fit the subject it describes. For instance, a report by a chemical engineer evaluating the effects of diesel fumes on the interior paint of bus garages would most likely be printed on standard bond paper, and its cover, if it had one, would be simple and functional. At the other end of the scale, a report by a firm of consulting engineers selecting a college site for a major city might be printed professionally and bound in an artistically designed book-type folder. But regardless of the appearance of a report, its internal arrangement will be basically the same.

Formal reports are made up of several standard parts, not all of which appear in every report. Each writer uses the parts that best suit the particular subject and the intended method of presentation. There are six major and several subsidiary parts on which to draw, as shown in Table 6-1. Opinions differ throughout industry as to which is the best arrangement of these parts. The two arrangements suggested in The Complete Formal Report later in this chapter, and illustrated in the mini-reports in Figures 6-3 and 6-8, are those most frequently encountered. Your knowledge of these parts and their two basic arrangements will help you adapt quickly to any variation in format preferred by a future employer.

Table 6-1 Traditional arrangement of formal report parts.

Cover or Jacket
Title Page
Summary
Table of Contents
Introduction
Discussion
Conclusions
Recommendations
References or Bibliography
Appendixes

Notes:

1. **Major Parts** are in boldfaced type.
2. A Cover Letter or Executive Summary normally accompanies a formal report.

The acronym SIDCRA is formed from the first letter of each major report part

Major Parts

Six major parts form the central structure of every formal report. In the traditional arrangement they are known by the acronym SIDCRA.

Summary

The summary is a brief synopsis that tells readers quickly what the report is all about. Normally it appears immediately after the title page, where it can be found easily. It identifies the purpose and most important features of the report, states the main conclusion, and sometimes makes a recommendation. It does this in as few words as possible, condensing the narrative of the report to a handful of succinct sentences. It also has to be written so interestingly—so enthusiastically—that it encourages readers to read further.

The summary is considered by many to be the most important part of a report and the most difficult to write. It has to be informative, yet brief. It has to attract the reader's attention, but must be written in simple, nontechnical terms. It has to be directed to the executive reader, yet be readily understood by almost any reader.

The criteria for a Summary are difficult to meet

Generally, the first person in an organization who sees a report is a senior executive, who may have time to read only the summary. If the executive's interest is aroused, he or she will pass the report down to the technical staff to read in detail. But if the summary is unconvincing, the executive may think the report is unimportant and put it aside; if this happens, the report may never be read.

Always write the summary last, after you have written the remainder of your report. Only then will you be fully aware of the report's highlights, main conclusions, and recommendations, so you can draw on them to form your words.

A summary needs to tell a story: it should have a beginning, in which it states why the project was carried out and the report was written; a middle, in which it highlights the most important features of the whole report; and an end, in which it reaches a conclusion and possibly makes a recommendation. The example below illustrates how the interest is maintained in an informative summary:

An informative summary answers readers' questions...

Informative Summary

We have tested a specimen of steel to determine whether a job lot owned by Northern Railways could be used as structural members for a short-span bridge to be built at Peele Bay in northern Alaska. The sample proved to be G40.12 structural steel, which is a good steel for general construction but subject to brittle failure at very low temperatures.

Although the steel could be used for the bridge, we consider there is too narrow a safety margin between the –51°C temperature at which failure can occur, and the –47°C minimum temperature occasionally recorded at Peele Bay. A safer choice would be G40.8C structural steel, which has a minimum failure temperature of –62°C.

Other informative summaries preface the two formal reports presented in this chapter and the semiformal evaluation report in Figure 5-6 (page 112).

Some writers prefer to write a topical summary for reports that describe history or events, or that do not draw conclusions or make recommendations. As its name implies, a topical summary simply describes the topics covered in the report without attempting to draw inferences or captivate the reader's interest:

...whereas a topical summary only suggests what can be found within a report

Topical Summary

Construction of the Minnowin Point Generating Station was initiated in 2000, and first power from the 1340 MW plant is scheduled for 2007. A general description of the structures and problems peculiar to the construction of this large development in an arctic climate is presented. The river diversion program, permafrost foundation conditions, and major equipments are described. The latter include the 16 propeller Turbines, among the largest yet installed, each rated at 160,000 horsepower.

Because they are less results-oriented, topical summaries are *not* recommended for most formal reports.

In a formal report, the summary should have a page to itself, be centered on the page, and be prefaced by the word "Summary." If it is very short, it may be indented equally on both sides to form a roughly square block of information. For examples, see pages 156 and 172.

Introduction

The introduction begins the major narrative of the report by preparing readers for the discussion that follows. It orients them to the purpose and scope of the report and provides sufficient background information to place them mentally in the picture before they tangle with technical data. A well-written introduction contains exactly the correct amount of detail to lead readers quickly into the major narrative.

The length of an introduction and its depth of detail depend mostly on the reader's knowledge of the topic. If you know that the ultimate reader is technically knowledgeable, but at the same time you have to cater to the executive reader who is probably only partly technical, write the introduction (and conclusions and recommendations) in semitechnical language. This permits semitechnical executives to gain a reasonably comprehensive understanding of the report without devoting time and attention to the technical details contained in the discussion.

Knowing your reader influences the depth of detail required

Technical Report Writing
www.sti.nasa.gov
Scientists at NASA's Lewis Research Center must write reports that are both technically correct and easy to read. This NASA guide was written to make writing reports easier. Separate chapters deal with the stages of report preparation, report style, the introduction, experiment and analysis descriptions, results and discussions, concluding and supporting sections, reviewing reports, and references. An author's checklist and reporting aids provide quick guidelines for technical report writers.

Most introductions contain three parts: **purpose**, **scope**, and **background information**. Frequently the parts overlap, and occasionally one of them may be omitted simply because there is no reason for its inclusion. Normally, the introduction is a straightforward narrative of one or more consecutive paragraphs; only rarely is it divided into distinct sections preceded by headings. It always starts on a new page (normally identified as page 1 of the report) and is preceded by the report's full title. The title is followed by the single word "Introduction," which can be either a center heading or a side heading, as shown in Figure 6-1.

The **purpose** explains why the project was carried out and the report is being written. It may indicate that the project has been authorized to investigate a problem and recommend a solution, or it may describe a new concept or method of work improvement that the report writer believes should be brought to the reader's attention.

The **scope** defines the parameters of the report. It describes the ground covered by the report and outlines the method of investigation used in the

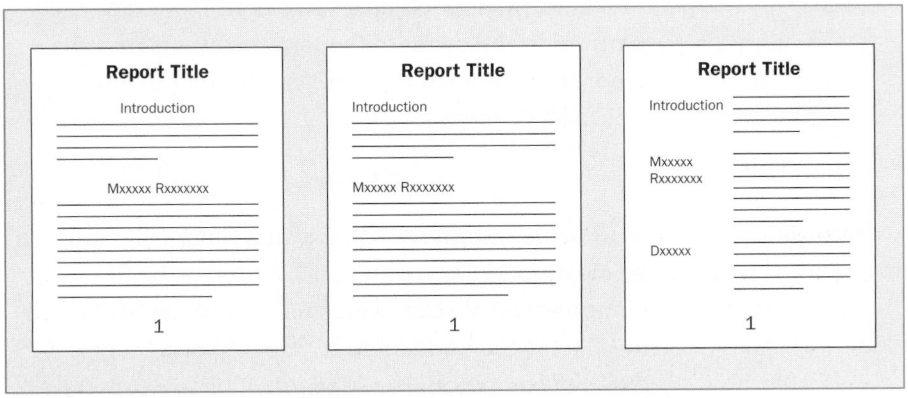

Figure 6-1 Different ways to integrate headings and text.

Place a short glossary in the introduction, a longer glossary in an appendix

project. If there are limiting factors, it identifies them. For example, if 18 methods for improving packaging are investigated in a project but only 4 are discussed in the report, the scope indicates which factors (such as cost, delivery time, and availability of space) limited the selection. Sometimes the scope may include a short glossary of terms that need to be defined before the reader starts to read the discussion.

A good introduction "sets the scene"

Background information comprises facts readers must know if they are to fully understand the discussion that follows. Facts may include descriptions of conditions or events that caused the project to be authorized, and details of previous investigations or reports on the same or a closely related subject. In a highly technical report, or when a significant time lapse has occurred between it and previous reports, background information may also provide a theory review and references to other documents. If the theory review is lengthy and there is a long list of documents, they are often placed in appendixes, with only a brief summary of the theory and a quick reference to the list appearing in the introduction.

The introduction shown here, plus those forming part of the two sample reports later in this chapter, represents the many ways a writer can introduce a topic.

Introduction 1

Background Northern Railways plans to build a short-span bridge 1 mile north of Lake Peele in northern Alaska and has a job lot of steel the company wants to use for constructing the bridge. In letter *Purpose* NR-70/LM dated March 20, 2003, Mr David L. Harkness, Northern Area Manager, requested that H. L. Winman and Associates test a sample of this steel to determine its properties and to assess its suitability for use as structural members for a bridge in a very low temperature environment. *Scope* Two Charpy impact tests were performed, one parallel to the grain of the test specimen, and the other transverse to the grain, at 10° increments from +22°C down to –50°C.

Introduction 2

Say why you are writing...

Purpose H. L. Winman and Associates was commissioned by Ms. Rita M. Durand, president and general manager of Auto Drive-Inns Inc. of Dallas, Texas, to select a Cleveland site for the first of a proposed chain of computerized, automatic drive-in grocery outlets to be built in the northeast. This area was chosen as the test site because it represents an average community in which to assess customer acceptance of such a service.

Background

Aside from exterior façade and foundation details, Auto Drive-Inns are built to a standard 80 ft by 30 ft pattern with an order window at one end of the longer wall, and a delivery window at the other end. Auto Drive-Inns carry only a limited selection of groceries, milk and fruit, but boast 60-second service from the time an order is placed to its delivery at the other end of the building. For this reason, Auto Drive-Inns attract people hurrying home from work, the impulse buyer, and the late-night traveler, rather than the selective buyer. The Drive-Inns depend more on the volume of customers than on the volume of goods sold to each individual.

...describe what has gone before, and...

Scope

The chief consideration in selecting a site must therefore be a location on the homeward-bound side of a main trunk road serving a large residential area. The site must have quick and easy entry onto and exit from this road, even during peak rush-hour traffic. The residential area should be occupied mainly by single persons and younger families in which both parents work. And there should be little competition from walk-in grocery stores.

...explain factors affecting both the study and the report

Discussion

The discussion, which normally is the longest part of a report, presents all the evidence (facts, arguments, details, data, and results of tests) that readers need to understand the subject. The writer must organize this evidence logically to avoid confusing readers, and present it imaginatively to hold their interest.

There are three ways you can build the discussion section of a report:

By **chronological development**—in which you present information in the sequence that the events occurred.

By **subject development**—in which you arrange information by subjects, grouped in a predetermined order.

By **concept development**—in which you organize information by concept, presenting it as a series of ideas that imaginatively and coherently reveal how you reasoned your way to a logical conclusion.

Reports using the chronological or subject method offer less room for imaginative development than those using the concept method, mainly because they depend on a straightforward presentation of information. The concept method can be very persuasive. Identifying and describing your ideas and thought processes helps your readers organize their thoughts along the same lines.

As a report writer, you must decide early in the planning stages which method you intend to use, basing your choice on which is most suitable for the evidence you have to present. Use the following notes as a guide.

Chronological Development

A discussion that uses the chronological method of development is simple to organize and write. Planning is minimal: you simply arrange the major topics in the order they occurred, and eliminate irrelevant topics as you go along. You can use it for very short reports, for laboratory reports showing changes in a specimen, for progress reports showing cumulative effects or describing advances made by a project group, and for reports of investigations that cover a long time and require visits to many locations to collect evidence.

Chronological development is essentially factual...

But the simplicity of the chronological method is offset by some major disadvantages. Because it reports events sequentially it tends to give equal emphasis to each event regardless of its importance, which may cause readers to lose interest. If you read a report of five astronauts' third day in orbit, you do not want to read about every event in exact order. It may be chronologically true to report that they were wakened at 7:15, breakfasted at 7:55, sighted the second stage of their rocket at 9:23, carried out metabolism tests from 9:40 to 10:50, extinguished a cabin fire at 11:02, passed directly over Houston at 11:43, and so on, until they retired for the night. But it can make dull reading. Even the exciting moments of a cabin fire lose impact when they are sandwiched between routine occurrences.

When using chronological development, if you are to hold your readers' attention you must still manipulate events. You must emphasize the most interesting items by positioning them where they will be noticed, and deemphasize less important details. This has been done in the following passage, which groups the previous events in descending order of interest and importance:

...but there still is room for some orchestration

> The highlight of the astronauts' third day in orbit was a cabin fire at 11:02. Rapid action on their part brought the fire, caused by a short circuit behind panel C, under control in 38 seconds. Their work for the day consisted mainly of metabolism tests and...
>
> They sighted the first stage of their rocket on three separate occasions, first at 9:23, then at...and..., when it passed directly over Houston. Their meals were similar to those of the previous day.

Over a five-year period H. L. Winman and Associates has been investigating the effects of salt on concrete pavement. Technical editor Anna King has suggested that the final report should have chronological development, because the investigation recorded the extent of concrete erosion at specific intervals. She wanted the engineer writing the report to describe how the erosion increased annually in direct relation to the amount of salt used to melt snow each year, and for the final conclusion to demonstrate the cumulative effect that salt had on the concrete.

Subject Development

If the previous investigation had been broadened to include tests on different types of concrete pavements, or if both pure salt and various mixtures of salt and sand had been used, then the emphasis would have shifted to an analysis of erosion on different surfaces or caused by various salt/sand mixtures, rather than a direct description of the cumulative effects of pure salt. For this type of report Anna King would have suggested arranging the topics in *subject* order.

The subject order could be based on different concentrations of salt and sand. The engineer would first analyze the effects of a 100% concentration of salt, then continue with salt/sand ratios of 90/10, 80/20, 70/30, and so on, describing the results obtained with each mixture. Alternatively, the engineer could select the different types of pavement as the subjects, arranging them in a specific order and describing the effects of different salt/sand concentrations on each surface.

Subject development sorts information into groups

Barry Brewster, who is head of H. L. Winman and Associates's design and drafting department, has been investigating high-speed color printers for use with the company's CAD system, and plans to recommend the most suitable model for installation in his department. He writes his report using the subject method. First, he establishes selection criteria, defining what he needs in an ideal printer, such as its speed, purchase price, economy of operation, and quality of printout. Then, as his tests have already determined the most suitable printer, he discusses the best machine either first or last. If he chooses to describe it first, he can state immediately that it is the best printer, and say why, by comparing it to the selection criteria. He can then discuss the remaining printers in decreasing order of suitability, also comparing each to the selection criteria to show why it is less suitable. If he prefers to describe the best printer last, he can discuss the printers in increasing order of suitability, again comparing each against the selection criteria and then stating why he has rejected each one, before describing the next.

The subject method of development permits report writers to hide their personal preferences until almost the end of their reports or, as Barry has done, to let their preferences show all the way through. These alternative approaches are illustrated in Figure 5-3 on page 106.

Concept Development

By far the most interesting reports are those using the concept method of development. They need to be organized more carefully than reports using either of the previous methods, but they give the writer a tremendous opportunity to devise an imaginative arrangement of the topic.

They can also be very persuasive. Because the report is organized in the order in which you reasoned your way through the investigation, your readers will much more readily appreciate the difficulties you encountered, and will frequently draw the correct conclusion even before they read it. This helps readers feel they are personally involved in the project.

Concept development has the greatest potential for orchestrating information

You can apply the concept approach to your reports by thinking of each project as a logical but forceful procession of ideas. If you are personally convinced that the results of your investigation are valid, and remember to explain in your report *how* you reached the results and *why* they are valid, then you will probably be using the concept method properly. Always anticipate reader reaction. If you are presenting a concept (an idea, plan, method, or proposal) that readers are likely to accept, then use a straightforward four-step approach:

1. Describe your concept in a brief overview statement.
2. Discuss how and why your concept is valid; offer strong arguments in its favor, starting with the most important and working down to the least important.
3. Introduce negative aspects, and discuss how and why each can be overcome or is of limited importance.
4. Close with a restatement of your concept, its validity, and its usefulness.

But if you are presenting a controversial concept, or need to overcome reader bias, then modify your approach. Try to overcome objections by carefully establishing a strong case for your concept *before* you discuss it in detail.

Andy Rittman used this approach in a report he prepared for Mark Dobrin, owner/manager of a company making extruded plastic and metal parts. Manufacturing costs had risen steeply over the past two years and Mark's prices had become uncompetitive. Mark thought he should replace some of his older, less efficient equipment, so he asked Andy to evaluate his needs.

Andy quickly realized that if Mark was to avoid going out of business he would have to replace much of his equipment with microprocessor-controlled machines, and do it soon. Because the cost would be high, he would have to lease, rather than buy the new equipment. Here, Andy had a problem. Mark was as old-fashioned as some of his extruders and shapers, and throughout his life he had steadfastly refused to purchase anything that he could not buy outright. He was unlikely to change now.

In his report, Andy used a carefully reasoned argument to prove to Mark that he needed a lot of new equipment and that the only feasible way he could acquire it would be to lease it. Throughout, Andy wrote objectively but sincerely of his findings, hoping that the logic of his argument would swing Mark around to accepting his recommendation. Very briefly, here is the step-by-step approach Andy used:

The concept method leads the reader to the right answer...

- He opened with a summary that told Mark that to avoid bankruptcy he would have to invest in a lot of expensive equipment and make extensive changes in his operating methods.
- Andy then produced financial projections to prove his opening statement, and discussed the productivity and profitability necessary for Mark to remain in business.
- He discussed why Mark's equipment and methods were inefficient, introduced the changes Mark would have to make, established why each change was necessary, and demonstrated how each would improve productivity. (Andy referred Mark to an appendix containing equipment descriptions, justifications, and costs.)
- Andy then introduced two sets of cost figures: one for making the minimum changes necessary for Mark's business to survive, and the second for more comprehensive changes that would ensure a sound operating basis for the future. He commented that both would require capital purchases likely to be beyond Mark's financial resources.
- He outlined alternative financing methods available to Mark, the implications and limitations of each, and the financial effect each would have on Mark's business. (Although he introduced leasing, Andy made no attempt to persuade Mark that he would have to lease; he let the figures speak for themselves.)

...carefully tracing a logical, persuasive, and sometimes intricate path

- Andy concluded by summarizing the main points he had made: that new equipment *must* be acquired; that to buy even the minimum equipment was beyond Mark's financial resources; and that, of the financing methods available, leasing was the most feasible.
- In his recommendation, Andy suggested that Mark should make comprehensive changes and lease the new equipment. (By then, Andy had become so involved with Mark's predicament that he wrote strongly and sincerely.)

Even though the concept method challenges a writer to fashion interesting reports, it is not always the best reporting medium. For instance, the concept method could possibly have been used for Barry Brewster's printer report mentioned earlier, but it is doubtful whether the topic would have warranted full analysis of the author's ideas. When a topic is fairly clear-cut, there is no need to lead the reader through a lengthy "this is how I thought it out" discussion. Reserve the concept method for topics that are controversial, difficult to understand, or likely to meet reader resistance.

Keep the narrative simple and uncluttered

Whichever method you use, avoid cluttering the discussion with detailed supporting information. Unless tables, graphs, illustrations, photographs, statistics, and test results are essential for reader understanding while the report is being read, banish them to an appendix. But always refer to them in the discussion, like this:

> The test results attached as Appendix C show that aircraft on a bearing of 265°T experienced considerably weaker reception than aircraft on any other bearing. This was attributed to...

Readers *interested only in results* will consider that this statement tells them enough and continue reading the report. Readers interested in *knowing how the results were obtained*—who want to see the overall picture—will turn to Appendix C to find out how the report writer went about performing the tests.

If an illustration or table is essential, extract the key points from an appendix and use them to create a miniature illustration or table that can be inserted beside or embedded in the narrative without impeding reading continuity.

Unless the discussion is very short, divide it into a series of sections that are each preceded by an informative heading. After each heading, start the section with an overview statement to describe what the section is about and suggest what conclusion will be drawn from it. Overview statements are miniature summaries that direct a reader's attention to the point you want to make. If the section is short, the overview statement may be a single sentence; if the section is long, it will probably be a short paragraph.

At the end of each major section, insert a concluding statement that summarizes the result of the discussion within that section. From these section conclusions you can later draw your main conclusions.

Conclusions

There should be no surprises here

Conclusions briefly state the major inferences that can be drawn from the discussion. You must base them entirely on previously stated information. *Never* introduce new material or evidence to support your argument. If there is more than one conclusion, state the main conclusion first and follow it with the remaining conclusions in decreasing order of importance. This is shown in the two examples below, which present the same conclusions in both narrative and tabular form.

Narrative Conclusion

> If we upgrade to version 4.1 of the *Mosaic* software, we will also have to upgrade our operating system. The one-time cost will be $7600, but we will increase our multimedia production capability by 65%.

Tabular Conclusion

If we upgrade to version 4.1 of the *Mosaic* software, we will

- experience a 65% increase in our multimedia production capability,
- have to upgrade our operating system, and
- incur a one-time cost of $7600.

Because conclusions are *opinions* (based on the evidence presented in the discussion), they must never tell the reader what to do. This task must always be left to the recommendations.

Recommendations

Now you can let your voice be heard

Recommendations appear in a report when the discussion and conclusions indicate that further work needs to be done, or when you have described several ways to resolve a problem or improve a situation and want to identify which is best. Write recommendations in strong, definite terms to convince readers that the course of action you advocate is valid. Use the first person and active verbs, as has been done here:

Strong I recommend that we build a five-station prototype of the Microvar system and test it operationally.

Compare this with the same recommendation written in the third person, using passive verbs:

Weak It is recommended that a five-station prototype of the Microvar system be built and tested under operational conditions.

If you feel you cannot use the personal "I," try using the plural "we," to indicate that the recommendations represent the company's viewpoint. For example:

Strong We recommend building a five-station prototype of the Microvar system. We also recommend that you

1. install the prototype in Railton High School,
2. commission a physics teacher experienced in writing programmed instruction manuals to write the first programs, and
3. test the system operationally for three months.

And no surprises here, either!

Because recommendations must be based solidly on the evidence presented in the discussion and conclusions, they must *never* introduce new evidence or new ideas.

Appendixes

Related data not necessary to an immediate understanding of the discussion should be placed further back in the report, in the appendixes. The data can vary from a complicated table of electrical test results to a simple photograph of a blown transistor. The appendixes are a suitable place for manufacturers' specifications, graphs, analytical data, drawings,

The appendix is *not* a storage place for information that only *might* be useful

sketches, excerpts from other reports or books, cost analyses, and correspondence. There is no limit to what can be placed in the appendixes, providing it is relevant and reference is made to it in the discussion.

The importance of an appendix has no bearing on its position in the report. Whichever set of data is mentioned first in the discussion becomes Appendix A, the next set becomes Appendix B, and so on. Each appendix is considered a separate document complete in itself and is paginated separately, with its front page labeled 1. Examples of appendixes appear at the end of Report 1, later in this chapter.

Subsidiary Parts

In addition to the six major parts of a formal report, there are several additional parts that perform more routine functions. Although referred to here as "subsidiary" parts, they nevertheless contribute much to a report's effectiveness. Because they directly affect the image conveyed of both you and your company, they must be prepared no less carefully than the rest of the report.

Cover

The cover should reflect the company's image

Almost every major formal report has a cover. It may be made of glossy cardboard printed in multiple colors and bound with a dressy plastic binding, or it may be only a light cover of colored fiber material stapled on the left side. The cover not only informs readers of the report's main topic but also conveys an image of the company that originated it. This "matching" of subject matter and company image plays an important part in setting the correct tone.

The cover should contain the report title, the name of the originating company and, perhaps, the name of its author. The title should be set in bold letters well-balanced on the page and separated from any other information.

The choice of title is particularly important. It should be short yet informative, implying that the report has a worthwhile story to tell. Compare the vague title below with the more informative version beneath it.

Original vague title

Radome Leakage

Revised informative title

Porosity of Fiberglass Causes Radome Leakage

Technical officers at remote radar sites would glance at the first title with only a muttered, "Looks like somebody else has the same problem we've got." But the second title would encourage them to stop and read the report. They would recognize that someone may have found the cause of (and, perhaps, a remedy for) a trouble spot that has bothered them for some time.

Note how the following titles each *tell* something about the reports they precede:

> **Reducing Ambient Noise in Air Traffic Control Centres**
> **Effects of Diesel Exhaust on Latex Paints**
> **Salt Erosion of Concrete Pavements**

The report title should capture readers' interest

These titles may not attract every potential reader who comes across them, but they will certainly gain the attention of anyone interested in the topics they describe.

Title Page

The title page normally carries four main pieces of information: the report title (the same title that appears on the cover); the name of the person, company, or organization for whom the report has been prepared; the name of the company originating the report (sometimes with the author's name); and the date the report was completed. It may also contain the contract number, a report number, a security classification such as CONFIDENTIAL or SECRET, and a copy number (important reports given only limited distribution are sometimes assigned copy numbers to control and document their issue). All this information must be tastefully arranged on the page, as has been done in the full report later in this chapter.

Table of Contents

All but very short reports contain a table of contents (T of C). The T of C not only lists the report's contents, but also shows how the report has been arranged. Just as a prospective book buyer will scan the contents page to discover what a book contains and whether it will be of interest, potential readers scan a report's T of C to find out how the author has organized the work.

If the T of C seems sparse or disjointed, check that the report narrative has sufficient *descriptive* headings

The pleasing arrangement for a T of C on page 157 uses the single word **Contents** rather than "Table of Contents;" we recommend the shorter title. The contents page also contains a list of appendixes, with each identified by its full title. In long reports you may also insert a list of illustrations and their page numbers between the T of C and the list of appendixes.

The introductory pages to a report (i.e. the Summary and Contents pages) are numbered using lower case roman numerals. All other parts of the report are numbered with arabic numerals, starting with the **Introduction**, which becomes page 1.

References (Endnotes), Bibliography, and Footnotes

Accurate documentation of information sources is essential

A report writer who refers to another document, such as a textbook, journal article, report, or correspondence, or to other persons' data or even a conversation, must identify the source of this information in the report. To avoid cluttering the report narrative with extensive cross-references, the reference details are placed in a storage area at the end of the report or at the foot of the page. This storage area is known as a list of references or a bibliography. Specifically:

Bibliography Styles Handbook
www.english.uiuc.edu/cws/wworkshop/bibliography.html
The Bibliography Styles Handbook, one section of the University of Illinois's Writers' Workshop, provides information about the bibliographic styles of the American Psychological Association (APA) and the Modern Languages Association (MLA), and about the old MLA style.

A **List of References** is the most convenient and popular way to list source documents. The references are typed as a sequentially numbered list at the end of the narrative sections of the report (usually immediately ahead of the appendixes). Such numbered references are sometimes referred to as ***Endnotes***. For a short example, see page 167. (Footnotes, which are printed at the foot of the page on which the particular reference appears, are seldom used in contemporary reports.)

A **Bibliography** is an alphabetical listing of the documents used to research and conduct a project. The documents are listed in alphabetical order of authors' surnames, and the list is placed at the end of the report narrative. A bibliography may list many more documents than are referred to in the report. An example appears on page 145.

By adopting the styles recommended by organizations such as the Modern Languages Association (MLA) and the American Psychological Association (APA), source references can be written many ways. The composite method style we show here most closely parallels the MLA approach, but with some differences that have become entrenched in the technical documentation field. If you are writing in industry, the suggested approach will work well for you. If you are writing in an academic situation or for a technical journal, you should enquire which style the professors or the journal editor prefer.

Until the mid-90s, most source referencing was to printed documents. Today, however, you will often be gaining information from an electronic resource, which may be by email or searching on the Web. Because Internet sources may only be transitory, you will need to quote more source information whenever possible. How to do this will be shown on the following pages.

Preparing a List of References

References should contain specific information, arranged in this sequence:

(a) Author's name (or authors' names).
(b) Title of document (article, book, paper, report).
(c) Identification details, such as:

For a book: city and state or country of publication, publisher's name, and year of publication.

For a magazine article or technical paper: name of magazine or journal; volume and issue number; date of issue.

For a report: report number; name and location of issuing organization; date of issue.

For correspondence: name and location of issuing organization; name and location of receiving organization; the letter's date.

For a conversation or speech: name and location of speaker's organization; name, identification, and location of listener; the date.

For an excerpt from a Web page: the name of the author/designer (if known); the title of the page or source; the name of the organization that owns the site; the latest update; the date the information was accessed; and the URL.

(d) The page number (if applicable) on which the referenced item appears or starts.

Each reference source must be recorded exactly as it appears on the original document

Referencing a Book

If you are referring to information in a book authored by only one person, the entry in your list of references should contain:

(a) Author's name (in natural order: first name and/or initials, and then surname).
(b) Book title (set in italic type).
(c) City of publication, publisher's name, and year of publication (all within one set of parentheses).
(d) Page number (the first page of the referenced pages).

If it is your first reference, and you are referring to an item on page 174 of the book, your entry would look like this:

1. Laurinda K. Wicherly, *Fiberoptic Modes of Communication* (New York: The Moderate Press Inc., 2002), p. 174.

Each reference entry is given a sequential number

If a book has two authors, both are named:

2. David B. Shaver and John D. Williams, *Management Techniques for a Research Environment* (Boulder, Colorado: Witney Publications, 2003), p. 215.

But if there are three or more authors, only the first-named author is listed and the remaining names replaced by "and others":

3. Donald R. Kavanagh and others,... *(etc).*

If a book is a second or subsequent edition (as this book is), the edition number is entered immediately after the book title:

4. Ron Blicq and Lisa Moretto, *Technically-Write!* 6th ed. (Englewood Cliffs, NJ: Pearson Education, 2004).

Some books contain sections written by several authors, each of whom is named within the book, with the whole book edited by another person. If your reference is to the whole book, identify it by the editor's name:

5. Donna R. Linwood, ed., *Seven Ways to Make Better Technical Presentations* (Portland, Oregon: Bonus Books Inc., 2004).

But if your reference is to a particular section of the book, identify it by the specific author, enclose the section title in quotation marks, set the book title in italics, then name the editor:

External document titles are listed in italic type; internal section titles are listed within quotation marks

6. Kevin G. Wilson, "Preparation: The Key to a Good Talk," *Seven Ways to Make Better Technical Presentations*, ed. Donna R. Linwood (Portland, Oregon: Bonus Books Inc., 2004), p. 71.

(In examples 5 and 6, "ed." means "editor" or "edited by.")

Referencing a Magazine or Journal Article

Similarly, if you are referring to an article in a magazine or journal, list these details:

(a) Author(s)'s name(s) (in natural order).
(b) Title of article (always in quotation marks).
(c) Title of journal or magazine (set in italics).
(d) Volume and issue numbers (shown as two numbers separated by a colon).
(e) Journal or magazine issue date.
(f) Page on which article or excerpt starts (optional entry).

If an article is your seventh reference, it would look like this:

7. Lilita Rodman, "You-attitude: A Linguistic Perspective," in *Technostyle*, 17:2, Winter 2002, p. 55.

If a magazine article does not show an author's name, then the entry starts with the title of the article:

8. "Selling to the EEC: Challenge of the New Millennium," *Technical Marketing*, 11:5, May 2003, p. 113.

Reports, memos, and email messages can be listed as reference sources...

Referencing a Report

To refer to a report written by yourself or another person, list this information:

(a) Author's name (or authors' names), if the author is identified on the report.
(b) Title of report, in italics.
(c) Report number, or other identification (if any).
(d) Name and location (city and state) of organization issuing report.
(e) Report date.
(f) Page number (if a specific part of the report is being referenced).

Here is an example:

9. Derek A. Lloyd, *Effective Communication and Its Importance in Management Consulting*. Report No. 61, Smyrna Development Corporation, Atlanta, Georgia, 18 February 2004.

Referencing an Email Message, Letter, or Memo

For an email message, the entry should look like this:

10. Christine Lamont (c.lamont@macroeng.com), "Replacing Vancourt Meters." Email to Wayne Kominsky (kominsky@7designgrp.net), 31 October 2003.

For a letter or memo, the email references are replaced by company name and location, the title often is omitted, and the word "email" is replaced by "letter" or "memo"):

11. Christine Lamont, Macro Engineering Inc., Phoenix, Arizona. Letter to Wayne Kominsky, No. 7 Design Group, Dallas, Texas, 31 October 2003.

Referencing a Conversation or Speech

For a conversation or speech follow these examples:

12. David R. Phillips, Lakeside Power and Light Company, Montrose, Ohio, in conversation with Anna King, H. L. Winman and Associates, Cleveland, Ohio, January 7, 2004.

...as can talks and telephone or face-to-face conversations

13. Francis R. Cairns, Elwood Martens and Associates, San Diego, California, speaking to the 8th Symposium on Videodisk Technology, Chicago, Ilinois, September 16, 2003.

Referencing an Excerpt from a Web Page

If information is available only on a Web page, and is *not* printed elsewhere, list:

(a) author's name (if an author is identified),
(b) title of the specific piece of information (enclosed within quotation marks),
(c) the title of the page or source (in italics),
(d) the name of the institute or organization that owns the Web page,
(e) the date the information was entered, or the latest update, using day (numeral), month (first three letters, spelled out), year (numeral),
(f) the date the information was accessed, and
(g) the Web identification (the URL, within angle brackets).

For example:

14. Göran Nordlund, "Documentation for Medical Equipment – a Real Cross-Cultural Challenge" in *Forum 2003 Preliminary Programme*, 19 Nov 2002, retrieved 2 Mar 2003 **<http://www.intecom.org/Forum 2003 Preliminary Programme>**.

However, if the information has also been published in print form, the entry should refer to the original document *and* the Web site. List:

(a) the full printed identification (for a book, article, technical paper, etc),
(b) the date the information was entered on the Web site (day, month, year), and
(c) the Web identification (within angle brackets).

Here is an example:

15. "Keeping Track of Your Performance" in *RGI News*, No. 13, Winter 2001-2002, 18 Mar 2002, retrieved 16 Jul 2003 **<http://www.rgilearning.com/newsletters>.**

Referencing an Excerpt from an Online Book

The entry should contain this information:
(a) Author's name
(b) Chapter title (in quotation marks)
(c) Book title (in brackets)
(d) Book identification (city of publication, publisher, copyright date), if available
(e) Date of electronic publication, or latest update
(f) Name of the organization responsible for the Web site
(g) Date the information was accessed
(h) URL identification.

Here is an example:

16. Marvin LeTouche, "Maintaining Quality Levels," *Quality Control in the Mining Industry*. Chicago: Bronzeline Publishers, 2003. 15 Oct 2003: Mainstream Mining Inc; retrieved 8 Jan 2004 **<http://www.mainstream.ca/qc>.**

Referencing an Excerpt from an Online Magazine Article

The following information should be recorded:
(a) Author's name (if author is identified)
(b) Title of article (in quotation marks)
(c) Title of magazine or journal (in italics)
(d) Issue number plus year of publication (in parentheses)
(e) Page number
(f) Date accessed
(g) URL identification.

For example:

17. Margery Leduc, "Are Handheld Computers Taking Over from Laptops?" *Computers Unlimited*, 8.3 (2003). Retrieved 23 Oct 2003 **<http://www.bearskincollege/lib/online.html>**.

Referencing Information in an Online Database

The information to be recorded is as follows:
(a) Author name (if identified)

(b) Title of information (in quotation marks)
(c) Name of database (in italics)
(d) Name of organization owning site (if available)
(e) Date accessed
(f) URL identification.

Here is a typical example:

18. "Radiant Heat in Tomorrow's Homes", Heating Industry Standards Institute. Heating Industry Institute. Retrieved 10 May 2003 **<http://www.heating.standards.com/market/2455/616.html>**.

Additional Factors to Consider

Remember that when a magazine article, technical paper, or report is published as one of several documents bound into a volume, then it is listed within quotation marks (only the title of the volume is set in italics). But if the article, technical paper, or report is published as a *separate* document, the quotation marks are omitted and the title of the article, paper, or report is set in italic type, as in entry No. 9.

Every entry in a list of references must have a corresponding reference to it in the Discussion section of your report. At an appropriate place in the narrative you should insert a superscript (raised) number to identify the particular reference. It should look like this:

Earlier tests[3] showed that speeds higher than 2680 rpm were impractical.
(*Alternatively, the raised 3 could go here.*)

Numbering each entry simplifies cross-referencing between the text and the list of references

If you refer to the same document several times, your list of references needs to show full details for that document only the first time you refer to it. Subsequent references can be shown in a shortened form containing only the author's name (or authors' names) and the page number. For example, if the first reference you make is to an item on page 48 of the particular book described below, the entry in the list of references would be

1. Wayne D Barrett, *Management in a Technical Domain* (San Francisco, California: Martin-Baisley Books, 2004), p. 48.

Now suppose that your second and third sources are other documents, but for your fourth source you again refer to *Management in a Technical Domain*, this time quoting from page 159. Now you need list only the author's surname and the new page number.

4. Barrett, p. 159.

Do the same for each future reference to the same document, simply changing the page number each time. (Note that the Latin terms *ibid.* and *op. cit.* are no longer used.) You can even make repeated references to several different documents by the same author by simply inserting the year of publication for the particular document between the author's name and the page number:

9. Barrett, 2004, p. 159.

Preparing a Bibliography

A bibliography lists not only the documents to which you make direct reference, but also many other documents that deal with the topic. The major differences between a list of references and a bibliography are:

In a bibliography, authors' surnames are used for easy cross-referencing

- Bibliography entries are *not* numbered 1, 2, 3, etc.
- The name of the *first-named* author for each entry is reversed, so that the author's surname becomes the first word in the entry. (If there is a second-named author, his or her name is *not* reversed.)
- The *first* line of each bibliography entry is extended about one-third inch to the left of all other lines (see Figure 6-2).
- The entries are arranged in alphabetical sequence of first-named authors.
- Punctuation of individual entries is significantly different, with each entry being divided into three compartments separated by periods: (1) author identification (name, etc); (2) title of book or specific article; and (3) publishing details. (Positions of periods are shown in Figure 6-2.)
- Page numbers are usually omitted, since generally the bibliography refers to the whole document. (Reference to a specific page is made within the narrative of the report.)

Most bibliography entries contain three distinct groups of information

Figure 6-2 shows how to list bibliography entries from various sources. The entries for this bibliography have been created from some of the "reference" entries listed earlier. The number to the right of each entry is cross-referenced to the explanatory list below:

1. Book by one author.
2. Book by two authors, sixth edition.
3. Conference speech.
4. Web source.
5. Email.
6. Report.
7. Conversation.

Bibliography

1. Barrett, Wayne D. *Management in a Technical Domain*. San Francisco, California: Martin-Baisley Books, 2004.

2. Blicq, Ron, and Lisa Moretto. *Technically-Write!*, 6th ed. Englewood Cliffs, NJ: Pearson Education, 2004.

3. Cairns, Frances R., Elwood Martens and Associates, San Diego, California. Speaker at the 8th Symposium on Videodisk Technology, Chicago, Illinois, 16 September 2003.

4. "Keeping Track of Your Performance." *RGI News*, No. 13, Winter 2001–2002. 23 June 2002. Retrieved 30 November 2003 <http://www.rgilearning.com/newsletters>.

5. Lamont, Christine, Macro Engineering Inc, Phoenix, Arizona. Email to Wayne Kominsky, No. 7 Design Group, Dallas, Texas, 12 December 2003.

6. Lloyd, Derek A. *Effective Communication and Its Importance in Management Consulting*. Report No. 61, Smyrna Development Corporation, Atlanta, Georgia, 18 February 2004.

7. Phillips, David R., Lakeside Power and Light Company, Montrose, Ohio. Conversation with Anna King, H. L. Winman and Associates, Cleveland, Ohio, 7 January 2004

8. Rodman, Lilita. "You-attitude: A Linguistic Perspective." *Technostyle*, 17:2, Winter 2002.

9. "Selling to the EEC: Challenge of the New Millenium." *Technical Marketing*, 18:5, May 2003.

10. Shaver, Donald B., and John D. Williams. *Management Techniques for a Research Environment*. Boulder, Colorado: Witney Publications, 2003.

11. Wilson, Kevin G. "Preparation: The Key to a Good Talk." *Seven Ways to Make Better Technical Presentations*, ed. Donna R. Linwood. Portland, Oregon: Bonus Books Ltd., 2004.

Extending the first line of each entry to the left helps readers find specific source references

Figure 6-2 A typical bibliogaphy.

8. Journal/magazine article.

9. Magazine article, with no author identification.

10. Book by two authors.

11. Section of book with section written by one author and whole book edited by another.

The text states author's name and relevant page number

Because a bibliography is not numbered, you cannot cross-refer directly to it simply by inserting a superscript number in the report narrative, as can be done with a list of references. The most common method is to insert a parenthetical reference in the narrative that includes the author's name (or authors' names) and the page number:

> Although the tests conducted in the Northwest Territories (Faversham, p. 261) showed only moderate decomposition...

The full descriptive listing for Faversham's book or report would be carried in the bibliography.

If several publications by the same author are listed in the bibliography, then the date of the particular publication is included as a parenthetical reference to identify which document is being mentioned:

> The most significant tests were those conducted 22 kilometres south of Old Crow, in the Yukon (Crosby, 2001, p. 17), which showed that...

Cover Letter

A cover letter is a brief letter that identifies the report and states why it is being forwarded to the addressee. The following letter accompanied the second report in this chapter (see pages 172 to 176).

A cover letter is often only one short paragraph...

Dear Mr. Merrywell:

We enclose our report No. 8-23, "Selecting New Elevators for the Merrywell Building," which has been prepared in response to your letter LDR/71/007 dated April 27, 2003.

If you would like us to submit a design for the enlarged elevator shaft, or to manage the installation project on your behalf, we shall be glad to be of service.

Sincerely,

Barry V. Kingsley

H. L. Winman and Associates

Some cover letters include comments that draw attention to key factors described in the report or that evolve from it, and sometimes summarize or interpret the report's main findings. This is done in the cover letter preceding the first report in this chapter (see Figure 6-5).

Executive Summary

An executive summary is an analytical summary of the purpose of the report, its main findings and conclusions, and the author's recommendations. Unlike the normal report summary prepared for all readers, the executive summary can present detailed information on aspects of particular concern to senior executives, and often may discuss financial implications. An executive summary can be presented in two ways: externally, as a letter pinned to the front of the report, or internally, as an integral part of the report.

...whereas an executive summary usually has two or more paragraphs

If the executive summary is attached to the report like a cover letter, the recipient may remove it before circulating the report to other readers. Hence, an external executive summary is written like a letter and is addressed to a specific reader, or group of readers, which permits its author to make comments that are intended only for that reader's eyes, or to deal with sensitive issues that for political reasons should not be discussed in the body of the report. The executive summary preceding Report No. 1 serves this purpose (see Figure 6-5 on page 154).

If, however, the executive summary is bound within the report so that it will be read by everyone, its purpose becomes more general and its author simply summarizes and perhaps comments on the report's major findings. Rather than being written like a letter, the words **Executive Summary** are centered at the top of the page and the summary is presented like a semiformal report. Such an executive summary may be positioned immediately inside the report cover, after the title page (in which case it replaces the normal short summary), or after the table of contents.

An integral executive summary rarely contains sensitive or confidential information

When an executive summary is bound inside the report, a brief cover letter may also be prepared as a transmittal document and attached to the front of the report.

An executive summary may often precede a major technical proposal, in which a company describes how it can successfully tackle a task at an economical price for the government or another company.

The Complete Formal Report

The Main Parts

Two formal reports are included in this chapter, each typical of the quality of writing and presentation that the technical business industry expects

The same information, but two ways to arrange it

of engineering, science, and computer graduates. The first report is presented in the traditional arrangement discussed so far, and the second in an alternative, pyramidal arrangement. In each case the parts of the report remain the same, but their sequence changes. Capsule descriptions of the main parts are listed in Table 6-2.

The text preceding each report discusses the report's sequence, identifies how the project was initiated and the report came to be written, and comments on both the report and the author's approach. The reports are typed single-spaced with a clear line between paragraphs, which is the style preferred by industry. In comparison, academic institutions tend to prefer double spacing throughout.

Traditional Arrangement of Report Parts

(Conclusions and Recommendations *after* the Discussion)

In the traditional arrangement there is a logical flow of information: the **introduction** leads into the **discussion**, from which the writer draws **conclusions** and makes **recommendations** (the two latter parts are sometimes referred to jointly as the **terminal summary**), as illustrated in the mini-report in Figure 6-3. This arrangement is used for most technical and business reports.

Formal Report 1: Installing a Radiant Energy Heating System for Hartwell Enterprises Inc.

The author of this report (Figures 6-5 and 6-6) is Karen Woodhouse, an engineer working in the Minneapolis, Minnesota, branch of H. L. Winman and Associates. She typed and edited the report on her desktop computer, and configured the word-processing program to automatically justify the right margin (make it vertically straight), insert page numbers, and create the contents page.

Comments on the Report

To understand the circumstances leading up to the report, first read Mark Hesseltine's letter of authorization in Figure 6-4. Mark is an architect with No. 5 Design Group in Duluth, Minnesota. He has designed a new building for Hartwell Enterprises Inc., a specialty manufacturer in Duluth, and his client has expressed interest in having radiant energy heating installed in the new building. Because No. 5 Design Group does not have radiant energy expertise in-house, Mark contracts with H. L. Winman and Associates to evaluate the practicability and cost of installing such a system.

Check before you place your hands on the keyboard: the person you are writing to may not be your primary reader

Karen admitted that at first that she had difficulty focusing her report: "The problem was that initially Mark seemed to be my primary reader: after all, he commissioned the report. Yet when I realized that he would

Table 6-2 The main parts of a formal report.

Cover:	Jacket of report; contains title of report and name of originating company; its quality and use of color reflect company image.
Title Page:	First page of report; contains title of report, name of addressee or recipient, author's name and company, date, and sometimes a report number.
Summary:	An abridged version of whole report, written in nontechnical terms; *very* short and informative; normally describes salient features of report, draws a main conclusion, and makes a recommendation; always written last, after remainder of report has been written.
Table of Contents:	Shows contents and arrangement of report; includes a list of appendixes and, sometimes, a list of illustrations.
Introduction:	Prepares reader for discussion to come, indicates purpose and scope of report, and provides background information so reader can read discussion intelligently.
Discussion:	A narrative that provides all the details, evidence, and data needed by the reader to understand what the author was trying to do, what the author actually did and found out, and what the author thinks should be done next.
Conclusions:	A summary of the major conclusions or milestones reached in the discussion; conclusions are only opinions, so can never advocate action.
Recommendations:	If the discussion and conclusions suggest that specific action needs to be taken, the recommendations state categorically what must be done.
References:	A list of reference documents that were used to conduct the project and that the author considers will be useful to the reader; contains sufficient information for the reader to correctly identify and order the documents.
Appendixes:	A storage area at the back of the report that contains supporting data (such as charts, tables, photographs, specifications, and test results) that rightly belong in the discussion but, if included with it, would disrupt and clutter the major narrative.

The traditional arrangement provides a continuous narrative

Sometimes the appendixes are bulkier than the rest of the report

want to attach my report to his proposal to Hartwell Enterprises, then I had to recognize that *they* would be my primary readers: they would be the people who will be deciding whether to install a radiant energy heating system." Consequently, Karen tailored her approach to suit nontechnical readers (the management at Hartwell Enterprises) and included more information than she would have done if she had been writing solely for No. 5 Design Group. This is evident in her report, which can be understood by any businessperson. **Clearly identifying the reader is an essential first step when writing a comprehensive report.**

"I also wrote the report *backwards*," she explained. "Doing so really helped me organize the information." This was her approach:

1. First, she collected data on the three heating systems she would be evaluating, and researched comparative costs with experts in the radiant energy field: Darryl Berkowski in Winnipeg, Manitoba, and Vincent Harding in St. Cloud, Minnesota.

A birdseye view shows report parts

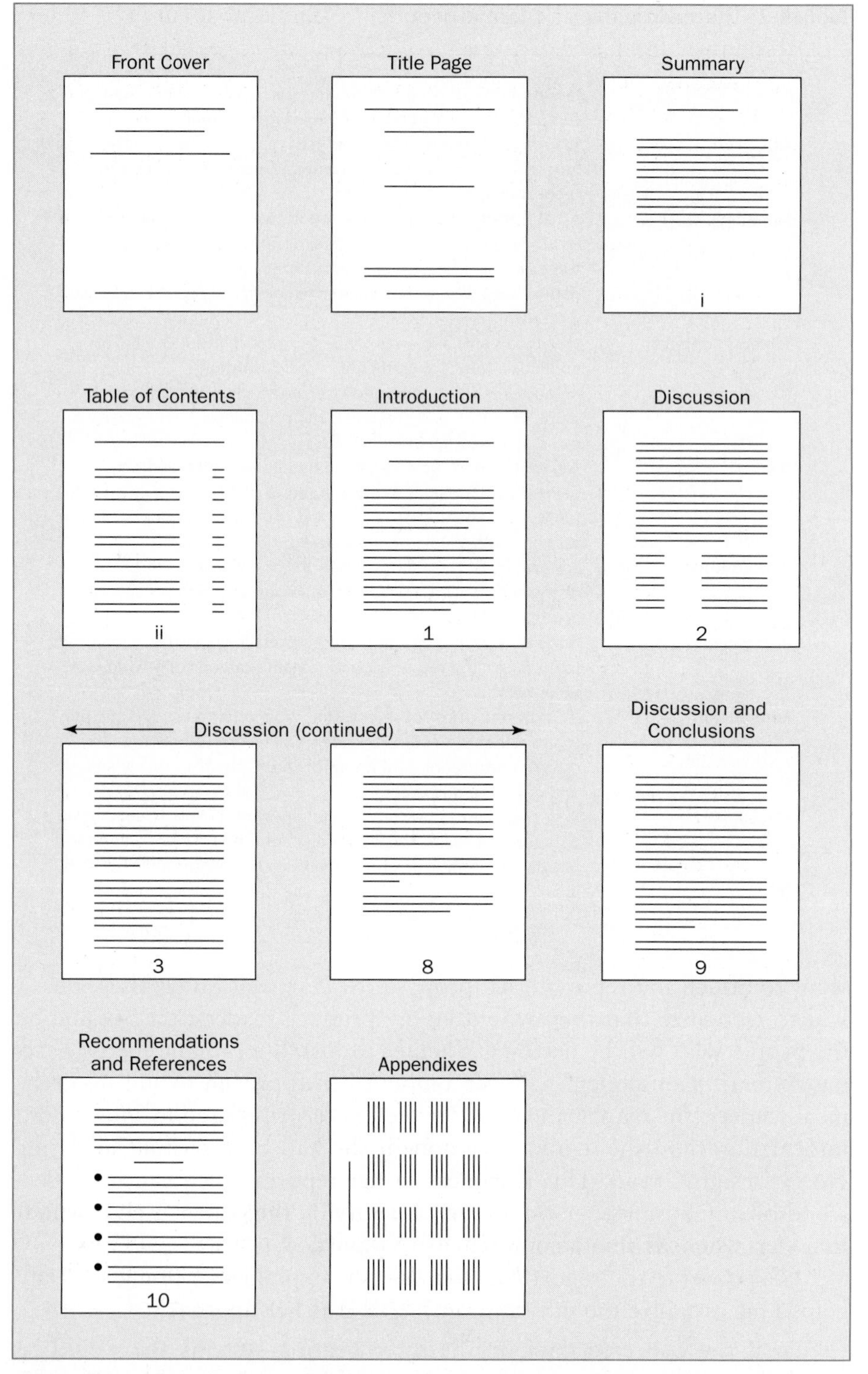

Figure 6-3 Formal report—traditional arrangement. (These are the individual pages of Formal Report 1; to conserve space, some of the discussion pages have been omitted.)

2. Then Karen created a table showing the relative costs of each system (this table became the report's Appendix, on page 168). From this table she also developed smaller tables identifying specific cost factors (these appear in the report's Discussion, on report page 8).
3. Next, she created two sets of illustrations depicting installation and operating costs at one, three, and five years: first a series of bar charts and then a series of graphs. From these she chose the graph in Figure 3 (on report page 9) as the most descriptive and simplest to read.
4. Karen's fourth step was to write the Introduction, to "set the scene." Here, she drew on information in No. 5 Design Group's letter of authorization to establish the background to the report, and its purpose and scope. She knew her primary readers would not have seen the letter.
5. Next, Karen wrote a preliminary outline, which was really an early version of the Table of Contents on page 157. She used this as a loose guide for structuring the report, making changes to it as she wrote. "Organizing the outline was surprisingly straightforward," she said, "once I had done my initial research, developed the cost analysis tables, and made the charts. All I had to do was identify what preliminary information my readers would need before they got into the system comparisons and cost analyses."
6. That preliminary information became the first four pages of the Discussion, in which she defined what a radiant energy heating system is like and how it is installed (see report pages 1 to 4).
7. Next she wrote the comparisons and analyses. "These were easy to write," she said, "because my charts and tables gave me a clear direction to take."
8. Now Karen wrote the Conclusions. "They fell naturally into place," she explained. "I went back to the Introduction and identified the three factors we were particularly asked to describe, and then provided brief answers for them."
9. She deliberated whether to write a Recommendation. The authorizing letter did not specifically ask for a recommendation yet she felt that, because she had researched the information and as such was the local expert, she should identify what route she felt Hartwell Enterprises should take.
10. And *last*, Karen wrote the Summary, drawing principally on the Conclusions to compose it. "If I had tried to write it first, before writing the report," she said, "I would have had much more difficulty writing it. Probably I would have had to go back and rewrite most of it, after the remainder of the report was finished."

Remember Karen's "backwards" approach when you have to write your next long report. By documenting all the details *first*, you will find you can organize your ideas more easily and write more fluidly.

Writing in reverse order may seem unnatural, yet it results in a better report

Sometimes inserting a recommendation is optional

The four bulleted items became Karen's project criteria

No. 5 Design Group

240 Victoria Drive – Suite 300
Duluth, Minnesota 50166
Tel: 218 234 1786; Fax: 234 1807
email: 5design@aol.com

November 27, 2003

Vern Rogers, Manager
H. L. Winman and Associates
970 Birchmount Road
Minneapolis MN 51023

Dear Mr. Rogers:

As we discussed by telephone earlier today, we are commissioning H. L. Winman and Associates to prepare a report on the efficacy of installing a radiant heating system in the new office and assembly plant we are designing for Hartwell Enterprises of Duluth. The plant is to be built at the northwest corner of the intersection of Seymour Drive and Graveley Street, with construction starting on April 1, 2004 (see attached preliminary design).

Our client has indicated interest in radiant heating but needs substantive information before deciding on installing such a system. Consequently, in your report will you please describe

- how radiant heat works and how it differs from traditional heating methods,
- the advantages of installing and using radiant heat,
- the cost to install radiant heat, compared to traditional heating systems, and
- the savings to be accrued by Hartwell Enterprises over, say, a five-year period.

Your contact at Hartwell Enterprises will be operations manager Vincent Correlli. He is aware that you are preparing a study for us and will be ready to answer questions about their operation.

I would appreciate receiving your report by January 8, 2004, because we will be submitting our design to Hartwell Enterprises on January 15. Please call me if you have any questions.

Sincerely,

Mark Hesseltine

Mark Hesseltine
Design Associate
No. 5 Design Group

Figure 6-4 Letter authorizing the report in Figure 6-6.

Comments on additional aspects of Karen's report follow, with specific references to the individual pages of the report in Figure 6-6.

- Karen's cover letter in Figure 6-5 is equivalent to an executive summary because she comments on the report's contents. Because she knows the cover letter will be seen only by Mark Hesseltine, she uses a friendly tone and the first person "I."
- A quick glance at the Contents (page ii) tells Karen's readers that she has organized her information into a logical, coherent flow, and that there are three main components: background information on radiant energy systems; a plan for installing a radiant energy heating system in the client's building; and a cost projection and analysis.
- In the Introduction (report page 1), the Background is in paragraph 1, the Purpose is in paragraph 2, and the Scope is in the three bulleted points.
- For the Discussion, Karen has adopted an overall "concept" arrangement of information, but internally it breaks into a subject arrangement when describing the proposed installation and some of the costs.
- She has written the entire report in the first person plural. "When I am presenting the results of a study I have done personally," she said, "and writing directly to my client, then normally I would use the first person singular: 'I.' But when I am writing on behalf of the company, and simultaneously writing for my client's client, and don't know the client personally, then I use the first person plural: 'We'."
- The Conclusions, on report pages 9 and 10, are longer than normal, but necessarily so to cover all the points the client requested. They show the advantages and disadvantages of radiant energy heating but do *not* advocate what action should be taken. Karen has taken care not to introduce new information into her Conclusions.
- The Recommendation advocates action and does so in strong definite terms, using "We recommend..." rather than "It is recommended...."
- The Appendix is a "landscape-view" page, and as such has been turned correctly so that it is read from the right (see page 168).

Writing in the first person means making a decision: "I" or "we"?

Pyramidal Arrangement of Report Parts

(Conclusions and Recommendations *before* the Discussion)

In recent years, more and more report writers have altered the organization of their reports so they more effectively meet their readers' needs. The pyramidal arrangement brings the conclusions and recommendations forward, positioning them immediately after the introduction so that executive readers do not have to leaf through the report to find the terminal

A way to address three different levels of reader within a single document

H. L. WINMAN AND ASSOCIATES

970 Birchmount Road, Minneapolis MN 51023
email: kwoodhouse@winman.com

January 7, 2004

Mark Hesseltine
Design Associate
No. 5 Design Group
240 Victoria Drive, Suite 300
Duluth MN 50166

Dear Mark:

I am enclosing our report, *Installing a Radiant Energy Heating System for Hartwell Enterprises Inc.*, as requested in your letter of November 22, 2003. The report shows that in the long term radiant energy will be the most economical heating system for Hartwell Enterprises' new building.

Inserting this second paragraph converted Karen's cover letter into an executive summary

To some extent I am concerned that Hartwell Enterprises may hesitate when they see the high front-end cost, particularly in comparison to electric heat. Hence, I have taken care to include a graph which shows clearly that radiant energy heating will be particularly efficient from a cost viewpoint after the fourth year. If the graph on page 9 were to be extended for another five years, the considerably lower operating cost of radiant energy heating would be even more noticeable. You may want to address this factor in your proposal.

Please call me if you need further information on any of the points addressed by the report. I'll be glad to supply it.

Sincerely,

Karen Woodhouse

Karen Woodhouse, P.E.
enc

Figure 6-5 Cover letter accompanying the formal report in Figure 6-6. This cover letter also is an executive summary.

H. L. WINMAN AND ASSOCIATES

The appearance of the title page subtly comments on the quality of the information you are presenting

Installing a Radiant Energy Heating System for Hartwell Enterprises Inc.

Prepared for

No. 5 Design Group
Duluth, Minnesota

Prepared by

Karen Woodhouse, P.E.
H. L. Winman and Associates
Minneapolis, Minnesota

January 7, 2004

Figure 6-6 Formal Report 1: traditional arrangement (14 pages).

The Summary: the full story in a capsule—difficult to write!

Summary

We have evaluated three methods for heating the proposed Hartwell Enterprises Inc. plant designed by No. 5 Design Group for construction in Duluth, Minnesota. Electric heat is the least expensive to install but the most expensive to operate. Gas-fired forced hot air is moderately expensive to install and moderately expensive to operate. Radiant energy heating is the most expensive to install and the least expensive to operate. Long term, however, radiant energy offers the most efficient and cost-effective method.

Installing a radiant energy heating system in a new building means the system can be incorporated into the overall design so that it becomes unobtrusive as well as efficient and cost-effective. It also provides more gentle warming than the other two methods, and the temperature in each area of the building can be controlled separately. The operating cost will be 39% less than for electric heating, and 30% less than for gas-fired hot air heating.

i

Contents

The preliminary pages bear roman page numbers; all other pages bear arabic numbers

Appendix

If more than one appendix, the title changes to "Appendixes" and each appendix is identified by a letter: "A," "B," "C," etc.

ii

Installing a Radiant Energy Heating System for Hartwell Enterprises Inc.

Introduction

The Introduction establishes why the project was undertaken and the report has been written

Hartwell Enterprises Inc. assembles and packages modules and specialty products for public and private organizations such as the Department of Defense, Department of Transport, Multiple Industries, Inc. and Northern Paging and Cellular Systems. The company does no original manufacturing itself, confining its role to assembling components supplied by carefully chosen manufacturers. Hartwell Enterprises Inc. has built a solid reputation as a fast, high-quality producer of specialty systems, and the company's business has increased steadily since its inception in 1988. Today, it needs a larger building, but research for suitable accommodation among Duluth area properties has failed to find a building that can be adapted to the company's special needs.

In August 2003, Hartwell Enterprises Inc. commissioned No. 5 Design Group of Duluth, Minnesota, to design a building that will meet the company's particular requirements, and to find a site on which to place it. No. 5 Design Group has, in turn, asked H. L. Winman and Associates to evaluate the efficacy and cost to install a radiant heating system in the new building, rather than a more traditional heating method. Specifically, they asked us to describe

These requirements were copied almost verbatim from the client's letter of authorization

- how radiant heat works and how it differs from traditional heating methods,
- the advantages of and the cost to install a radiant heating system, and
- the savings to be accrued by Hartwell Enterprises Inc. over the first five years.

Radiant Heating vs Traditional Heating

A traditional heating system warms air directly, which we feel on our skin as immediate heat, but its impact is transitory. Turn off the source of the heat, and the space being warmed immediately starts to cool. On a winter's day, for example, a residential furnace pumps hot air into the rooms until a preset temperature is reached, then the thermostat switches off the furnace. The

1

warming effect stops immediately and the air temperature, influenced by cooler windows and walls, begins to drop.

A radiant heating system, rather than warming air directly, radiates heat outwards in all directions until the rays contact another surface. If the surface is cooler than the radiant panel, the surface begins to warm up and the air near to it also warms, but gently, and so our skin feels the warmth as a gentle, comfortable heating. On a cool winter's day, a furnace pumps heat into the radiant panels as hot water or they are heated electrically until a preset room temperature is reached, when the source of the heat is switched off. The warming effect, however, does not stop immediately because the radiant panel continues to radiate residual heat for a considerable time. Consequently, the air in the room cools much more slowly than with hot air heating.

Compare the difference in heat produced by a gas ring and an electric hotplate. The gas ring provides immediate heat to the surrounding air, but the heat stops immediately when the gas is switched off. The electric hotplate builds up its heat more slowly, but *continues to radiate heat* for 10 to 15 minutes after the electricity is switched off.

Comparing a complex technical concept to a familiar everyday event helps reader understanding

In locations that are unlikely to be affected by external sources introducing sudden changes in temperature, a 100% radiant heating system is ideal. Even in areas that are subjected to marked changes in temperature, such as an aircraft hangar, radiant heat also is effective. Lawrence Drake, executive director of the Radiant Panel Association, explains that

> Thermal mass in a heated shop or hangar floor responds rapidly to the change of air temperature when a big overhead door is opened. All the heat that has been "trickled" into the slab over time is released quickly to combat the cold air rolling in over the floor. This happens because of the sudden, dramatic increase in temperature difference between the slab and the new air. Once the door is closed the building returns to its normal comfort setting almost immediately.[1]

He adds, however, that under such conditions a combination of radiant energy and a back-up hot air heating system can be even more effective, because occupants of the space immediately feel the heated air.

2

A statement made in an email message or during a conversation may be documented as a source reference

Darryl Berkowski, who installs radiant heating systems in Manitoba and Northwestern Ontario, points out that electric or hot water baseboard heaters may *appear* to produce radiant heat, but in effect they release only a small amount of radiant energy. Primarily, they heat the air.[2]

A particular advantage of radiant heating is that *it can be controlled locally*. In a traditional heating system, hot air is supplied to vents, or hot water to radiators, from a central furnace, the operation of which is controlled by a single thermostat mounted on a wall of only one of the rooms. Thus, the temperature in the other rooms cannot be controlled separately. If, for example, one of the rooms tends to be cooler than the others because it has large windows on a north wall, the heating system is unable to compensate for the variation. Conversely, in a radiant heating system, the heat supplied to the radiant panels can be controlled separately in each room. This is true of radiant panels heated either electrically or by hot water.

Radiant Panel Installations

Radiant heating panels may be installed in the wall, ceiling or floor. Because they heat all objects within their line of sight, floor panels heat the ceiling and walls, wall panels heat the opposite wall, and ceiling panels heat the floor and walls. Similarly, all three heat furniture they can "see" and, of course, the people in the room. The radiant energy is felt as a very gentle, subtle warming, never as a searing blast.

Generally, wall panels tend to be smaller than ceiling panels, and ceiling panels tend to be smaller than floor panels, which often use the whole floor as their radiating surface. Also generally, the smaller the panel the greater the temperature at which it must operate to gain a measurable effect. Consequently, wall panels may be as hot as 145°F, ceiling panels as warm as 110°F, but floor panels rarely exceed 80°F. Lawrence Drake writes that

Introducing excerpts from an existing document adds credibility and reduces writing effort

> A heated floor normally "feels" neutral. Its surface temperature is usually less than our body temperature, although the overall sensation is one of comfort. Only on very cold days when the floor is called on for maximum output will it actually "feel" warm.
>
> Heat coming from a wall radiator can be felt the closer you get to it because its surface is much warmer than your body. Radiant ceiling

3

panels are also generally warmer than your body so you will feel some warmth on your head and shoulders.[3]

Wall and ceiling panels usually come preassembled and are fixed onto the surface of the wall or ceiling. Floor panels normally are embedded in the floor, as part of the floor construction, and then are covered with a layer of concrete or similar floor material. Consequently the whole floor becomes a single, large radiant panel.

When installed as an integral part of a new installation, either electric elements or water pipes are laid in continuous parallel rows in the floor (see Figure 1). The pipes are made of a strong, durable, light, cross-linked polyethylene (PEX). Connectors are made of noncorroding copper, brass, or plastic.

Although it's possible to install a radiant heating system as a retrofit in an existing building, the ideal arrangement is to design a radiant heating system specifically for a new building. It can then be installed as an integral part of the new structure.

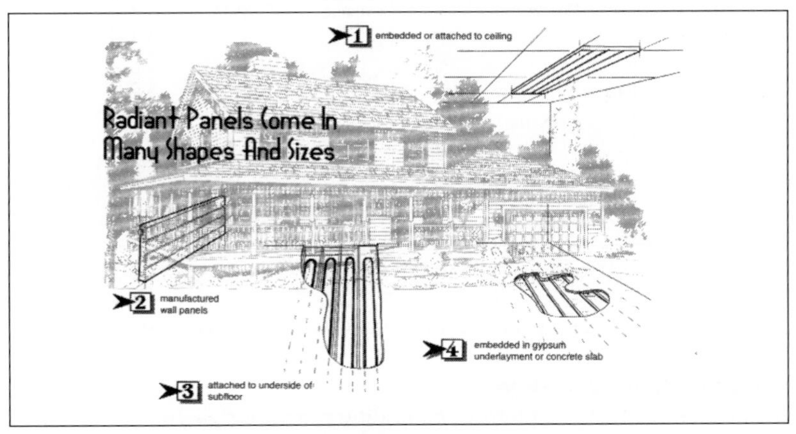

Figure 1. Various types of radiant energy panels
(Illustration courtesy of Radiant Panel Association)

A cutaway illustration like this helps readers visualize the technology

Research into a client's business practices can help you focus information accurately

Suggested Installation Plan for the Proposed Hartwell Enterprises Building

Hartwell Enterprises' operation is unique, in that it works on a just-in-time method of delivery for the components that are assembled into products. The company's contracts with its suppliers stipulate that components are to be delivered in small lots only a few hours before they are to be assembled. This has two major effects:

1. Because components are moved rapidly from the delivery semitrailer to the assembly line, and then to the shipping area for packing and storing in a second semitrailer ready for shipping, Hartwell Enterprises requires only a small warehouse area.

2. Because the loading bay doors have to be opened frequently, the loading bay demands special attention from a heating viewpoint.

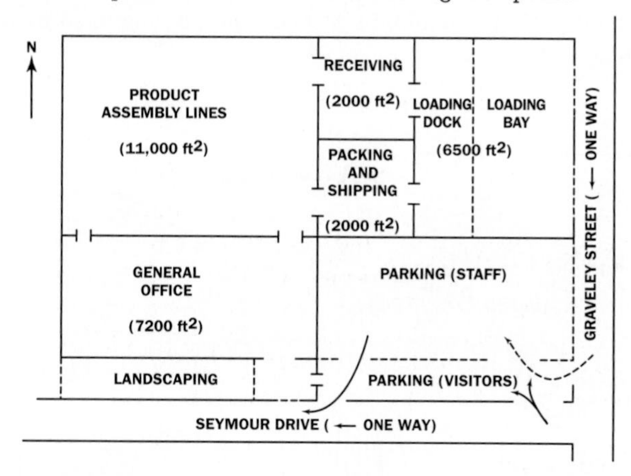

Figure 2. Design of proposed new building for Hartwell Enterprises Inc.

Proposed Heating System

The scene has been set: now for the plan!

The Proposed Hartwell Enterprises building particularly lends itself to heating by radiant energy. The diverse activities in the various parts of the building call for different types and methods of heating. As Figure 2 shows, there will be five main areas of activity, each requiring a different level and form of heating, and each with its own thermostat. The floor in each area will be concrete, but there will be some variations:

5

- The general office will have an 8 inch thick slab with its upper surface 14 inches above grade. It will be covered by Orlando carpet and underlay.

- The product assembly, receiving, and shipping areas will be set on an 8 inch thick slab with the floor level at 42 inches above grade. There will be basement under these three areas, with a 7 ft high ceiling.

- The unloading and loading bays will be at grade level, built on a 14 inch reinforced concrete slab.

As the primary source of heat for these areas, we propose installing hot-water pipes embedded in the concrete with a separate circulation system and thermostat for each area. The mechanical room for the hot water boiler and the controls will be set up in the basement. (We are not recommending electrically heated panels, because in Minnesota it is on average 20% more expensive to heat the radiant panels by electricity than by natural-gas-fired hot water.[4])

The product assembly, receiving, and shipping areas will require no supplementary heating. However, the general office and the loading bays will. For these areas we propose the following additional heating arrangements.

The main paragraphs contain the plan...

- The ceiling in the general office will slope upward, toward the north (the back of the office), which will tend to draw warm air aloft, into the higher part of the ceiling and away from the south side of the area. To maintain an even warmth on the south side, we propose installing two 50 in. × 24 in. horizontal radiant panels under the windows along the south wall. These will be heated by hot water and will be controlled by a separate thermostat.

- The loading bays will cool rapidly in midwinter when the doors are opened to admit and exit semitrailers. Here we propose installing two overhead unit heaters, one above each door. They also will be hot-water heaters, and will be triggered to start up when a door opens and to shut down when a predetermined ambient temperature is reached. (In effect, the radiant energy from the floor panels will restore the temperature quite quickly. The unit heaters will provide a supplementary, readily noticeable, immediate source of heat during extreme cold conditions.)

...while the bulleted subparagraphs introduce variations and exceptions

6

Proposed Cooling System

Although radiant energy panels can provide moderate cooling in the summer, particularly in dry climates, in moist or semimoist conditions they tend to be less efficient.[5] Consequently we propose that all cooling be carried out through a separate system, comprising

- a Hyperion Model 2000 air conditioning unit for cooling the general office, product assembly, and receiving and shipping areas, and
- a Hyperion Model 2720 air conditioning unit for cooling the loading bay.

The two air conditioners will be located in the basement, with the model 2000 at the west end and the model 2720 at the east end. The model 2000 will feed cooled air through ductwork concealed in the ceiling of the office, assembly, and receiving/shipping areas. The model 2720, which will be a fast-response unit designed to recover quickly from rapid changes in temperature, will feed cooled air through ductwork under the roof of the loading bay.

Costs: Radiant vs Traditional Heating

Two cost factors have to be considered when comparing a radiant energy heating system with a traditional heating system: the cost of installation and the cost of operation. The cost of installing and operating air conditioning from May to September also has to be taken into account.

A section opening paragraph should act like a minisummary, identifying aspects to be discussed

For this study, we have examined the cost of installing and operating three systems:

1. Radiant energy heating, plus a separate air-conditioning system.
2. All-electric heating, plus a separate air-conditioning system.
3. Forced hot-air heating fueled by natural gas, with integral air-conditioning.

Installation Costs

The costs to install these three systems in the proposed Hartwell Enterprises building are listed in Table 1, which at first glance shows that electric heating is the least expensive and radiant energy is the most expensive. However, when air conditioning is included, gas-fired hot air becomes the least expensive.

7

Table 1. Installation Costs

System	System ($)	Heating Conditioning ($)	Air- Total ($)
Radiant energy	47,200	21,600	68,800
All-electric plus air-conditioning	26,600	21,600	48,200
Gas-fired hot air with integral air-conditioning	38,700	(incl)	38,700

The table lists key cost figures, provides a ready comparison

Annual Operating Costs

To calculate potential operating costs, we referred to a 2002 study carried out by Vincent Harding of V. Harding Associates, in which he averaged the annual heating costs of 30 industrial buildings in Minnesota, Michigan, and Ontario for the years 1996 to 2001.[6] From these we culled the heating costs for eight light-industry buildings in Minnesota, each of a similar size to the proposed 3200 m^2 Hartwell Enterprises Building. Two are heated by electricity, three by gas-fired forced hot air, and three by radiant energy. The results are summarized in Table 2, which shows that radiant energy heating has the lowest annual operating cost and all-electric heating has the highest annual operating cost. A detailed breakdown is shown in the Appendix.

Table 2. Average Annual Operating Costs

System	Heating Only ($)	With Air-Conditioning ($)
Radiant energy	17,300	20,800
All-electric	31,700	35,200
Gas-fired forced hot air	27,600	31,100

A simple table with key cost figures can be embedded conveniently into the narrative

Projected Costs Over Five Years

In Figure 3, the combined installation and operating costs for both heating and cooling are shown for year one, and then the projected operating costs for heating and cooling are shown cumulatively for years two through five. The graph shows that in the first year the installation and operating cost for electric heating is the least expensive, and radiant energy is the most expensive. However, after 1 year and 5 months the positions are reversed, with the cumulative costs of installing and operating radiant energy heating becoming less expensive than electric heating.

Installing and operating a gas-fired forced-air heating and air-conditioning system remains less expensive than radiant energy for the first 3 years and 2 months, after which the cumulative cost of radiant energy heating plus a companion air-conditioning system becomes more economical.

Both this graph and the two previous tables are supported by the detailed cost comparison in the appendix

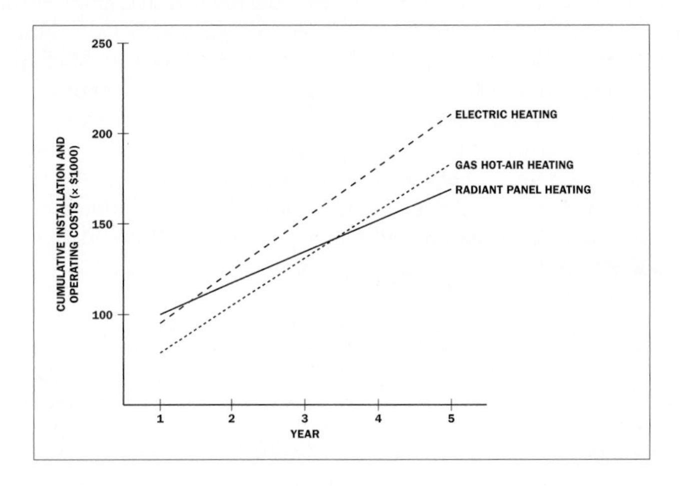

Figure 3. Comparison of installation and operating costs over five years, for radiant energy, hot air, and electric heating systems

Conclusions

Conclusions sum up key outcomes, never specifically advocate action

For the proposed Hartwell Enterprises Inc. office and product assembly plant planned for 1650 Seymour Drive in Duluth, Minnesota, a radiant energy heating system offers several advantages:

9

- Its operating cost will be 39% less expensive than for electric heating, and 30% less expensive than for gas-fired forced hot-air heating.

- It will provide a softer, less obtrusive, and more stable source of warmth than either electric or hot-air heating.

- It can be controlled separately for each area of the building.

- Because it will be a new building, it can be designed and installed as an integral part of the structure and thus be less obtrusive.

Its chief disadvantage is that its installation cost will be 77% higher than for an electric heating system, and 22% higher than for a gas-fired hot-air system.

However, in the long term, the combined installation and operating cost of radiant energy heating becomes less than that of electric heating in 17 months, and less than that of forced hot-air heating in 38 months.

Conclusions may, however, *imply* the course to be taken

Recommendation

Viewing Hartwell Enterprises' move into a new building as a long-term venture, we recommend installing a radiant energy heating system to take advantage of the long-term low operating expenses it will incur.

References

The reference section lists all written and spoken information sources

1. Lawrence V. Drake, *Radiant Panel Heating and Cooling*. Report: Radiant Panel Association, Hyrum, Utah, 1995, p. 3.
2. Darryl Berkowski, D & M Innovators, Winnipeg, Manitoba. Email to Karen Woodhouse, H. L. Winman and Associates, December 10, 2003.
3. Drake, p. 2.
4. Berkowski, p. 2.
5. Drake, p. 4.
6. Vincent Harding, *Comparison of Costs: Electric, Gas, and Radiant Energy Heating in 30 Industrial Buildings, 1996–2001*. Report: V. Harding Associates, St. Cloud, Minnesota, February 23, 2002.

10

Placing a detailed table like this within the report narrative would distract a reader's attention

Appendix

Comparison of Costs: Three Heating/Cooling Systems for Hartwell Enterprises Inc.

Heating/Cooling System	*Installation Cost ($)*	*Annual Operating Cost ($)*	*First Year Costs ($)*	*Cumulative Three-Year Costs ($)*	*Cumulative Five-Year Costs ($)*
Radiant Energy:					
• Heating	47,200	17,300			
• Air-Conditioning	21,600	3,500			
Combined Systems:	***68,800***	***20,800***	***89,600***	***131,200***	***17, 800***
Electric Heat:					
• Heating	26,600	31,700			
• Air-Conditioning	21,600	3,500			
Combined Systems:	***48,200***	***35,200***	***83,400***	***153,800***	***224,200***
Gas-fired Forced Hot Air:					
• Heating	} 38,700	27,600			
• Air-Conditioning		3,500			
Combined Systems:	***38,700***	***31,100***	***68,700***	***130,900***	***193,100***

summary (the report's outcome). The advantages of the pyramidal approach are immediately evident: busy readers have only to read the initial pages to learn the main points contained in the report, and the writer can help them along by gradually increasing the technical content of the report, catering to semitechnical executive readers up to the end of the recommendations, and to fully technical readers in the discussion and appendix. Although the natural flow of information that occurs in the traditional arrangement is disrupted, Figure 6-7 shows there is now a reader-oriented flow, with the three compartments each containing progressively more technical details.

This gradually increasing development of the topic in three separate stages is similar to the newspaper technique described earlier, in which the first one or two paragraphs contain a capsule description of the whole story, the next three or four paragraphs contain a slightly more detailed description, and the final eight or nine paragraphs repeat the same story, but this time with more names, more peripheral information, and more details. Newspapers cater to both the busy reader who may not have time to read more than the opening synopsis, and the leisurely reader who wants to read all the available information.

A minireport showing the pyramidal arrangement of report parts is illustrated in Figure 6-8. Segments of a sample report written using the pyramidal approach are shown in Figure 6-9.

Excerpts from Formal Report 2: Selecting New Elevators for the Merrywell Building

Before reading these excerpts, read the client's letter authorizing H. L. Winman and Associates to initiate an engineering investigation:

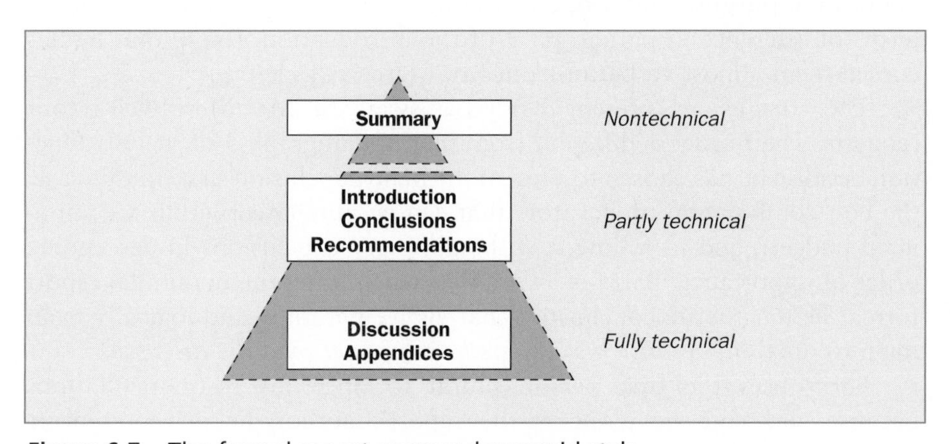

Figure 6-7 The formal report arranged pyramid style.

Three reports in one, each a complete story in itself

Dear Mr. Bailey:

The elevators in the Merrywell Building are showing their age. Recently we have experienced frequent breakdowns and, even when the elevators are operating properly, it has become increasingly evident that they do not provide adequate service at the start of work, at noon, and at the end of the working day. I have therefore decided to install a complete range of new elevators, with work starting in mid-August.

The report writer used some of these words in the Introduction on page 173

Before I proceed any further, I would like you to conduct an engineering investigation for me. Specifically, I want you to evaluate the structural condition of my building, assess the elevator requirements of the building's occupants, investigate the types of elevators available, and recommend the best type or combination of elevators that can be purchased and installed within a proposed budget of $950,000.

Please use this letter as your authority to proceed with the investigation. I would appreciate receiving your report by the end of June.

Regards,

David P. Merrywell, President

Merrywell Enterprises Inc.

By comparing this letter with the conclusions and recommendations, you can assess how thoroughly Barry Kingsley (the report's author) has answered the client's requests.

Comments on the Report

The summary is short and direct because it is written primarily for one reader: the president of Merrywell Enterprises Inc. It encourages him to read the report immediately, and to accept its recommendations by offering the opportunity to save $60,000.

A report may be directed to one reader, but also must consider other readers who may see it

Although the background information contained in the first two paragraphs of the introduction seems to repeat details the client already knows, Barry recognizes he must satisfy the needs of other readers who may not be fully aware of the situation in the Merrywell Building. He then defines the purpose and scope of the investigation by stating the client's terms of reference in paragraph 3 of the introduction. (Note that he has copied them almost verbatim from Mr. Merrywell's letter.)

The conclusions present Barry's answers to Mr. Merrywell's four requests. Their order is different from that in paragraph 3 of the introduction because he has chosen to present the main conclusion first (in this case, the best combination of elevators that can be purchased within the stipulated budget), and to follow it with subsidiary conclusions in descending order of importance. Barry is aware that when using the pyramidal report format he must write conclusions that evolve naturally and logically from the introduction, *because his readers have not yet read the discussion.*

Barry uses the first person plural to open his recommendations because, although he alone is the report's author, he is representing H. L. Winman and Associates' views to the client.

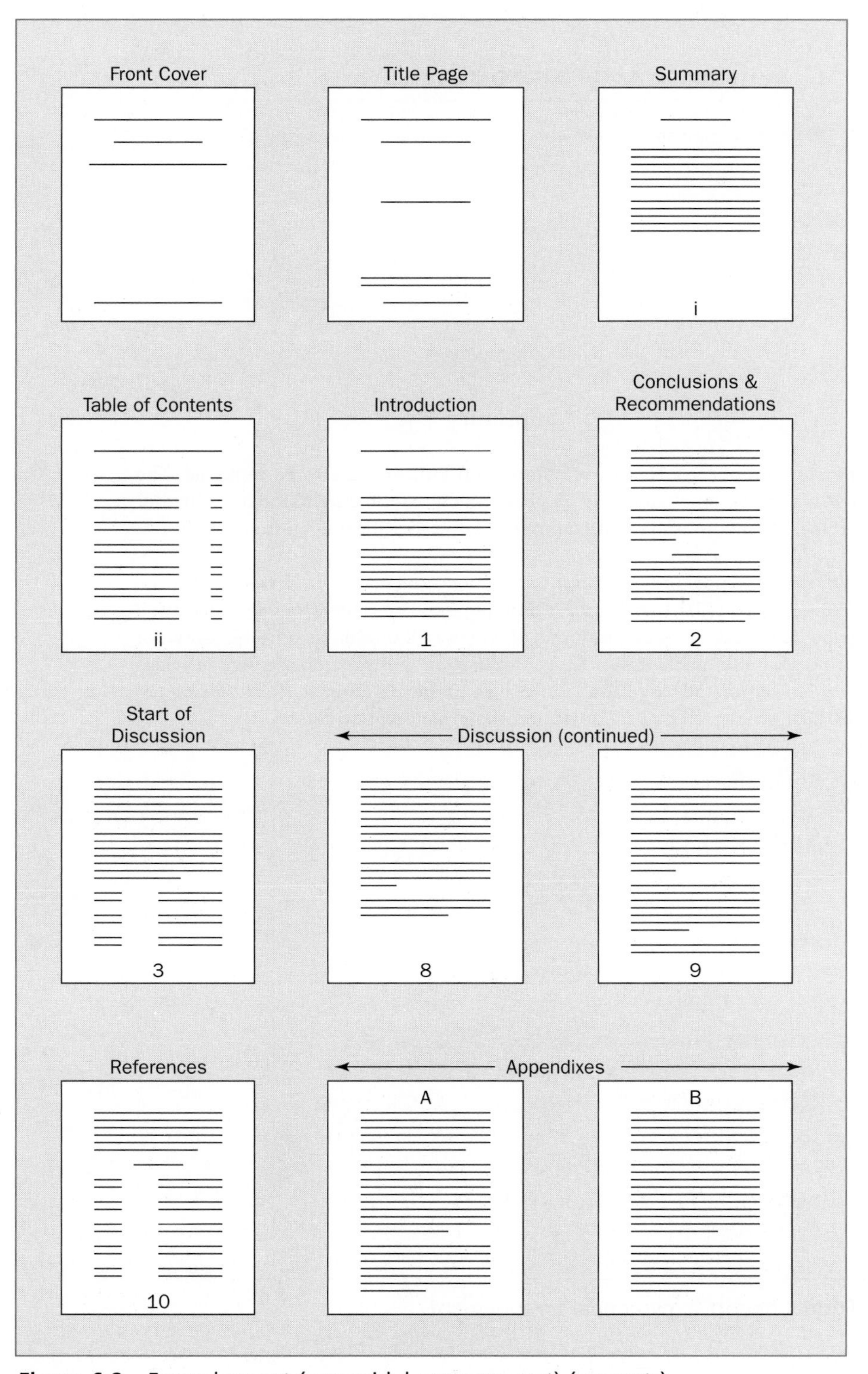

Figure 6-8 Formal report (pyramidal arrangement) (excerpts).

Compare this arrangement with the birdseye view in Figure 6-3

H. L. WINMAN AND ASSOCIATES

A simple, straightforward summary that answers the reader's most immediate question

Summary

The elevators in the 71-year-old Merrywell Building are to be replaced. The new elevators must not only improve the present unsatisfactory elevator service, but must do so within a purchase and installation budget of $950,000.

Of the many types and combinations of elevators considered, the most satisfactory proved to be four 8 ft x 7 ft deluxe passenger elevators manufactured by the YoYo Elevator Company, one of which will double as a freight elevator during off-peak traffic times. This combination will provide the fast, efficient service requested by the building's tenants for a total price of $890,000, which will be 6.3% less than the projected budget.

Figure 6-9 Formal Report 2: pyramidal arrangement.

Selecting New Elevators for the Merrywell Building

Introduction

When in 1970 Merrywell Enterprises Inc. purchased the Wescon property in Montrose, Ohio, they renamed it "The Merrywell Building" and renovated the entire exterior and part of the interior. The building's two manually operated passenger elevators and a freight elevator were left intact, although it was recognized that eventually they would have to be replaced.

Recently the elevators have been showing their age. There have been frequent breakdowns and passengers have become increasingly dissatisfied with the inadequate service provided at peak traffic hours.

In a letter dated April 27, 2003, to H. L. Winman and Associates, the president of Merrywell Enterprises Inc. stated his company's intention to purchase new elevators. He authorized us to evaluate the structural condition of the building, to assess the elevator requirements of the building's occupants, to investigate the types of elevators that are available, and to recommend the best type or combination of elevators that can be purchased and installed within the proposed budget of $950,000.

The Introduction, Conclusions, and Recommendations stand alone as a composite section

Conclusions

The best combination of elevators that can be installed in the Merrywell Building will be four deluxe 8 ft × 7 ft passenger models, one of which will serve as a dual-purpose passenger/freight elevator. This selection will provide the fast, efficient service desired by the building's tenants, and will be able to contend with any foreseeable increase in traffic. Its price at $890,000 will be 6.3% less than the proposed budget.

The primary conclusion comes first, followed by subsidiary conclusions

Installation of special elevators requested by some tenants, such as a full-size freight elevator and a small but speedy executive elevator, would be feasible but costly. A freight elevator would restrict passenger-carrying capability, while an executive elevator would elevate the total price to at least 20% above the proposed budget.

The quality and basic prices of elevators built by the major manufacturers are similar. The YoYo Elevator Company has the most attractive quantity price structure and provides the best maintenance service.

1

The building is structurally sound, although it will require some minor modifications before the new elevators can be installed.

Recommendations

Recommendations should be written in the first person, singular or plural

We recommend that four Model C deluxe 8 ft × 7 ft passenger elevators manufactured by the YoYo Elevator Company be installed in the Merrywell Building. We further recommend that one of these elevators be programmed to provide express passenger service to the top four floors during peak traffic hours, and to serve as a freight elevator at other times.

2

Evaluating Building Condition

We have evaluated the condition of the Merrywell Building and find it to be structurally sound. The underpinning done in 1971 by the previous owner was completely successful and there still are no cracks or signs of further settling. Some additional shoring will be required at the head of the elevator shaft immediately above the 9th floor, but this will be routine work that the elevator manufacturer would expect to do in an old building.

The Discussion tells the reader the building is in good shape...

The existing elevator shaft is only 24 feet wide by 8 feet deep, which is unlikely to be large enough for the new elevators. We have therefore investigated relocating the elevators to a different part of the building, or enlarging the existing shaft. Relocation, though possible, would entail major structural alterations and would be very expensive. Enlarging the elevator shaft could be done economically by removing a staircase that runs up the center of the building immediately east of the shaft. This staircase is used very little and its removal would not conflict with fire regulations. Removal of the staircase will widen the elevator shaft by 11 feet, which will provide sufficient space for the new elevators.

Establishing Tenants' Needs

To establish the elevator requirements of the building's tenants, we asked a senior executive of each company to answer the questionnaire attached as Appendix A. When we had correlated the answers to all the questionnaires, we identified five significant factors that would have to be considered before selecting the new elevators. (There were also several minor exceptions that we did not include in our analysis, either because they were impractical or because they would have been too costly to incorporate.) The five major factors were:

- Every tenant stated that the new elevators must eliminate the lengthy waits that now occur. We carried out a survey at peak travel times and established that passengers waited for elevators for as much as 70 seconds. Since passengers start becoming impatient after 32 seconds,[1] we estimated that at least three, and probably four, faster passenger elevators would have to be installed to contend with peak-hour traffic.

...and then goes on to assess what needs to be done

- Although all tenants occasionally carry light freight up to their offices, only Rad-Art Graphics and Design Consultants Inc. considered that a freight elevator would be essential. However, both agreed that a separate

3

freight elevator would not be necessary if one of the new passenger elevators is large enough to carry their displays. They initially quoted 9 feet as the minimum width they would require, but later conceded that with other modifications they could reduce the length of their displays to 7 feet, 6 inches. All tenants agreed that if a passenger elevator is to double as a freight elevator, they would restrict freight movements to non-peak travel times.

- The three companies occupying the top four floors of the building requested that one elevator be classified as an express elevator serving only the ground floor and floors 6, 7, 8, and 9. Because these companies represent more than 50% of the building's tenants, we considered their request should be entertained.
- Three companies expressed a preference for deluxe elevators. Rothesay Mutual Insurance Company, Design Consultants Inc., and Rad-Art Graphics all stated that they had to create an impression of business solidarity in the eyes of their clients, and felt that deluxe elevators would help convey this image.
- The managements of Rothesay Mutual Insurance Company and Vulcan Oil and Fuel Corporation requested that a small key-operated executive elevator be included in our selection for the sole use of top executives of the building's major tenants. We asked other companies to express their views but received only marginal interest. The consensus seemed to be that an executive elevator would have only limited use and the privilege would too easily be abused. However, we retained the idea for further evaluation, even though we recognized that an executive elevator would prove costly in relation to passenger usage.[2]

Identifying users' needs helps establish clear criteria

We decided that the first two of these factors are requirements that must be implemented, while the latter three are preferences that should be incorporated if at all feasible. The controlling influence would be the budget allocation of $950,000 stipulated by the landlord, Merrywell Enterprises Inc. In decreasing order of importance, the requirements are:

1. Passenger waiting time must be no longer than 32 seconds.
2. At least one elevator must be able to accept freight up to 7 feet, 6 inches long.
3. An express elevator should serve the top four floors.
4. The elevators should be deluxe models.
5. A small private elevator should be provided for company executives.

4

We have included the first two pages of the discussion to show that, early in his report, Barry establishes criteria that will subsequently influence how he selects a combination of elevators that will best meet his client's needs. By carefully identifying the five criteria and describing why each is valid, he shows his readers the direction his report will take. (In later sections of his report—not included in the sample pages—he identifies various combinations of elevators that could be installed, and demonstrates which ones meet the criteria, until he finally reaches an optimum configuration.)

ASSIGNMENTS

Project 6.1: Testing Highway Marking Paints

For the past six years the Highways Department in your state has used "Centrex CL" for marking highway pavement centerlines and lanes. Recent advances in paint technology, however, have brought several new products onto the market, which their manufacturers claim are better than Centrex CL. To meet this challenge, Centrex Inc. has developed a new paint ("TL") and has recommended that the Highways Department use it in place of CL.

In a letter dated March 18 of this year, Senior Highways Engineer Morris Hordern commissioned you to carry out independent tests of the new paints. (You own a home-based consulting company known as Pro-Active Consultants Inc.) You start your project by obtaining samples of white and yellow highway paint from six manufacturers, transferring the samples into unmarked cans and then coding the cans like this:

New technology creates new products for evaluation

		Paint Coding	
	Manufacturer	**White**	**Yellow**
1.	Centrex Inc., Hartford, Connecticut, Paint type: CL (the "old" paint)	WA	YL
2.	Novell Paint Ltd, Moorstown, New Jersey, Paint type: 707	WB	YM
3.	Hi-Liner Products, Rockford, Illinois, Paint type: HILITE	WC	YN
4.	Multiple Industries Corporation, Cleveland, Ohio, Paint type: MICA	WD	YO
5.	Wishart Incorporated, Utica, New York, Paint type: ROADMARK 8	WE	YP
6.	Provincial Paint Company, Pittsburgh, Pennsylvania, Paint Type: 81-234	WF	YQ
7.	Centrex Inc, Hartford, Connecticut, Paint type: TL (their "new" paint)	WG	YR

You then place the coding list into a sealed envelope, and lock it away in a safety deposit box at a local bank.

You decide to paint sample stripes on two regularly traveled stretches of highway and to assess the samples in four ways:

1. Spraying characteristics.
2. Drying time.
3. Visibility after three months.
4. Visibility after six months.

Some factors demand personal judgment, others are measurable

You assess spraying characteristics as excellent, very good, good, fair, and poor. The ratings are:

Very good:	WA,	WB,	WC,	WD,	WF,	YL,	YR
Good:	WE,	WG,	YM,	YN,	YO,	YQ	
Fair:	YP						

You assess drying time in minutes:

WA:16	WC:18	WE:14	WG:19	YM:26	YO:14	YQ:12
WB:33	WD:11	WF:13	YL:13	YN:18	YP:10	YR:15

After three months you assess visibility by day and by night. You use five drivers (you are one) to rate the stripes independently and to place the stripes' visibility on a scale of 1 to 10. You then average the five assessments (night readings are taken with headlights at high beam).

Paint	**Concrete Pavement**		**Asphalt Pavement**	
Code	**Day**	**Night**	**Day**	**Night**
WA	8	8	8	9
WB	7	8	7	7
WC	8	9	10	9
WD	9	9	7	8
WE	7	6	7	8
WF	8	9	9	10
WG	7	7	7	8
YL	8	9	9	9
YM	7	7	7	9
YN	7	9	7	8
YO	8	8	7	8
YP	7	8	6	8
YQ	5	6	4	6
YR	9	10	9	9

After another three months the same five drivers again assess stripe visibility, with these results:

Paint Code	Concrete Pavement		Asphalt Pavement	
	Day	Night	Day	Night
WA	6	6	5	7
WB	4	5	5	5
WC	8	8	9	9
WD	6	7	6	8
WE	6	5	6	7
WF	5	7	5	6
WG	6	7	6	7
YL	6	7	7	8
YM	6	7	6	8
YN	6	7	6	7
YO	6	7	6	8
YP	3	4	4	5
YQ	2	3	3	4
YR	8	9	8	9

You consolidate all your results into two tables, one for white paint, one for yellow paint, and then

You will need to create two comparison tables before writing your report

- reject any unacceptable paints (see guidelines below),
- rank acceptable paints in order of suitability,
- identify the best paint(s) to use for highway marking,
- retrieve the paint coding list from the bank deposit box, and
- write your report.

Some factors you use to conduct your study and to write your report are:

1. Senior Highways Engineer Morris Hordern's office address is 416 Inkster Building, 2035 Perimeter Road of your city.
2. The paint stripes were painted on two stretches of highway:
   2.1 Highway 17 (concrete surface), 1.5 miles north of the intersection with Highway 43.
   2.2 Highway 43 (asphalt surface), 1 mile west of the intersection with Highway 17.
3. You are unable to borrow the regular highway paint stripe applicator from the Highways Department. Instead, you mount an applicator on a small garden tractor. The paint stripes are applied at night, between midnight and 6:00 a.m.
4. Paint Manufacturers' Association specification PMA-02-28H states that spraying characteristics for fast-drying highway paint should be at least "Good," and preferably "Very Good." To achieve "Very

Good" the paint must flow smoothly and evenly without forming globules or dripping from the nozzle.

These factors help establish acceptability criteria

5. You refer to specification ASTM D-711 to establish the maximum acceptable paint drying time, which is 20 minutes.
6. Guidelines you give to the drivers assessing paint visibility are:

	Distance Visible	
Rating	**Day** ***(yards)***	**Night** ***(yards)***
10	500	200
8	400	160
6	300	120
4	200	80
2	100	40

You then average the five assessments.

7. You establish minimum acceptable visibility levels for the paints to be:
   After three months' traffic wear: 7
   After six months' traffic wear: 6

Note: Calculate real dates for each stage of the study and quote them in your report.

Your report should not only present the results of your tests, but also analyze them, draw conclusions, and make a recommendation.

Project 6.2: Correcting a Noise Problem

Assume that today is March 3. This morning you receive a letter from Trudy Parsenon, the area manager of Mirabel Realty. (You are the owner/manager of Pro-Active Consultants Inc., which you operate from your home.)

A letter confirming a telephone request

Dear (you):

As I mentioned when I telephoned last week, my staff have been complaining for the past three months that the noise level in our office is too high. They claim it is affecting their work and causing fatigue. I have noticed, too, that staff turnover has been higher lately.

Will you please look into the problem for me to determine whether their complaints are justified. If they are, will you suggest what can be done to remedy the problem, recommend the most suitable method, and include a cost estimate.

Sincerely,

Trudy G. Parsenon

Area Manager, Mirabel Realty

Part 1

At 4:00 p.m. on March 3 you visit Mirabel Realty (the office is in room 210, on the second floor of the Fermore Building at 381 Conway Avenue, Montrose, Ohio). You notice a background hum, which you consider to be caused by motors in the computers and printers. You are still there when the office staff quits at 4:30. After they go, you notice you can still hear the hum, but at a lower level.

You walk around the office with Trudy, who plagues you with questions. "What do you think?" she asks. "Seems like the same noise level you get in any business office, don't you think?"

You suspect she is hoping for a good report from you, which she can use to prove to her staff that their complaints are imaginary.

"I can't tell you without taking readings," you hedge. "Noise is a pretty tricky thing. What some people think is too noisy, others hardly notice."

But you do notice that the hum gets significantly louder near the north wall of the office. Then suddenly it stops, or rather, dies away. The time is 4:45.

"What's on the other side of this wall?" you ask.

An initial visit shows there may well be a problem

"Oh, that's Superior Giftware," Trudy replies. "They distribute cheap imports—that sort of thing."

"Have you talked to them about the hum?"

"Yes. I asked Saul Ferguson about it—he's the manager next door. Pretty hostile, he was."

"And what time do they quit work?" you ask.

"Right now," Trudy replies. "You can always tell, because they switch their machines off."

You arrange with Trudy that you will take sound-level measurements one week from today. You want to find out how much of the noise in the realty office is generated by normal office activity and how much by the company next door.

You consider that a visit to Superior Giftware is essential, since you suspect that the machines Trudy mentioned may be the problem. You want to know the sound levels on both sides of the wall between the two companies, to assess the extent of soundproofing you may want to recommend.

Write to Saul Ferguson, manager of Superior Giftware, to ask permission to carry out sound-level measurements in his offices one week from today. His business is in room 208, 381 Conway Avenue.

Part 2

It is now March 10, one week later. You take a Nabuchi Model 1300 sound-level meter with you and return to 381 Conway Avenue. You plan to measure sound levels at various locations in the Mirabel Realty office under four conditions:

- When both businesses are empty.
- When only Superior Giftware is working (4:30–4:45).
- When only Mirabel Realty is working (8:00–8:15).
- When both businesses are working.

You also plan to take readings in Superior Giftware's office.

Table 6-3 Average sound levels second floor, 381 Conway Avenue, Montrose, Ohio.

Location	Both Offices Working (dB)	Only Superior Working (dB)	Only Mirabel Working (dB)	No One* Working (dB)
Mirabel Realty				
A	74	73	48	27
B	71	69	51	27
C	66	65	50	27
D	64	61	52	26
E	63	59	51	28
F	59	53	49	26
G	54	49	44	28
Superior Giftware				
H	86	—	—	25
I	83	—	—	26

*Mostly air-conditioner noise.

Note: Measurements made with Nabuchi Model 1300 Sound-Level Meter set to "A" scale.

You suspect the next door neighbor is being defensive

Since Mr. Ferguson has not replied to your letter, you telephoned him yesterday afternoon to ask if you could come in today to take the measurements. He said he was "terribly busy" and that it was "damned inconvenient," but he somewhat reluctantly agreed. (It was worthwhile being persistent. When you visit Superior Giftware, almost right away you notice a packaging and sealing machine only 7 feet from the wall separating the two business offices.)

You record the measurements you take (see Table 6-3) and compare them to the general ratings for office noise, which you obtain from City of Montrose standard SL2020, dated January 20, 2001. The recommended sound levels for an urban office are:

Quiet office: 30–40 dB
Average office: 40–55 dB
Noisy office: 55–75 dB

You note that the sound level in Mirabel Realty's office increases as you move toward the dividing wall between the two offices (see Figure 6-10).

You also notice there seem to be two components of noise in Mirabel Realty's office, some being transmitted through the air and some being transmitted through the structure (from Superior Giftware's machines, through the floor). When you place a hand on the walls or floor, you can feel the vibration. Floors in both offices are tiled.

Before leaving, you tell Trudy Parsenon there seems to be a noise problem, but it can be corrected. You warn her, however, that it may prove expensive. She says it will be difficult to justify the costs to her head office.

You prepare your client for news she does not want to hear

On returning to your office you summarize your findings in a brief progress report, which you mail to Trudy Parsenon.

Part 3

It's now March 14 and you are considering possible ways to reduce the sound levels in Mirabel Realty's office:

1. You could erect a false wall, insulated internally with Corrugon, from floor to ceiling on Mirabel Realty's side of the wall between the two companies.
2. Black cork panels, 18 mm thick, could be glued on the Mirabel Realty side of the wall.
3. You could install carpeting throughout Mirabel Realty's office.
4. Superior Giftware's machine could be mounted on Vib-o-Rug (insulating rubber that eliminates transmission of vibration from machine to building structure).

You recognize that remedies 1 and 2 are alternatives (they both deal with sound transmitted through the air). Remedies 3 and 4 are also alternatives (they both dampen vibrations and sound carried through the structure). Remedy 3 also quite effectively dampens internal office noise.

You consider the approximate costs:

Remedy 1 – $8700
Remedy 2 – $1600
Remedy 3 – $14,900
Remedy 4 – $950

You consider possible problems each remedy may present:

You have several options, some of which can be used in tandem

Remedy 1. Corrugon is in short supply; delivery time would be a minimum of 3 months.

The packaging machine is the problem, but difficult to remedy at source

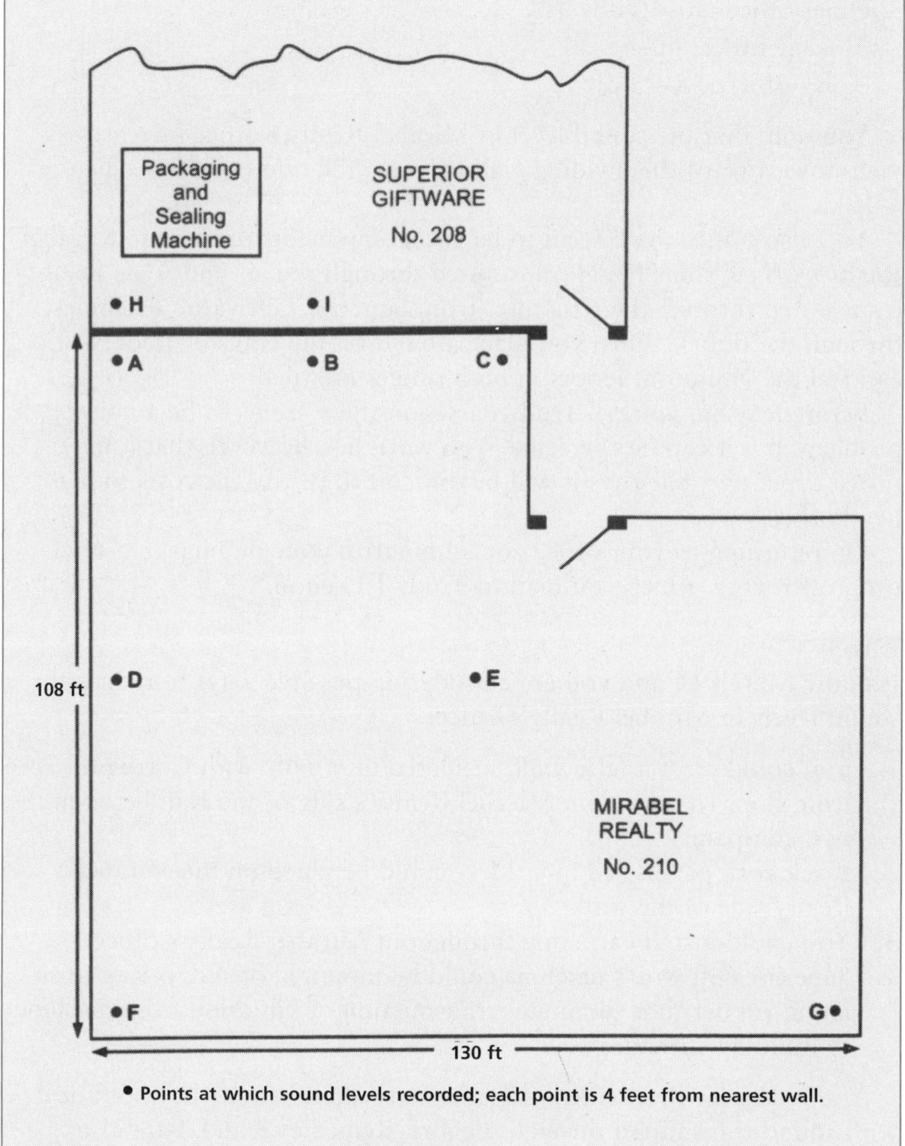

Figure 6-10 Plan of Mirabel Realty's office.

Remedy 2. To some people, cork has an offensive smell; this can be partly corrected by treating the cork with polymethynol.

Remedy 3. The carpet must be dense and have a good quality rubber underlay (included in the approximate cost above).

Remedy 4. Depends on cooperation of Superior Giftware's manager.

You calculate probable noise reductions for each method:

Unfortunately, predicted sound reductions cannot be added together

Remedy 1 – 6 to 10 dB
Remedy 2 – 4 to 7 dB
Remedy 3 – 8 to 12 dB
Remedy 4 – 3 to 5 dB

(These anticipated reductions apply only to the Mirabel Realty office, when both businesses are working.)

You consider which alternatives to recommend to Mirabel Realty, and then write an investigation report describing your findings and suggested corrective measures. You also write a brief cover letter to Trudy Parsenon summarizing your findings.

Note: You prepare a *formal* report because Trudy has mentioned she might have difficulty convincing her head office executives that they *must* authorize the cost of the remedial measures. And, because she knows little about noise and its effects, you decide to include some explanatory information. (You would also be wise to research and document such information at a library, to establish positive evidence for the statements you make in your report.)

Chapter 7 Technical Proposals

In an informal proposal, you can comfortably use the first person: "I"

Proposals
www.io.com/~hcexres/tcm1603/acchtml/props.html
This document is one chapter from the online textbook used in Austin Community College's online course, Online Technical Writing (www.io.com/~hcexres/tcm1603/acchtml/acctoc.html). It describes types of proposals, their organization and format, and the common sections in a proposal. Included are several sample proposals and a revision checklist.

Here, "We" is more prevalent

A technical proposal often seems like a technical report, but there is one major difference. A report usually identifies a situation that needs to be improved or a problem that needs to be resolved, describes ways of correcting the situation or problem, and recommends what action needs to be taken. As a result, it is primarily a "tell" document. A technical proposal might also describe a situation or problem and describe how it can be resolved, but its main purpose is to *convince* or *persuade* the reader to take a certain course of action. As a result, it is primarily a "sell" document.

At various times during their careers, scientists, engineers, technologists, technicians, and their managers and supervisors are called upon to write a proposal. The proposals they write fall into three categories:

An **Informal Proposal** offers an idea and discusses why it should be implemented. Most often it is circulated only within the company, and is usually written as an email or a memo. In each case the writer believes there is a better way to do something and proposes that this idea be implemented. Typical informal proposals might be

- a plan to introduce a new software-driven electronic calibration system throughout the company,
- a proposal to research local resources for replacement equipment components, rather than importing them, or
- a request to attend a conference (such a request is a proposal).

A **Semiformal Proposal** can range from one page to 30 pages or more and may be sent from one company to another, or to senior management within a large company. Short semiformal proposals are often written as letters. Longer semiformal proposals may stand as a separate document with a title page, and be preceded by a cover letter or executive summary. They may suggest ways to increase productivity, provide a service, conduct research, or resolve a problem. Typical examples might be

- a proposal to research new office space to alleviate crowded conditions,

- a proposal to provide specialist consulting services for a potential client,
- a proposal to amalgamate company departments, to provide a more efficient and cost-effective management structure, or
- a proposal to provide portable computers with built-in wireless transmission capabilities for field crews.

A **Formal Proposal** normally is a large, often multiple-volume document designed to impress the government or a major organization that the proposing company has the capability to carry out an important, usually multi-million-dollar task or project. Such proposals are substantial because they describe in detail what will be done, how it will be done, who will be responsible for specific aspects of the work, and why the proposing company has the potential to complete the project on time, within budget, and to the client's satisfaction.

"We" is also common here, to maintain a confident active voice

Formal proposals are usually prepared in response to a Request for Proposal (RFP) that defines exactly how they are to be organized and what must be covered in the proposal. They are always accompanied by a cover letter or letter of transmittal, which often acts as an executive summary. Typical examples are

- a proposal to develop a deep-water holding pond for a city that regularly experiences an overloaded draining system and flooding during heavy rainfalls,
- a proposal to a bank to research ways to improve automatic teller services for customers, or
- a proposal to refurbish mobile communication systems for the Department of Highways.

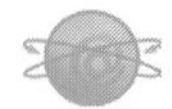

Writing Research Proposals
www.cpsc.ucalgary.ca/Research/grouplab/699/research_proposal.html
This site shows scientists, engineers and technologists how to write research proposals.

In this chapter we will focus on writing informal and semiformal proposals, which are the types you are most likely to encounter in industry.

Overall Writing Plan

All proposals, regardless of their length, contain the following parts:

- A **Summary** that describes briefly what is being proposed and identifies any significant factors (such as cost).
- **Background** information that outlines the circumstances that have caused the proposal to be prepared.
- Definitive **Details** that describe what needs to be done, how it will be done, what the results will be, and why the proposing company is capable of doing the job. This is the body of the proposal.
- An **Action Statement** that requests approval to go ahead (for an in-house proposal), or make a decision (for a client who will buy the services being offered).

The overall writing plan is similar to that for semiformal reports

- **Attachments** or **Appendixes** that contain detailed evidence to support statements made in the body of the proposal (appropriate for most semiformal proposals; not always present in informal proposals).

On the following pages we will demonstrate how to alter these five writing compartments to suit different proposal configurations.

Short Informal Proposal

Marina writes with confidence, and it shows

The plan in Figure 7-1 was used by Marina Albrecht to organize the proposal in Figure 7-2. It is an in-house proposal because Marina is writing only to her manager, Karen LePage. We have inserted the label for each writing compartment beside the proposal to demonstrate how it was constructed; the labels were not shown beside the original document. Sometimes a very short proposal like this does not need supporting evidence, which is why they were omitted from Marina's proposal.

Longer Informal or Short Semiformal Proposal

Marina has written what is known as a single-solution proposal, a proposal that offers only one way to do something. However, there are times when you may want to describe alternative solutions, to demonstrate to the reader that you have considered a number of options, only one of which you propose should be adopted.

Summary	A brief statement that describes what you want to do, or what you want done.
Introduction	The circumstances leading up to the situation that caused you to write the proposal (the **Background** and **Reason**).
Proposal Details	A carefully developed description, in two parts: • **Suggestion:** The proposed changes or improvements, why they are necessary, and what they will cost. • **Evaluation:** An assessment of the viability of the proposed changes and the effect they will have, including any problems that will evolve and how they will be overcome.
Action	A firm statement identifying what you want done, when, and by whom.
Attachments	Supporting data, such as drawings, plans, cost estimates, and spreadsheets. *(This compartment is optional.)*

Figure 7-1 Writing plan for a short informal proposal.

Rossmore Environmental Consultants

To: Karen LePage, Office Manager
From: Marina Albrecht, Project Engineer
Date: March 5, 2004
Re: Proposal to Change to Recycled Copy Paper

Summary Statement When the current supply of regular office copier paper is exhausted, I propose that we change to recycled paper. The cost will be marginally higher, but our company will be seen to be following the advice we give our customers.

Reason/ Background This is exactly the right moment to make the change. The American public has become increasingly sensitive to the damage being done to the environment by extensive use of paper manufactured from the country's timber resources. We will not only make a contribution by using recycled paper, but also can use that fact when proposing that other companies do the same. Coincidentally, we will be changing to a locally made product.

Strong, definite statements...

Details: Suggestion The paper we have used for the past four years is 20 lb Westburn stock, which is imported by Manor Industries Inc. of Dayton, Ohio. (Our two other offices have similar arrangements with local distributors of imported paper products.) The recycled paper I am proposing is 20 lb Environ stock manufactured by Schultz Industries Inc. in Rossmore, Connecticut.

Details: Evaluation I bought 1000 sheets of Environ stock and tested them on a trial basis. I found the following:

- The Environ paper fed as well as the Westburn paper and experienced no paper jams.
- The Environ paper appears very slightly coarser than the Westburn paper, and is slightly less white, but the print image is the same quality.
- 1000 sheets of Environ paper are 1.5 mm thicker than the same quantity of Westburn paper, but that does not affect printing or handling.
- The cost of the Environ paper is $66.95 per 5000 sheets, compared with $59.95 for the Westburn paper.

...and the first-person active voice...

I have discussed the possibility of obtaining a discount from Schultz Industries, and they have agreed that, providing we contract to bulk-purchase all our copy paper from them for one year, for our offices in White Plains and Charlotte as well as Rossmore, they will give us a 10% discount. This will result in a purchase price of $60.26 per 5000 sheets, which is only $0.31 more per 5000 than we are currently paying.

Action Statement I propose that we use Environ recycled copy paper on a 12-month trial basis. May I have your approval by March 25 to place an order with Schultz Industries Inc. of Rossmore, for deliveries to start May 1, 2004.

...help convince the reader the idea is valid

Figure 7-2 A short informal proposal prepared for an in-house audience.

For example, Terence Watkinson is the Chief Executive Officer (CEO) of a successful and rapidly growing business that develops innovative computer software for controlling and routing shipments for the trucking industry. At a recent manager's meeting, it was decided that the company would have to find larger space. Terence instructed Wally Meyers, the company's office manager, to research a suitable building and prepare a proposal to present to the next management meeting.

Satisfy your readers' curiosity

Wally found three suitable locations in different parts of the city, each with different advantages. Although he could have selected what he considered the best site, and proposed just that one, he chose to present alternatives. This achieved two objectives: it demonstrated that he had done a thorough research job, and it satisfied some of the managers who, he knew, had fixed opinions on where the building should be. The writing plan Wally used to organize his proposal is shown in Figure 7-3.

Several factors affect how you write a proposal that offers alternative solutions:

Show you have anticipated your readers' questions

1. When you establish the criteria you will use to evaluate the different alternatives, you must "prove" any criterion the reader might question. If Wally writes, "We will need a minimum of "X" square feet of office space immediately, and another "Y" square feet within two years," he needs to recognize that not every member of the management committee may be aware of the space requirements. He must explain ("prove") why the figures are valid.
2. When you present your Proposed and Alternative Solutions, you must prevent your opinions from intruding. For example, Wally must present only *facts* about each property, and neither comment on its advantages or disadvantages (that will be done in the Evaluation section), nor compare it with other properties. His readers must feel he is completely objective.
3. When presenting Alternative Solutions, to avoid confusing your readers you need to present the facts about each alternative *in the same sequence* that you presented that information for the Proposed Solution. This means Wally must describe the availability, age, condition, accessibility, and cost of each property in exactly the same order.

Maintain your objectivity until you make your recommendation

4. When writing the Evaluation, take great care not to compare one solution with another. Compare the solutions *only* against the criteria.
5. Be positive when writing the Action Statement: use the active-voice expression, "I recommend..." or "We recommend...," rather than the passive-voice expression, "It is considered that..." or "It is recommended that...."

When Meridian Engineering Consultants of Minneapolis, Minnesota, decided to provide courses in technical writing for their staff, they inserted a Request for Proposal (RFP) as a display advertisement in *Midwest*

Summary	A synopsis of the proposal's key points, which identifies the proposal's purpose, main advantages, result, and cost.
Introduction	A description of the situation, condition, or problem that demands attention, and the circumstances leading up to it. This part represents the **Background** and **Reason**.
Proposal Details	The **Details** section is the body of the proposal. It should open with a brief statement that identifies the overall approach. It is then divided into four subcompartments:
Objective	• The **Objective** defines what needs to be achieved to improve the situation or condition, or resolve the problem, and establishes the **Criteria** that must be met.
Proposed Solution	• The **Proposed Solution** offers what the writer considers to be the best way to achieve the objective. It includes a full description of the solution, the expected result or improvement, its advantages, and its cost.
Alternative Solutions	• The **Alternative Solutions** section describes other ways that the objective can be met. Each alternative addresses the same topics as those covered for the proposed solution.
Evaluation	• The **Evaluation** analyzes each solution and compares it against the criteria for an optimum solution established in the Objective. The solutions should be compared only against the criteria, never against each other.
Action	The **Action Statement** recommends what action needs to be taken. It is often titled **Recommendation** and must be written in strong, confident terms.
Attachments	The **Evidence** or **Supporting Data** contains drawings, cost analyses, spreadsheets, etc., that establish the validity of statements made in the body of the proposal.

Avoid saying or implying that this is the best solution...

...let the facts speak for themselves

Figure 7-3 Writing plan for a longer informal proposal or short semiformal proposal.

Business News (see Figure 7-4). One of the companies submitting a proposal was Online Writing Trainers Inc. (OWTI) of Rochester, Minnesota. The proposal was written by Arlene Tetrault, OWTI's projects manager, and is shown in Figure 7-5.

Arlene started by listing the advantages of each of the three training methods she would present, and then made in-depth calculations of the cost for each. This showed her that, because the costs were so close, they would not be a governing factor in MEC's choice. She was then able to concentrate on the advantages that each method offered from the MEC learners' point of view.

MEC Call for Proposals

Provision of Training Services:
Writing Technical Letters, Email, Reports, and Proposals

MEC is soliciting proposals from innovative training consultants to provide courses in technical writing for our 120-person staff, 80% of whom are engineers and engineering technicians working primarily in Civil and Structural Engineering, Information Technology, and the Environmental Sciences. The training is to cover letter, email, report, and proposal writing, and include methods for sharpening individual writing style.

The training is to start January 15, 2004 and be completed by March 31, 2004. Vendors are to provide three copies of their technical and cost proposal by noon on Thursday, October 31, 2003, marked RFP 3/014. Late proposals will not be accepted.

Meridian Engineering Consultants Inc.
334 Willows Avenue, Minneapolis, MN 55261

Figure 7-4 The Request for Proposal (RFP) used by Online Writing Trainers Inc. to prepare the proposal in Figure 7-5.

Encourage readers so they *want* to read your words

The design of the proposal, with headings in a narrow column on the left and the text in a wider column on the right, is an effective application of information design principles. Readers can readily see how Arlene has structured her ideas, and can find information easily. The headings also parallel the labels in the writing compartments shown in Figure 7-3. Here are some additional comments on the proposal:

- The paragraph in the center of page 1 is Arlene's **Summary**, in which she identifies the preferred training method she recommends and lists its cost. Many people hesitate to state the cost in the Summary, fearing that readers may not continue reading if they feel the cost is too high. We believe it should be there, because it is the first question readers are likely to ask, so they will search for it and be irritated if they find it has been buried far down in the proposal.
- The first paragraph of the Introduction provides the **Background**, which sets the scene for the information to follow. Arlene draws much of this information from the RFP in Figure 7-4.
- The paragraph at the foot of page 1 provides a quick statement that identifies OWTI's capacity to handle the project. Arlene keeps it short, placing the detailed information in an attachment.
- By listing the **Objectives** (on page 2), Arlene identifies the factors she will use to evaluate the three methods. In Objective 1, she lists the

Online Writing Trainers Inc.
Suite 200 – 450 Bridgeview Road
Rochester MN 55952

Proposal to Provide Training Services:
Writing Technical Letters, Email, Reports, and Proposals

Prepared for
Meridian Engineering Consultants Inc.
Minneapolis, Minnesota

In response to
MEC RFP 3/014
Proposal prepared October 27, 2003

We have investigated three methods for providing training in letter, email, report, and proposal writing for Meridian Engineering Consultants Inc. The method we propose is a mix of web-based and traditional classroom-style learning. It will meet the needs of MEC staff who prefer electronic delivery and those who prefer more traditional instruction. The total cost at $51,400 is comparable to solely online or solely classroom instruction.

Indent the summary on both sides to catch readers' attention

Introduction

Meridian Engineering Consultants Inc. (MEC) plans to upgrade its technical staff's ability to write effective letters, email, reports, and proposals. Training is to be conducted between January 15 and March 31, 2004, and is to include approximately 96 technical professionals and 24 support staff. MEC published a Request for Proposals (MEC RFP 3/014) in the *Midwest Business News* on October 10, 2003, calling for interested training consultants to submit training and cost proposals for providing the appropriate services.

Online Writing Trainers Inc. (OWTI), of Rochester, Minnesota, has been providing onsite courses for engineering and other business organizations in the US and Canada since 1972, and in the UK since 1994 (see Attachment 1 for a detailed company description). To meet the growing demand by both North American and European businesses to access training over the Internet, in 2001 we converted our onsite courses for electronic delivery. They are now available online from our corporate website.

Highlight company experience only very briefly; focus on what your company can do for the reader

1

Figure 7-5 A short semiformal proposal offering alternative solutions.

Objectives

We established the following requirements that must be met:

1. The training is to cover eight main subjects:

   Foundation Topics

   - Getting to the point (identifying and placing key information for immediate access)
   - Organizing the details (developing the remainder of the document)
   - Writing effective email
   - Sharpening language skills

   Advanced Topics

   - Writing business letters and memos
   - Writing short reports
   - Planning and writing formal reports
   - Planning and writing business and technical proposals
2. Technical staff are to receive training on both foundation and advanced topics. Support staff are to receive training only on the foundation topics.
3. The cost for the training must be comparable, whether delivered onsite or online, or in a blended format.
4. The training must be completed within a 2.5 month period, between January 15 and March 31, 2004.
5. The training must accommodate the schedules of technical staff who travel frequently.

Draw on the client's requirements to write the Objectives

Proposed Delivery Method
Blended Training

Our proposal offers a combination of onsite and online courses under a "blended" arrangement, with some portions of the training being taught online and other portions being taught onsite. The costs for implementing blended training will be slightly less than for traditional classroom training.

In the following discussion, we have relabeled the eight topics as four foundation courses and four advanced courses. See Attachment 2 for course descriptions.

Cover new technology or methods in detail

The **online training** will be held first and will cover the four foundation courses, which will be taken by all staff. Because the program is maintained on OWTI's server, participants will not need to download the courses to their individual computers. They will also be able to access the courses from any computer at any location at any time. The system will record their progress and each time participants log on they will be taken immediately to the point where they stopped. There will be an examination at the end of each course, which will be evaluated electronically and the results reported to the participant.

The **onsite training** will cover the four advanced courses, which will be presented at spaced intervals, after each person has completed the foundation courses online. Support staff will not take the advanced courses, because they normally do not write technical correspondence, reports, and proposals.

2

Blended Training (continued)

Schedule

The **online segment**, comprising the four foundation courses, will be taken over a six-week period, between January 15 and February 24, 2004. The four courses require a total of 6 to 8 hours of study.

The **onsite segment**, comprising the four Advanced Courses, will be covered in a single 8-hour class, with a maximum of 12 participants in each class. To cover the 96 technical professionals, the classes will be held on eight separate days, four between March 1 and 4, and four between March 22 and 25. The spread of dates will allow for possible travel absences of engineering staff.

Explain how the system will work

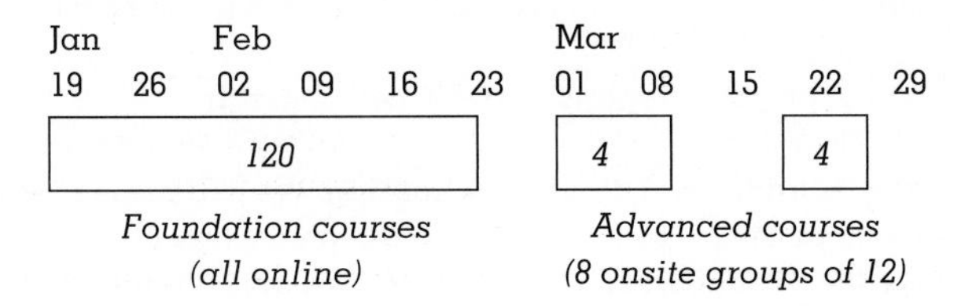

Cost

The cost for providing courses under the blended approach will be $51,400. The cost covers provision of

- 120 passwords and individual IDs for participants to access the four foundation courses,
- 8 one-day onsite training sessions, to cover the four advanced courses, with 12 staff members attending each session,
- 2 instructors for each onsite course,
- evaluation of 2 assignments written by each onsite course participant, and
- instructors' travel and accommodation expenses.

Identify total cost and what it covers in the proposal; put detailed costs in an attachment

Alternative Delivery Method *Online Training*

In the online training mode, all eight courses will be taken electronically. Participants will be able to access the courses from any computer at any location and at a time convenient to themselves. The system will record their progress and each time participants log on they will be taken immediately to the point where they stopped. There will be an examination at the end of each course.

The 96 technical professionals will register for all eight courses, which will require approximately 14–16 hours of study. They will also write four assignments and submit them electronically to an OWTI instructor, who will return them with feedback.

The 24 support staff will register for the four foundation courses, which will require approximately 8 hours of study.

All participants will receive a copy of the course textbook, which will become a permanent resource for future reference.

3

Online Training (continued)

Schedule

The courses will be taken between January 15 and March 31, 2004. OWTI will monitor course progress and submit a report to MEC every two weeks. The reports will list those who have started, how far each person has progressed, and those who have completed their courses.

Cost

The cost for providing training online will be $49,220, which will include

- 120 passwords and individual IDs for participants to access the four foundation courses,
- 120 course textbooks,
- evaluation of four assignments submitted by each person completing courses 5 to 8, and
- submission of progress reports at two-week intervals.

Alternative Delivery Method
Onsite Training

Onsite training is traditional classroom training. We will present eight two-day workshops for the 96 technical professionals, and two one-day workshops for the 24 support staff, with 12 participants attending each workshop. Topics to be covered will include the following:

2-day workshop: All 8 topics
1-day workshop: 4 foundation topics

All participants will receive a copy of the course textbook and approximately 30 pages of additional notes. The workshops will be held on MEC's premises.

Schedule

The workshops will be presented in four timeframes to accommodate staff absences while on field assignment:

A table summarizes key points and draws readers' attention

Dates	2-day Workshops	1-day Workshops
January 19–23	2	1
February 2–5	2	-
March 1–5	2	1
March 15–18	2	-

Cost

The cost for providing the 10 workshops will be $52,520, which will include

- 2 instructors for each workshop,
- evaluation of four assignments completed by participants attending the two-day workshops,
- 120 course textbooks and course notes,
- instructors' travel and accommodation expenses.

4

Evaluation of Alternative Methods

All three of the proposed methods will provide the required depth of training established in the Objectives, can be completed within the required time frame, and will accommodate the schedules of technical staff who travel. The costs also are comparable:

Blended training *(online and onsite)*	$51,400
Solely online training	$49,220
Solely onsite training	$52,520

The primary differences are in the delivery methods and individual participants' reaction to them. For a detailed cost analysis, see Attachment 3.

Identify where cost explanations can be found

Blended Training will meet the needs of both types of course participant: those who prefer electronic instruction and those who prefer the interactive classroom environment. The more basic foundation topics will be taught online. The more intense advanced topics will be presented in person, which will provide participants with personal instruction and the opportunity to ask questions.

Solely **Online Training** will please participants who prefer the privacy and ability to study at their own time, pace, and location. It will not, however, provide personalized instruction or the interactive environment that some participants prefer.

The Evaluation permits the writer to air her views

Solely **Onsite Training** will please participants who prefer to work face-to-face with an instructor and like the interactive environment in which they may ask questions and hear the questions of others. It will not, however, offer much flexibility because participants must attend at a fixed time.

Conclusions

Although all three methods will meet MEC's requirements, we consider that the blended training option will provide the flexibility MEC needs to train both technical and support staff, and will suit employees who often have to travel and work offsite.

Online Writing Trainers Inc.
27 October 2003

5

(Note: the three attachments are not printed here to conserve space in this edition of *Technically-Write!*)

topics that need to be taught. She divides them into two groups to suit the two different groups of employees, which she identifies in Objective 2. She drew Objectives 3, 4 and 5 from MEC's RFP.

Maintain your objectivity, almost to the end

- The **Proposed Solution** starts in the center of page 2 and continues to two-thirds of the way down page 3. Arlene presents only *facts*, without commenting on their value at this stage in the proposal.
- The two **Alternative Solutions** appear on pages 3 and 4. Each is shorter than the proposed solution, but the information is presented in the same sequence. Again, Arlene presents only facts.
- In the **Evaluation** of the three methods (page 5), Arlene compares each method against the Objectives she established on page 2. She starts by identifying the objectives met by all three methods, and then continues with comments on the advantages and disadvantages of each method. Here, she allows her (i.e. OWTL's) opinions to appear, for the first time in the whole proposal.
- The **Conclusions** present the outcome of the Evaluation, but Arlene only *suggests* which method MEC should select. Although Figure 7-3 labels this as an **Action Statement,** in which the writer normally makes a strong recommendation that the reader approve the proposal, Arlene chose to move her Action Statement into a cover letter to send with the proposal:

A cover letter is also known as a transmittal letter or an executive summary

Dear Contracts Manager:

I am enclosing Online Writing Trainers' proposal to present courses on writing technical letters, email, reports, and proposals to Meridian Engineering Consultants' staff, in response to MEC RFP 3/014. We recommend that Meridian Engineering Consultants adopt "Blended Instruction" as the preferred training method, which will be a combination of electronic and in-person delivery of the training. The cost will be $51,400, which is slightly less than for regular classroom-style training. Blended Instruction will also meet the needs of staff who prefer electronic delivery and those who prefer personal instruction.

Our corporate web site at www.owti.com provides a detailed description of Blended Instruction. Please call me at 507.488.1827 if you have further questions.

Sincerely,

Arlene Tetrault

Contracts Manager

Online Writing Trainers Inc.

Student Project Proposal

Many technical students nearing the end of their education have to undertake a technical term project, sometimes working alone but more often working in teams. Although the instructor may assign a project to each team, there are times when the instructor invites each team to identify a technical problem and then write a proposal identifying how the team will tackle it. You can use the writing plan in Figure 7-6 to help you.

This practical approach is particularly suited to college writing

The plan shows that you cannot simply decide, without considerable forethought, that "We'll put two computers at different ends of the building and work out whether there is less information loss between them, using fiberoptic cable compared with RS-232 wire cable." That would make a good project, but before writing it up as a proposal you need to work out the amount of cable you will need, how you will get the computers, what software you will be using, how you will measure information loss at different frequencies, how long all this will take you, and so on. Only when you have "done your homework" and have the facts at your fingertips, will you be ready to write the proposal.

Longer Semiformal Proposal: Single Solution

If a company is already working on a project with a client, and the client runs into a technical problem at a nearby location, the client is more

Summary	A brief outline of what you plan to do, and what doing it will achieve.
Background	Why the project needs to be tackled. Include historical information concerning the topic and identify the team members.
Proposal Plan	Describe how your project team will carry out the project. Provide the following information:
Details	• Your overall approach or plan. • Who (in your team) will be doing what. • Special equipment or parts you require (attach a list)
Project Schedule	• Identify the dates on which you plan to > complete your research, > finish the design, > complete product construction, and > complete testing and troubleshooting the product.
Reporting Schedule	• Identify the dates when you plan to > submit progress reports, > submit a topic outline for your project report, > submit first draft sections for evaluation, > complete the final project report, and > present your oral report.
Action	Request approval to go ahead with the project.
Evidence	Provide supporting information to validate your plan, plus a list of materials or parts you will need.

This writing plan parallels how project proposals are written in industry

Figure 7-6 Writing plan for a student project proposal.

likely to turn to the existing consultant and ask that they research the problem and propose a solution, rather than solicit bids from several consultants. This is known as a *solicited* proposal. (In an unsolicited proposal, the proposing company has not been asked to submit a proposal.)

The writing plan is similar to that for a short informal proposal, but is extended to include additional features. In the **Proposal Details** it will

Insert additional steps into the writing plan

- describe in-depth what can be done, and why,
- outline the gains that will be achieved,
- draw attention to any problems that may occur if the proposal is implemented, then explain what will be done to alleviate them, and
- calculate the cost.

There is also likely to be an additional section that describes the company's **Capability** to do the work if the proposal is approved.

These writing compartments are shown in Figure 7-7.

Longer Semiformal Proposal: Multiple Solutions

A longer semiformal proposal may also present and evaluate alternatives, just as the longer informal proposal does, but it will have much greater internal development and demand more attention to detail. A typical writing plan is shown in Figure 7-8.

A single-solution semiformal proposal can be quite straightforward

Summary	Very briefly describe what needs to be done, and why. Identify the overall advantage, what will be achieved, and the cost.
Introduction	Describe the situation, condition, or problem that demands attention, and the circumstances leading up to it. (The **Background** and **Reason**.)
Proposal Details	Open with an introductory statement that identifies the overall approach. Divide the information that follows into four subcompartments:
Description	• Fully describe what will be done and how the work will be implemented.
Gains	• Describe all the advantages and the effect each advantage will have.
Problems	• Identify any problems that implementing the proposal will create, and describe how the problems will be resolved or at least lessened.
Cost	• Provide definitive cost details to show how the overall cost mentioned in the Summary has been calculated.
Capability	Describe your company and the company's capability and experience in doing similar work.
Action	Recommend what action needs to be taken and ask for approval to implement the proposal. Use strong, positive terms.
Attachment	Attach drawings, sketches, cost analyses, spreadsheets, etc, to support the statements made in the Details section of the proposal.

Figure 7-7 Writing plan for a longer semiformal proposal offering a single solution.

A primary difference between this writing plan and that for a longer informal proposal (Figure 7-3) occurs during the evaluation of alternatives. The writing plan for an informal proposal suggests presenting all the alternatives before starting to evaluate them. The writing plan for the semiformal proposal suggests evaluating each alternative immediately after you have presented it. However, in both cases you evaluate the alternatives *only against the criteria*, never against each other. We are not suggesting that you can use each approach only for the writing plan you now

Still maintain your objectivity

Summary	Provide a synopsis of the proposal's main features and state the cost, possibly as a range, depending on which option is approved by the client.
Introduction	Describe the circumstances leading up to the problem or unsatisfactory condition to be corrected, and the proposal being written. (This is the **Background** and **Reason**.)
Proposal Details	As with the previous proposals, start the Proposal Details section with a short introduction to the six subsections that follow.
Approach	• Describe your overall approach (how you will investigate the problem/situation and what you anticipate the proposal will achieve).
Parameters	• Describe the factors that the client established for a satisfactory conclusion to the project, and include a budget and timeline. Also identify the criteria you will use to evaluate the effectiveness of the solution(s) being offered.
The Plan	• This will become your most detailed subsection. Describe > exactly what will be done (as a general statement here, and step-by-step in an Attachment), > how it will be done, > the advantages that will accrue by doing the work in the way you suggest, > how well the plan meets the criteria established earlier, and > the cost, broken out for each significant factor you discuss.
The Alternatives	• Describe what other options are available (anticipate that, on reading The Plan, your readers may say to themselves: "Yes, but didn't you also consider...?"). Evaluate each option separately against the criteria.
Cost	• Provide a cost calculation for each of the different configurations or alternatives listed earlier.
Conditions	• Establish contractual details, such as for how long the cost analysis is valid and who will be responsible for what activities if the proposal is approved.
Capability	Describe your company's capability to do the work and experience on similar projects.
Conclusions	Sum up the key features of the proposal, and restate the primary advantages it offers.
Attachments	Insert charts, drawings, sketches and spreadsheets to support earlier statements, and detailed steps and cost analyses.

A multiple-solution semiformal proposal can offer in-depth descriptions of alternatives

The recommendation can appear in an accompanying letter (like an executive summary: see page 147)

Figure 7-8 Writing plan for a longer semiformal proposal offering multiple solutions.

see. When you become an experienced proposal writer, it's acceptable to transfer each approach into either writing plan.

Here are further comments on the writing plan in Figure 7-8:

- When establishing the **Criteria** (in the **Parameters** section), ensure that you prove any criteria your readers might question. They must feel comfortable with the criteria before you start comparing your different plans against them.
- Know that there can be two types of **Alternatives**, and that normally you will present only one:
  1. You can offer different ways to correct a problem or improve a situation, only one of which will be selected. (When Wally Meyers presented alternative sites to the company management, he used this approach.)
  2. You can offer a basic plan as your **Proposed Solution**, and then present additional features as "add-ons" for the reader to consider (they supplement, rather than replace, the proposed solution).
- The **Conditions** are like insurance: they are there to protect you in case the reader assumes you will be performing certain tasks that you expect the client to perform. If you prepared your cost estimate on that assumption, you could be in for an expensive surprise. For example, Metronome Telecommunications proposed that H. L. Winman and Associates employ them to install Mercury 7.0 high-speed Internet connections in each of the company's engineering departments. Winman approved the proposal, but when Metronome started to install the equipment they discovered that Winman had assumed that Metronome would remove the existing system, whereas Metronome had assumed that Winman would do that before they started the installation work.

Anticipate and identify "who does what"

- The section where you describe your company's experience and capability is often mishandled. We have seen many proposals that start with from 10 to 40 pages describing how good the proposing company is, before their readers encounter any section that tells them *what they most want to hear*. A prominent manufacturer told us once: "I don't want to wade through pages of 'Look at who we are and what we have done for others in the past!'"
- As a result, the **Capability** section in Figure 7-8 is positioned after the **Proposal Details**. We suggest, however, that it's acceptable to insert a paragraph or two in the Introduction, to summarize your company's strengths, as a way of explaining why you are submitting the proposal.
- Previous proposal writing plans in this chapter have concluded with an **Action** statement, in which you make a recommendation and ask

for approval. In this longer, more complex proposal we suggest closing with a **Conclusion**, in which you sum up the key points (this is sometimes known as a Terminal Summary.) The proposal is then accompanied by an **Executive Summary**, which is a letter of transmittal that comments on any point of particular importance and ends with an Action Statement. For more information about Executive Summaries, see Chapter 6.

Writing Plan Flexibility

Design your proposal to suit the particular audience

We have shown you five writing plans for informal and semiformal proposals, each of varying length and complexity. They are typical of the many designs you may encounter. It's important to remember that the designs are not "written in stone." As a writer of proposals, you should always be ready to augment or modify the designs shown here to fit the particular situation that affects you, the information you have to convey, and the reader(s) you have to address.

The Language of Proposal Writing

There can be nothing wishy-washy about the language you use in a proposal. If you have organized your proposal using one of the writing plans shown in this chapter, you will provide a smooth flow of information. Now you must let your language convince your readers that you have a strong case to present. Here are four suggestions.

Types of Proposals
http://writing.colostate.edu/references/documents/proposal/pop2b.cfm
This site shows engineers how to write different types of proposals, from grants to bids.

1. Present Only Essential Information

Before writing, divide all your information into two parts: (1) information the reader *must have* to make a decision (we call this the "Need to Know" information); and (2) information that is of general interest but the reader *does not need* to make a decision (we call this the "Nice to Know" details). Often, because we know a project well, we tend to present everything we know because it interests us. Take a step back and look at your information from the potential readers' point of view.

2. Use the Active Voice

The active voice will make you sound firm and definite. Instead of writing:

> The two computers would be connected by means of a metal wire and a fiber-optic cable, whereas alternating from one cable to the other would be accomplished by a Model 1880 switching unit. *(Passive voice: 33 words)*

Write strongly and positively

Write:

> A wire and a fiberoptic cable will connect the two computers, while a Model 1880 switching unit will alternate between them. *(Active voice: 22 words)*

In addition to having 33% fewer words, the active voice makes the second writer sound much more confident and knowledgeable. (For more information on using the active voice, refer to Chapter 12.)

3. Avoid Wishy-washy Words

Replace weak words like *would, could*, and *should* with a strong word like *will*. In the first example about the two cables, the word *would* occurs twice and creates only a "soft" impression (the reader may comment: "Well, I guess that might be okay."). In the second example, the word *would* has been replaced with *will*, creating a much more confident impression (the reader will feel like commenting: "Now that makes sense!")

Use words that convey a strong image

You can make a similarly weak impression if you insert low-information-content expressions into your proposal. Examples are *bring to a conclusion* (use *concludes*), *in the direction of* (use *toward*), and *by means of* (use *by*, or change from passive voice to active voice, as in the two sample sentences, above). For an extensive list of such words, see the section on Low-Information-Content Expressions in Chapter 12.

A third damaging effect occurs if you write vague statements like *an adequate supply, got some help*, or *many tests will be attempted*. Whenever possible, use descriptive words that convey clear images, words such as *a three-week supply, two technicians helped us*, and *we will carry out 30 tests*.

4. Avoid Giving Opinions

Experienced proposal writers know just when to insert an opinion or a subjective statement. As a beginning proposal writer, you will be much safer if you withhold your opinions until the end of the proposal, when you make your recommendation.

ASSIGNMENTS

Project 7.1 Acquiring Handheld Computers

Assume that you are employed by the local branch of H. L. Winman and Associates (a nationwide consulting firm), in a department related to the discipline you are studying—e.g. civil engineering, mechanical engineering, biophysics, environmental science. The local branch has 130 employees, 27 of whom are in your department. In your work, you and your associates have to travel frequently. Most of you use laptops.

In the coming year's budget, the company has set $30,000 aside to purchase replacement laptops. However, you feel that a handheld computer would be more useful and convenient. You discuss the idea with your associates and nearly all agree with you. You describe your idea to Wilson Harcourt, your department manager, who says "Your idea has merit." He asks you to prepare a proposal he can take to the next Capital Budget meeting, and suggests that you describe

Overcome readers' resistance by anticipating their questions

- why handheld computers would be of value to departmental staff,
- what you can do with a handheld computer, compared to a laptop,
- the advantages of a handheld computer, and
- how many should be purchased.

He also suggests you identify several different types of handheld computers, evaluate them, and propose that the company buy a specific brand.

Project 7.2 Installing an Alternative Power Supply

You are the technician who experienced a 5 hour and 24 minute overnight power outage that destroyed the tests on electronic and mechanical switches for Terrapin Control Systems, and delayed the project (see Project 4.4 on page 96).

Today your manager (John Grayson) tells you that the company has received a second, even larger contract to test switches for Terrapin Control Systems. The tests will start on the 20th of next month and the scheduling will be very tight. As a result, there must be no delays.

To prevent a future power outage affecting the new set of tests, you recognize you must have a backup power supply. Investigate what power supplies are available and make an informal proposal to John Grayson recommending either the purchase or lease of a suitable power supply that can handle the current required for heat and cold chambers. Ideally, present several alternative power supplies, with different capabilities and prices, and recommend one of them.

Project 7.3 A Proposal to the Student Council

Write a proposal to the Student Council (or its equivalent) at your school or college, describing an innovative idea you have that you would like the council to implement. The idea may be one of the following:

Answer even more questions than are listed here

1. A plan to set up a 2–4 day skiing trip to one of the ski resorts nearest to your college, to be held during a mid-term or between-term break. Work out the details and use them to answer questions the Student Council is likely to ask, such as these:
   - When will it happen?
   - How much will it cost?
   - Where is the resort?
   - What lodgings are available, and at what cost?
   - What arrangements will be made to rent a coach, and at what cost?
   - Who will make all the arrangements?
   - How will it be marketed?
2. A plan to set up a money-making event that will generate funds for a charity (you choose which one). Identify an event that will be particularly visible and so promote the charity's need for funds. Typically, students taking part in the event obtain sponsors who promise to donate a specified amount if the student they sponsor completes an activity such as a 50-mile bicycle ride, a 20-lap swim, a 10-mile hike, and so on. Describe why the charity is worth supporting, how the event will be organized, who will do the organizing, and how the event will be publicized, either through the school's internal media or to the general public.
3. A plan to clean up the neighborhood around the school or college. From time to time there have been complaints from residents on neighboring streets that the students drop gum and candy wrappers, looseleaf pages, cigarette packs and butts, and so on, as they walk to and from the local bus stop or their cars. The residents complain that the students' debris lowers the quality of the neighborhood and reduces house values. Suggest that the Student Council set up clean-up crews who will regularly (once a week?) search the neighboring streets for rubbish and collect it in large garbage bags. A key factor in your proposal is that the Student Council should send out a news release to the local media, to demonstrate that the College's/School's students are very conscious of the image they create and that they want to contribute actively to the neighborhood's environment. Be ready to counter remarks from the Student Council that the local residents contribute much of the garbage they are complaining about.

Alternatively, if you have a different idea you feel the Student Council should address, you may select it as your topic. Whatever topic you select, you must research it in sufficient depth so that you can write a confident proposal.

Develop your own idea for a project!

Chapter 8
Other Technical Documents

This chapter describes how to write a user's manual, provides detailed guidelines for writing a technical instruction, offers suggestions for writing a scientific paper, and describes how to convert your knowledge of a process, equipment, or new technique into an interesting magazine article or technical paper.

User's Manual

In our technological era, with its increasingly complex range of hardware and software, there is a growing need for manufacturers to write clear technical manuals to accompany what they sell. Most manufacturers issue a user's manual with each of their products, which contains (1) a brief description of the product, (2) instructions on how to use it, and (3) suggestions for fixing problems that may occur. For qualified repair specialists they may also produce a set of maintenance instructions containing detailed service and repair procedures. Both publications perform the same task for different readers: user manuals assume the reader has only slight technical knowledge, while maintenance instructions assume the reader is a technical expert.

The suggestions that follow apply to any basic user manual. Whether you are writing a manual to accompany heavy construction equipment, a delicate instrument, or a new version of a software program, you must organize your description so that it follows a coherent pattern.

Identify the Audience

The problem with many user guides is that they are written from an engineer's or a technical person's point of view and are often too complex or incomplete from the user's point of view. To avoid this, you need to identify your audience before you start writing, and understand what level of knowledge and experience it has with the product. The language you use must be appropriate for the intended audience: be careful not to use jargon it is not familiar with.

We often suggest to engineers that they write a brief description of who the user is, so they can refer back to it as they write and keep the focus of their writing on that audience. This audience analysis often becomes the first section in the user's guide, called "Who Should Read This Document."

The first step in *any* writing situation: know your audience

Writing Plan

The writing plan for most user manuals has four compartments, as shown in Figure 8-1. The two top compartments describe the product, and the two lower compartments tell the reader how to use it.

The following sections demonstrate how these four compartments are used. The product in this instance is electronic mail software.

Describing the Product

The **Summary Statement** briefly describes the product and its main purpose:

> This electronic mail software allows you to communicate with other email users by sending and receiving messages. It allows you to connect to a remote computer, called a server, and access messages other people have sent to you. The server computer will also send any messages you have written to other people.

The **Product Description** identifies each part and describes its components. For example, a user manual accompanying a word-processing program would describe the program's contents and application. For equipment that has several discrete components or parts, the Product Description may be subdivided into two sections: Overview and Detailed Description.

- The Overview section simply lists the main components:

  The electronic mail program contains four components:

  - The In Tray
  - The Out Tray
  - The File Cabinet
  - The Address Book

List the components in the same sequence you will describe them

- The **Detailed Description** provides more specific information about each component, and in particular draws attention to items the user will operate or use. The parts must be described *in the same sequence* that they were presented in the Overview section:

> The In Tray is identified by a square icon with an arrow pointing down. This indicates that messages are directed to you. Any messages that are sent to you are automatically placed in this area. You can view your messages in the In Tray by pointing your mouse on the icon and clicking.
>
> The Out Tray is identified by a square icon with an arrow pointing up. This indicates that messages are from you to someone else. This is the area where messages you have written are stored until you are ready to send them.
>
> The File Cabinet is identified by an icon that looks like a traditional two- or three-drawer filing cabinet you might find in an office. This is the area where you can

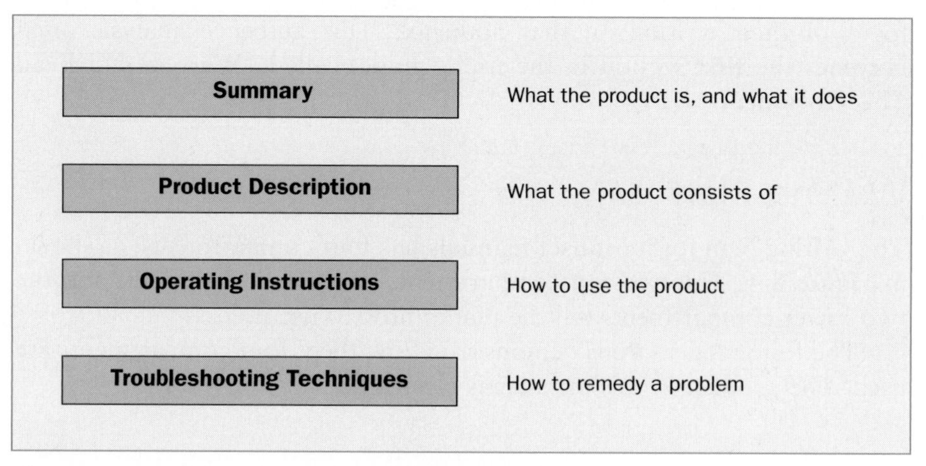

Figure 8-1 Writing plan for a user's manual.

Clear, simple language helps *all* readers understand

store or file messages that are important or that you want to keep. You can create different folders for different situations and thus create a filing system for your messages.

The Address Book is identified by an icon that looks like a small book. This is the area where you store the electronic addresses of people to whom you frequently send messages. When you write a message you can select the address of the person or persons you are sending it to directly from the Address Book.

Using the Product

The **Operating Instructions** provide step-by-step instructions for each task the product can do. One of the major problems with many user manuals is that they are not written from the user's perspective: instead of describing *how* to do something with the product, the manual describes *what* can be done with it. Too often, writers jump right into writing the steps without determining what the reader will need. If you follow these three steps, your User's Manual will be user-oriented:

1. Perform a Task Analysis.
2. Group and label the tasks.
3. Write the steps for each task.

This is similar to the writing process described in Chapter 2. See Figure 2-3.

Step 1—Perform a Task Analysis

When you have decided who your audience is, you need to list everything *they might want to do* with the product. Make sure you don't list everything *the product can do*. Instead, focus on the tasks the user will perform. For example, a task analysis for a simple electronic mailing package might look like this:

Tasks:
Sending messages
Receiving messages
Forwarding messages
Addressing messages
Printing messages
Adding names to an address book
Storing messages in folders
Creating folders
Downloading messages
Assigning a password
Deleting messages
Creating messages
Connecting to the server
Checking spelling
Attaching documents
Installing
Customizing
Calling manufacturer for support
Using the Help system

When writing a task analysis, use verbs that end with "...ing" (*sending, receiving*, etc.)

This list is developed while brainstorming and is not meant to be in any particular order. Later, you can add items; for now, just list the tasks.

Step 2—Group and Label Tasks

Looking at the above task list, you can see that certain items are related. For example, Receiving messages and Forwarding messages are related, just as Creating messages, Addressing messages and Checking spelling form a second group of related tasks or topics. We suggest grouping the related topics and giving them a letter to identify each group. For example, in the following revised list, each task has been assigned a letter (A, B, C, etc) to show which group it belongs to.

Identify which tasks seem related

Tasks:	
Sending messages	F
Receiving messages	A
Forwarding messages	A
Addressing messages	B
Printing messages	C
Adding names to an address book	B
Storing messages in folders	A
Creating folders	A
Downloading messages	A

Assigning a password	D
Deleting messages	A
Creating messages	B
Connecting to the server	F/A
Checking spelling	B
Attaching documents	F
Installing	D
Customizing	D
Calling manufacturer for support	E
Using the Help system	E

This becomes the first step toward organizing the information

So far, no organization has been done with the list of topics. The first organizational step is to assign a label to each group. Choose a label that *describes what the user is doing* or is trying to accomplish with the tasks. Use "ing" words whenever you can in your labels, because they *indicate an action performed by the user.* Here are labels for the above groups:

A – Handling Incoming Messages
B – Writing Messages
C – Printing Messages
D – Getting Started
E – Getting Additional Help
F – Sending Messages

An outline emerges naturally, almost painlessly...

Now is the time to organize the groups of topics into a logical sequence for the intended audience. User manuals are usually structured in a sequential arrangement listing what needs to be done first, or by introducing easy tasks first. In this example the structure might look like this:

Getting Started ***(D)***
- Installing Your Software
- Customizing Your Software
- Assigning a Password to the System

Writing Messages ***(B)***
- Creating Messages
- Addressing Messages
- Adding Names to an Address Book

Sending Messages ***(F)***
- Sending Messages You Have Written
- Connecting to the Server
- Attaching Documents

Handling Incoming Messages ***(A)***
- Connecting to the Server

Receiving Messages
Forwarding Messages
Creating Folders
Storing Messages in the Folders
Downloading Messages
Deleting Messages

Printing Messages *(C)*

Getting Additional Help *(E)*
Using the Help System
Calling the Manufacturer for Support

We now have a user-focused, task-oriented structure that describes how to use the product.

Step 3—Write the Steps

Each of the identified tasks now becomes a procedure, and you can write the steps it takes to accomplish each task. Here are two examples from the electronic mail tasks:

...from which a logically flowing procedure can be written

Installing Your Software

1. Unpack the contents of the box and make sure you have
   - this manual
   - six 3.5-inch disks
   - the software license.
2. Turn your computer on and start Windows.
3. Put the disk labeled Disk 1 into your A drive.
4. Click on the Windows Start menu.
5. Select Run...
6. Type **A:INSTALL**
7. Follow the directions on the screen.

Connecting to the Server

1. Click on the File menu.
2. Choose the Connect to Server command.
3. When a dialog box appears, enter your ID and password.
4. Click OK.
5. Wait while the system initiates the protocol sequence defined in the Server Settings dialog box.
6. When the message **You are Now Connected** appears on screen, you have successfully connected to the remote server.

You can now send and receive your electronic mail messages.

Note that each step is short, has a number, and uses verbs in the imperative mood. (For more information on writing instructional steps, see "Give Your Reader Confidence" on page 216 of this chapter.)

The **Troubleshooting Techniques** tell the reader what to do if, having followed the Operating Instructions correctly, the equipment does not work. It also has short, numbered steps and verbs in the imperative mood.

When all else fails, call for help

If the message **Server Connection Failed** appears, follow this procedure:

1. Click on the Settings Menu.
2. Choose Server Settings... command.
3. Make sure your ID, password, and IP address are correctly identified.
4. Click OK.
5. Try connecting to the server again.
6. If the problem continues, call the manufacturer for technical support.

Technical Instruction

When H. L. Winman and Associates' special project engineer, Andy Rittman, wants a job done, he issues instructions in clear, concise terms: "Take your crew over to the east end of the bridge and lay down control points 3, 4, and 7," he may say to the survey crew chief. If he fails to make himself clear, the crew chief has only to walk back across the bridge to ask questions. But Fred Stokes, chief engineer at Macro Engineering Inc., seldom gives spoken instructions to his electrical crews. Most of the time they work at remote sites and follow printed instructions, with no opportunity to walk across a project site to clarify an ambiguous order.

A technical instruction tells somebody to do something. It may be a simple one-sentence statement that defines what has to be done but leaves the time and the method to the reader. Or it may be a step-by-step procedure that describes exactly what has to be done and when and how. The latter type of technical instruction will be described here.

An instruction *must* be written from the reader's point of view

Before attempting to write an instruction, you must first define your readers, or establish their level of technical knowledge and familiarity with your subject. Only then can you decide the depth of detail you must provide. If they are familiar with a piece of equipment, you may assume that the simple statement "Open the cover plate" will not pose a problem. But if the equipment is new to them, you may have to broaden the statement to help them first identify and open the cover plate:

Find the hinged cover plate at the bottom rear of the cabinet. Open it by inserting a Robertson No. 2 screwdriver into the narrow slot just above the hinge and then rotating the screwdriver half a turn counterclockwise.

Start with a Plan

A clearly written instruction contains four main compartments, as shown in Figure 8-2. These compartments contain the following information:

- A **Summary Statement** outlining briefly what has to be done.

  The 28 Vancourt Model AL-8 overhead projectors in rooms A4 and A32 are to be bolted to their projection tables...

- The **Purpose** explaining why the work is necessary:

  ...to reduce the current high damage rate caused by projectors being accidentally knocked onto the floor.

  (A technician who understands *why* a job is necessary will much more readily follow an instruction.)

- A short paragraph or list describing the **Tools and Materials** that technicians will need to perform the task (they can use this as a checklist to ensure they have gathered everything they need before they start work).

  To carry out the modification you will require:

  - Modification kit OHP4, comprising

    1 template, OHP4-1

    4 bolts, flat head, 2 in. long, $\frac{1}{8}$ in. dia

    4 washers, 1 in. dia, with $\frac{5}{32}$ in. dia central hole
  - A $\frac{1}{4}$ in. drill with a $\frac{3}{16}$ in. drill bit
  - A Phillips No. 2 screwdriver
  - A slot-head No. 3 screwdriver
  - A sharp pencil

One barbecue manufacturer lists the assembly tools on the outside of the shipping carton

- The **Steps** that readers must follow to take them through the whole process.

  Proceed as follows:

  1. Disconnect the projector's power cord from the wall socket, then take the projector to a table and turn it on its side.
  2. Use a slot-head No. 3 screwdriver to unscrew the four bolts that hold the feet onto the base of the projector. Remove, but retain the bolts and feet for future use.
  3. Place template OHP4-1 onto the projection table and position it where the projector is to stand. Using a sharp pencil, mark the table through each of the four holes in the template.
  4. Drill four $\frac{3}{16}$ in. dia holes through the table top at the places marked in step 3.
  5. Place the overhead projector right side up and with the lens assembly facing the screen so that the four screwholes identified in step 2 coincide with the four holes in the table top.

Number the steps: show there is a sequence

6. From beneath the table, place a washer under each hole and insert a 2 in. flat head bolt up through the washer and hole until it engages the corresponding hole in the projector base. Tighten the four bolts in place, using a Phillips No. 2 screwdriver.

The writing plan embodying these four compartments, when combined with the suggestions below, will consistently ensure that any instructions you write will be clear, direct, and convincing.

Give Your Reader Confidence

A well-written technical instruction automatically instills confidence in its readers. They feel they have the ability to do the work even though it may be new to them and highly complex. Consider these examples:

Vague Before the trap is set, it is a good idea to place a small piece of cheese on the bait pan. If it is too small it may fall off and if it is too big it might not fit under the serrated edge, so make sure you get the right size.

Clear and Concise Cut a 17 in. length of 10-gauge wire and strip 1 inch of insulation from each end. Solder one end of the wire to terminal 7 and the other end to pin 49.

An instruction is not the place for weak, wishy-washy words

The first excerpt is much too ambiguous. It only suggests what should be done, it hints where it should instruct (almost inviting readers to nip their fingers), and despite using 31 explanatory words, it fails to define the size of a "small" piece of cheese. The second excerpt is assertive and keeps strictly to the point. The verbs *cut*, *strip*, and *solder* make readers feel they have no alternative but to follow the instructions. Such clear and authoritative writing immediately convinces them of the accuracy and validity of the steps they have to perform.

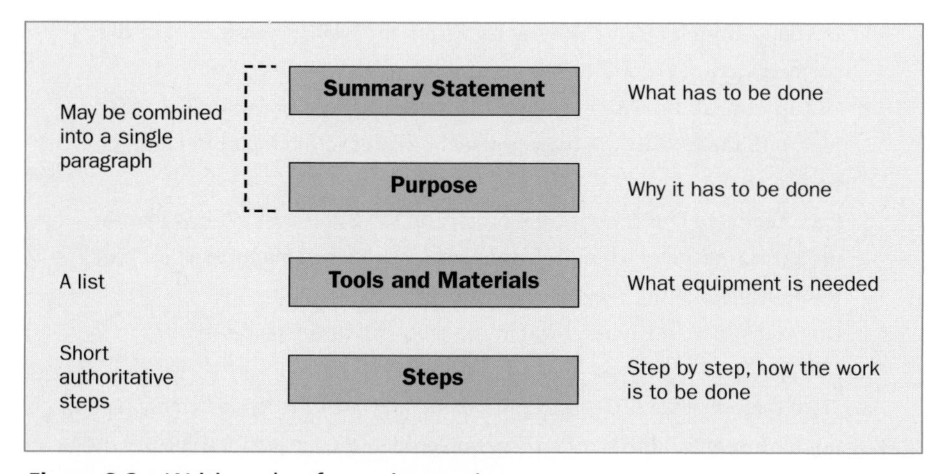

Figure 8-2 Writing plan for an instruction.

The best way to be authoritative is to write in the imperative mood. This means beginning each step with a strong verb, so that your instructions are commands:

Ignite the mixture...
Mount the transit on its tripod...
Apply the voltage to...
Cut a 2-inch wide strip of...
Connect the green wire...
Excavate 3 feet down...
Measure the current at...
Insert the PCMCIA card...

The imperative mood in the clear, concise excerpt quoted above keeps the instruction taut and definite. The vague excerpt would have been equally effective (and much shorter) if it had also been written in the imperative mood:

> Before setting the trap, wedge a $^3/_8$-inch cube of cheese firmly under the serrated edge of the bait pan.

(Note how the vague word "place" has been replaced by the image-conveying verb–adverb combination "wedge...firmly.")

The following two statements clearly show the difference between an instruction written in the imperative mood and one that is not:

Make each step a command, not a broad statement of intent

A. Disengage the gear, then start the engine. (***Definite: uses strong verbs***)
B. The gear should be disengaged before starting the engine. (***Indefinite: uses weaker verbs***)

Statement A is strong because it *tells readers to do something*. Sentence B is weak because it neither instructs nor insists that anything need be done ("should" implies it is only *preferable* that the gear be disengaged before the engine is started).

In the imperative mood, the first word in a sentence is almost always a strong verb:

> *Position* the pointer on File, then select Print.

Sometimes, however, the verb may be preceded by an introductory or conditional clause:

> Before connecting the meter to the power source, *set* all the switches to "zero."

The imperative mood is maintained here because the main verb starts the statement's primary clause (the clause that describes the action to be taken).

If you want to check whether a sentence you have written is in the imperative mood, ask yourself whether it *tells* the reader to *do* something. If it does, then you have written an *instruction*.

Avoid Ambiguity

There is no room for ambiguity in technical instructions. You have to assume that the person following your instructions cannot ask questions, so you must never write anything that could be interpreted in more than one way. The following statement is open to misinterpretation:

Align the trace so that it is inclined approximately 30° to the horizontal.

Write specific details; never generalize

Each technician will align the trace with a different degree of accuracy, depending on his or her interpretation of "approximately." How accurate does "approximately" require the technician to be? Within 5°? Within 2°? Within $\frac{1}{2}$°? Maybe even 10° either side of 30° is acceptable, but the reader does not know this and is left feeling doubtful. Worse still, the reader's confidence in the technical validity of the whole instruction is undermined. Replace such vague references with clearly stated tolerances:

Align the trace so that it is inclined 30° (±5°) to the horizontal.

More subtle, but equally open to misinterpretation, is this statement:

Adjust the capstan handle until the rotating head is close to the base.

Here the offending word is "close," and needs to be replaced by a specific distance:

Adjust the capstan handle until the distance between the rotating head and the base is 2.5 mm.

Never leave a statement open to misinterpretation

Similarly, replace vague references such as "*relatively* high," "*near* the top," and "*an adequate* supply" with clearly stated measurements, tolerances, and quantities.

Avoid weak words such as "should," "could," "would," "might," and "may," because they weaken the authority of an instruction and reduce the reader's confidence in the writer. For example:

Set the meter to the +300 V range. The needle should indicate 120 V (±2 V).

Here "should" implies that it would be nice if the needle indicated within 2 V of 120 V, but not essential! No doubt the writer meant it *must* read the specified voltage, but has failed to say so. Neither has the reader been told to note the reading. The writer has forgotten the cardinal rule of instruction writing: *Tell* the reader to *do* something. To be authoritative, the instruction needs very few changes:

Set the meter to the +300 V range, then check that its needle indicates 120 V (±2 V).

Notice that the steps in the sample instructions in Figure 8-3 are clear, concise, and definite. You need not be a specialist in the subject to recognize that they would be easy to follow.

Write Bite-Size Steps

Technicians working on complex equipment in cramped conditions need easy-to-follow instructions. You can help them by writing short paragraphs, each containing only one main step. If a step is complicated and its paragraph grows unwieldy, divide it into a major step and a series of substeps, numbering the paragraphs and subparagraphs:

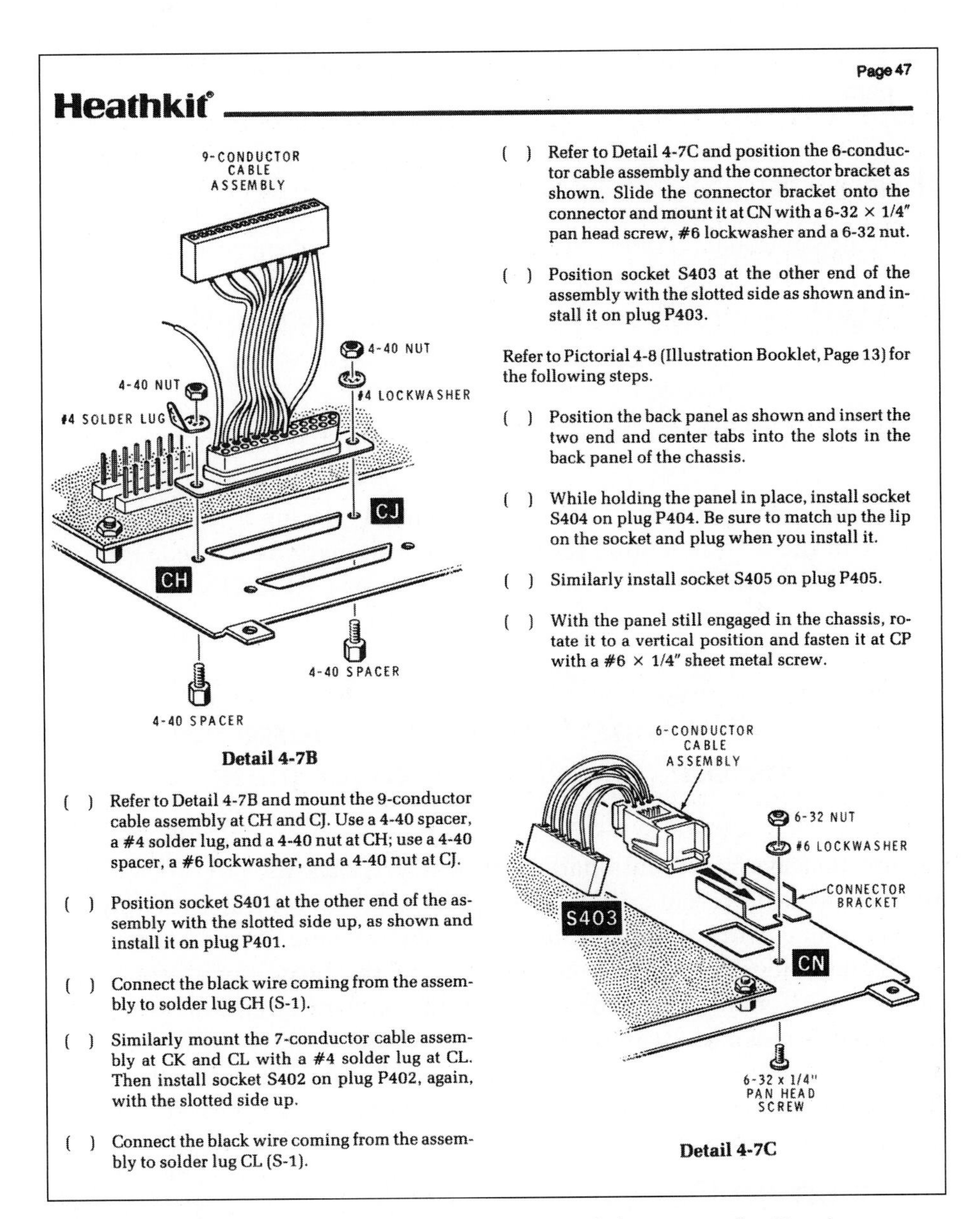

Page 47

Heathkit®

Detail 4-7B

( ) Refer to Detail 4-7B and mount the 9-conductor cable assembly at CH and CJ. Use a 4-40 spacer, a #4 solder lug, and a 4-40 nut at CH; use a 4-40 spacer, a #6 lockwasher, and a 4-40 nut at CJ.

( ) Position socket S401 at the other end of the assembly with the slotted side up, as shown and install it on plug P401.

( ) Connect the black wire coming from the assembly to solder lug CH (S-1).

( ) Similarly mount the 7-conductor cable assembly at CK and CL with a #4 solder lug at CL. Then install socket S402 on plug P402, again, with the slotted side up.

( ) Connect the black wire coming from the assembly to solder lug CL (S-1).

( ) Refer to Detail 4-7C and position the 6-conductor cable assembly and the connector bracket as shown. Slide the connector bracket onto the connector and mount it at CN with a 6-32 × 1/4" pan head screw, #6 lockwasher and a 6-32 nut.

( ) Position socket S403 at the other end of the assembly with the slotted side as shown and install it on plug P403.

Refer to Pictorial 4-8 (Illustration Booklet, Page 13) for the following steps.

( ) Position the back panel as shown and insert the two end and center tabs into the slots in the back panel of the chassis.

( ) While holding the panel in place, install socket S404 on plug P404. Be sure to match up the lip on the socket and plug when you install it.

( ) Similarly install socket S405 on plug P405.

( ) With the panel still engaged in the chassis, rotate it to a vertical position and fasten it at CP with a #6 × 1/4" sheet metal screw.

Detail 4-7C

Figure 8-3 Excerpts from an instruction manual. (Courtesy the Heath Company, Mississauga, Ont.)

Keep diagrams uncluttered...

...and shown from the reader's point of view

3. List the documentary evidence in block J of Form 658. Check that blocks A to G have been completed correctly, then sign the form and distribute copies as follows:
   - 3.1 Attach the documentary evidence to Copies 1 and 2 and mail them to the Chief Recording Clerk, Room 217, Civic Center, Montrose, Ohio.
   - 3.2 Mail Copy 3 to the Computer Data Center, using one of the special preaddressed envelopes.
   - 3.3 File Copy 4 in the "Hold—Pending Receipt" file.
   - 3.4 When Copy 2 is returned by the Chief Recording Clerk, attach it to Copy 4 and file them both in the "Action Complete" file.

Offer bite-size pieces of information

As each step is completed, the user inserts a check mark beside the appropriate paragraph.

Insert Fail-Safe Precautions

Insert precautionary comments into instructions whenever you need to warn readers of dangerous conditions, or of damage that may occur if they do not exercise care. There are two precautionary notices you can use:

Warning:	To alert readers to an element of personal danger (such as unprotected high voltage terminals).
Caution:	To tell readers when care is needed to prevent equipment damage.

Draw attention to a precautionary comment by placing it in a box in the middle of the text, indenting the box from both margins. Precede the cautionary note with the single word WARNING or CAUTION.

Use word-processing technology to design a simple but noticeable warning

WARNING

Disconnect the power source before removing the cover plate.

Ensure that every precautionary comment *precedes* the step to which it refers. This will prevent an absorbed reader, who concentrates on only one step at a time, from acting before reading the warning. Never assume that mechanical devices, such as indentation and the box drawn around the precautionary note, are enough to catch the reader's attention.

Use warnings and cautions sparingly. A single warning will catch a reader's attention. Too many will cause a reader to treat them all as comments rather than as important protective devices.

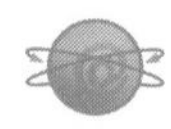

Usability Testing on documents
www.uwec.edu/jerzdg/orr/handouts/TW/proj/usability.htm
This document suggests some ways you might conduct tests to measure the usability of your technical documents. It covers the kind of data you should collect, how many test subjects you need, and how you should treat those subjects.

Insist on an Operational Check

The final test for any technical instruction is the reader's ease in following it. Usability testing is an essential part of instruction writing but, sadly, it's often overlooked in the rush to get a product out. Since you cannot always peer over a reader's shoulder to correct mistakes, you should find out whether users are likely to run into difficulty *before* you send an instruction out. To obtain an objective check, give the instruction to someone roughly equal to the people who will eventually be using it, and observe how well that person performs the task.

Note every time the user hesitates or has difficulty. When he or she has completed the task, ask if any parts need clarification. Rewrite ambiguous steps and then recheck your instruction with another person. Repeat these steps until you are confident your readers will be able to follow your instruction easily.

Scientific Paper

Earlier chapters described how technical reports should be planned, organized, written, and presented by engineers and computer specialists working for industry, business, and government. Within this context, one additional report remains to be described: the research report prepared by scientists and technologists working in industrial and university laboratories. Research reports are most often prepared and published as *scientific papers* and differ in style, organization, and emphasis from an investigation report, although their parts are similar.

A scientific paper either identifies and attempts to resolve a scientific problem, or it tests (validates) a scientific theory. It does so by describing the four main stages of the research:

1. Identifying the problem or theory.
2. Setting up and performing the tests.
3. Tabling the test results (the findings).
4. Analyzing and interpreting the findings.

A scientific paper is like a highly professional lab report

These four stages represent the major divisions of a scientific paper, with each stage preceded by a descriptive heading: **Introduction, Materials and Methods, Results,** and **Discussion**. These stages are similar to those used for the laboratory report in Chapter 4 and the investigation report in Chapter 5. There are, however, differences in a scientific paper's appearance and writing style.

Appearance

A scientific paper straddles the borderline between semiformal and formal presentation. Normally the title is centered about 2 to 3 inches from the top of the first page (see Figure 8-4). The author's name and the name of the company or organization the author works for can also be centered at the top of the page, about 1 inch below the title. Alternatively, the author's name and affiliation can be placed at the bottom left of the first page, or at the end of the paper.

The abstract (summary) appears next, and starts about $1\frac{1}{4}$ inches below the title or the author's name. It should be indented about 1 inch from both side margins.

The body of a scientific paper starts about $\frac{3}{4}$ inch beneath the abstract and looks much like the body of a semiformal report. Normally, a scientific paper is double-spaced throughout, including the abstract. Often, the first line of every paragraph is indented $\frac{1}{4}$ to $\frac{1}{2}$ inch from the left margin.

Writing Style

A slight bending of the rules!

The rules for brevity, clarity, and directness suggested in Chapters 3 and 12 for technical letters and reports apply equally to scientific papers. But there is one exception: In some technical disciplines it's more common to write scientific papers in the passive voice. For example, in a technical report we have advised you to write

> I placed the sample in the chamber...

or

> We placed the sample in the chamber...

or

> The technician placed the sample in the chamber...

However, in a scientific paper you will more likely be expected to write

> The sample was placed in the chamber...

In other words, scientific papers often conceal the identity of the "doer," and carefully avoid using the first person ("I" or "We").

Talk to the journal editor: determine the journal's guidelines

Unfortunately, we cannot give you definitive advice, because different organizations, academic institutions, and technical disciplines adhere to different guidelines. In papers on medical research, for example, scientists tend to write predominantly in the passive voice (although this tradition is gradually changing), whereas in papers on computer technology, authors tend to write in the active voice. If you are writing a paper that may be published in a peer-reviewed scientific journal, you will need to study previous issues—or write to the journal editor—to identify whether the active voice is acceptable.

Organization

A scientific paper has six main parts, which are described in detail below.

1. **Abstract** (Summarizes the paper, emphasizing the results)
2. **Introduction** (Provides background details and outlines the problem or theory tested)
3. **Materials and Methods** (Describes how the tests were performed)
4. **Results** (States the findings)
5. **Discussion** (Analyzes and interprets the findings)
6. **References** (Lists the documents consulted)

Sometimes you may receive significant assistance or guidance from other people during your research and may want to acknowledge their help. Such acknowledgments are normally placed after the discussion but before the references.

If you are accustomed to writing laboratory reports or investigation reports, you have probably noticed that the "Conclusions" heading has been omitted from this list. Where the conclusions in a laboratory or investigation report serve as a *separate* terminal summary (a summing up of the results and their analysis), in a scientific paper they are more often embedded as a closing statement at the end of the Discussion.

Abstract

Write an informative rather than a topical abstract

The rules for writing an abstract are almost identical to those for writing the summary of an investigation report. In an abstract you (1) outline the problem and the purpose of your investigation, (2) mention very briefly how you conducted the investigation or tests, (3) describe your main findings, and (4) summarize the conclusions you have drawn. All this must be done in as few words as possible; ideally, your abstract will be about 125 words long and never more than 250. A typical abstract is shown in Figure 8-4.

Writing Report Abstracts http://owl.english.purdue.edu/workshops/hypertext/reportW/abstract.html
The Online Writing Lab at Purdue University has information about writing report abstracts. Included are sections on types of abstracts, qualities of a good abstract, and steps for writing effective abstracts.

From the abstract, readers must be able to decide whether the information you provide in the remainder of the paper is of particular interest to them and whether they should read further. Because a scientific paper is addressed to readers who are generally familiar with your technical or scientific discipline, you may use technical terminology in the abstract. (This is the major difference between a report summary and a scientific abstract.) You should write the abstract last, after you have written the remainder of the paper, so that you can *abstract* the brief details you need from what you have already written.

Introduction

In the introduction you prepare readers so that they will readily understand the technical details in the remainder of the paper. The introduction contains four main pieces of information:

1. A definition of the problem and the specific purpose of the investigation or tests you conducted. (This should be a much more detailed definition than appears in the abstract.)

Conduct a detailed literature review

2. Presentation of background information that will enable the reader to fully understand and evaluate the results. Often it will include—or sometimes may consist entirely of—a review of previous scientific papers, journal articles, books, and reports on the same subject. (This is known as a literature survey.) Rather than simply list the pertinent documents, you are expected to summarize the main findings of each and their relevance to your investigation. These documents should be cross-referenced to your list of references at the back of the report. Here is an example:

   Previous measurements of acid rain in Montrose were recorded in 2001 by Gershwain (3), who reported an average acidity level of xxxx, and...

This is the only time we suggest you write a summary in the passive voice!

Acid Rain Testing in the City of Feldspar, Ohio, 2004

Corrine L. Danzig and Mark M. Weaver
University of Feldspar Research Laboratory

Abstract

Acid rain is a growing concern in the United States, with its effects becoming increasingly noticeable south of Lakes Erie and Ontario. To determine what increases have occurred in the City of Feldspar over the past nine years, acidity levels were measured and compared to measurements recorded in 1995, when the average pH level was YYY. The current tests showed that the average pH now is ZZZ, but that the acidity levels are not uniform. The greatest toxicity was found to be at the University of Feldspar, in the southeastern area of the city, with the lowest toxicity in the northwestern area of the city. The change was attributed primarily to the increase in fossil-fuel-burning heavy industries in the Poplar Heights industrial park, which is 3 miles northwest (generally upwind) of the principally affected areas.

Introduction

Over the past 20 years, observers in the counties adjacent to Lake Ontario have reported an increased incidence of crop spoilage and tree defoliation, which has been attributed to acid rain created primarily by fossil-fuel-burning industries and automobile emissions. Tests were taken initially in 1995 to assess the pH levels of the precipitation falling in and around the City of Feldspar, Ontario.[1] These were compared with the measurements recorded in 2000, 2002, and 2003 at the exact same locations. For consistency purposes, the test sites were secured to keep the variables stable.

1

Figure 8-4 The first page of a scientific paper.

3. A short description of your approach and why you chose that particular method for your investigation.
4. A concluding statement that outlines your main findings.

Notice that the introduction of a scientific paper *includes a brief summary of the results*, which is much less common in a semiformal investigation report and rare in a laboratory report. Yet here there is a parallel with the alternative method of presenting a formal report, in which the writer presents the results three times (see Figure 6-7). In a scientific paper the results are also presented three times: very briefly in the abstract and introduction, and fully in the results section.

Materials and Methods

Be detailed and specific: help the reader understand *exactly* what you did

This section has to be thoroughly prepared and presented because readers must be able to replicate (perform) an identical investigation or series of tests from what you write here. The materials list must include *all* equipment used and specimens or samples tested, which may range from a whole-body nuclear radiation counter to a tiny microorganism. For ease of reference they should be listed in representative groups, such as:

Equipment	Plants
Instruments	Animals
Chemicals	Birds or fishes
Specimens	Humans

Precede the specific methods with the subheading "Method." Describe the tests chronologically in paragraph and subparagraph form, using a main paragraph to introduce a test or part of a test and short, numbered subparagraphs to describe the specific steps you took. Avoid writing in the imperative mood, so that you do not inadvertently start writing an instruction. For example:

Write this: The test unit was connected to the X-Y terminals of the recorder.

Or this: We connected the test unit to the X-Y terminals of the recorder.

But not this: Connect the test unit to the X-Y terminals of the recorder.

Results

The results section is often the shortest part of your paper. If your methods section has described clearly how the investigation or tests were conducted, the results section has only to state the result:

Present the facts; avoid offering opinions

> Acid rain is above average in the southeastern part of the city, below average in the northwestern part, and average in the southwestern and northeastern parts. Tables 1 through 8 show the measurements recorded at the eight metering stations.

(Tables and charts depicting your findings often will be a major part of the results section.)

Never comment on the results. Analysis and interpretation belong *only* in the discussion.

Discussion

Your readers now expect you to analyze and interpret your findings. You will be expected to discuss

- the results you obtained, compared to the results obtained by previous researchers,
- any significant correlation, or lack of correlation, between parts of your own findings,
- factors that may have caused the differences,
- any trends that seem to be evident or to be developing (ideally, by referring to graphs and charts you have presented in the results), and
- the conclusions you draw from your analysis and interpretation.

Embed the conclusions in the Discussion, or introduce them with a heading

The conclusions should show how you have responded to the problem stated at the start of your introduction, and identify clearly what your investigation, tests, and analyses show, thus forming a fitting close to the narrative portion of your research paper.

References

The final section of your paper is a list of the documents you have referred to earlier or from which you have extracted information. Chapter 6 provides general instructions for preparing a list of references, which is the preferred method for most technical reports, and a bibliography. You can use Chapter 6 as a guideline, but you should check first with the editor of the journal in which your scientific paper is likely to be published to determine (1) whether a list of references or a bibliography is preferred, and (2) the exact format the journal uses for listing authors' names, book and journal titles, publisher details, and so on.

Technical Papers and Articles

The likelihood that one day you might be asked to write a technical paper for publication, or even want to do so, might seem so remote to you now that you feel justified in skipping this section. Yet this is something you should think about, for getting one's name into print is one of the fastest ways to obtain recognition. Suddenly you become an expert in your field and are of more value to your employer, who is happy because the company's name appears in print beneath yours. You become of more value to prospective employers, who rate authors of technical papers more highly than equally qualified persons who have not published. And you have positive proof of your competence, and sometimes a few extra dollars from the publisher.

Writing and presenting a paper enhances your professional reputation

Mickey Wendell has an interesting topic to write about: As a senior technologist with H. L. Winman and Associates' materials testing laboratory, he has been testing concretes with various additives to find a grout that can be installed in frozen soil during the winter. One mixture he analyzed but discarded contained a new product known as Aluminum KL. As a byproduct of his tests he has discovered that mixing Aluminum KL with cement in the right proportions results in a concrete with very high salt resistance. He reasons that such concrete could prove invaluable to builders of concrete pavements in snow-affected areas of the United States and Canada, where salt mixtures are applied in winter to melt the snow.

Mickey has been thinking about publishing this particular aspect of his findings and has jotted down a few headings as a preliminary outline. Here are the four steps he must take before his ideas appear in print:

Step 1: Solicit Company Approval

Most companies encourage their employees to write for publication, and some even offer incentives such as cash awards to those who do get into print. However, they expect prospective authors to ask for permission before they submit their manuscripts.

To obtain permission, Mickey must write a brief email outlining his ideas to John Wood, his department head. He should ask for approval to submit a paper, explain what he wants to write about and why he thinks the information should be published, and outline where he intends to send it. His proposal is shown in Figure 8-5. John Wood will discuss the matter at management level, then signify the company's approval or denial in writing. Mickey knows he must have *written* consent before he can publish his findings.

Have a clear, well-thought-out idea before requesting approval

Step 2: Consider the Market

Mickey must decide early where he will try to place his paper. If he prefers to present his findings as a technical paper before a society meeting, he will be writing for a limited audience with specialized interests. If he decides to publish in the journal of a technical society, he will be writing for a larger audience, but still within a limited field. If he plans to publish in a technical magazine, he will be appealing to a wide readership with a broad range of technical knowledge. His approach must differ, therefore, depending on the type of publication and level of reader.

A guiding factor may be Mickey's writing ability. A paper to be published by a technical society requires high-quality writing because the editor of a society journal does not normally do much prepublication editing, other than making minor changes to suit the format and style of the society's publications. A technical magazine article, however, will be edited—sometimes quite fiercely—by a professional editor who knows the exact style that readers expect. Such an editor prefers authors to approximate

Articles written for a professional journal are usually sent out for peer review

that style and expects them to organize their work well and to write coherently; but he or she is always ready to prune or graft, and sometimes even completely rewrite portions of a manuscript. Hence, the pressure on authors of magazine articles to be good writers is not as great.

Perhaps the most important factor for Mickey is identifying a potential audience for his information. The same people may read society journals and technical magazines, but they expect different information coverage in a technical paper than they do in a magazine article.

If you are describing a scientific breakthrough, wait until you are ready to send in the whole paper

Technical Paper

Readers of society journals are looking for facts. They neither expect nor want an explanation of basic theories, and they can accept a strongly technical vocabulary. A technical paper can be very specific. It can describe a minute aspect of a large project without seeming incomplete, or it can outline in bold terms the findings of a major experiment. No topic is too large or too small, too specialized or too complete, to be published as a technical paper.

Technical Article

Most readers of magazine articles are looking for information that will keep them up-to-date on new developments. Some will have definite interest in a specific topic and would welcome a lot of technical skills. Others will be looking mainly for general information, with no more than just the highlights of a new idea. Magazine authors must therefore appeal to a maximum number of readers. Their articles should be of general interest, their style can be brief and informal, their vocabulary must be understandable, and they should include some background details for readers who possess only marginal technical knowledge.

Step 3: Write an Abstract and Outline

Many editors prefer to read either a summary of a proposed paper or an abstract and outline before the author submits the complete manuscript. They may want to suggest a change in emphasis to suit editorial policy, or even decline to print an interesting paper because someone else is working on a similar topic.

Writing Abstracts
http://writing.colostate.edu/references/documents/abstract/index.cfm
This site covers the various aspects of writing abstracts from the definition and purpose of abstracts to writing abstracts for specific disciplines.

This type of summary is much longer than the summary at the head of a technical report; the abstract, however, is usually quite short. The summary contains a condensed version of the full paper in about 500 to 1000 words. An abstract contains only very brief highlights and the main conclusion (rather like the summary of a report), since it is supported by a comprehensive topic outline.

Some authors write the complete first draft of the paper before attempting to write a summary or abstract, then leave the revising and final polishing until after the paper has been accepted by an editor. Others prepare a fairly comprehensive outline, often using the freewheeling approach suggested in Chapter 2, and leave the writing until after acceptance. Both methods leave room for the author to incorporate changes before writing the final manuscript.

Since the summary or abstract and outline have to "sell" an editor on the newsworthiness of his topic, Mickey Wendell must make sure that the material he submits is complete and informative. In addition he must indicate clearly

- why the topic will be of interest to readers,
- how deeply the topic will be covered,
- how the article or paper will be organized,
- how long it will be (in words), and
- his ability to write it.

Details like these demonstrate you have developed your idea in depth

Mickey can cover the first four items in a single paragraph. The fifth he will have to prove in two ways. He can prove his technical ability by mentioning his involvement in the topic and experience in similar projects. And he can demonstrate his writing ability by submitting a clear, well-written summary or abstract.

If Mickey later decides to prepare his paper for presentation at a conference, he will have to prepare a summary in response to a "call for papers" sent out by the society, and submit it to the chairperson of the papers selection committee. If his paper is to be accepted, his summary has to convince the committee that the subject is original, topical, and interesting, and that he has the capability to write and present the paper.

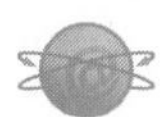

How to Have Your Abstract Rejected
www.eecs.harvard.edu/cs245/liptonadvice.html
This is a tongue-in-cheek guide describing how to guarantee the rejection of an abstract submitted for a conference. The comments can be generalized to technical papers of any length.

Step 4: Write the Article or Paper

A good technical paper is written in an interesting narrative style that combines storytelling with factual reporting. Articles published in general interest magazines tend to be written like feature newspaper stories, whereas technical papers more nearly resemble formal reports. If the article deals with a factual or established topic, the writing is likely to be crisp, definite, and authoritative. If it deals with the development of a new idea or concept, the narrative will generally be more persuasive, since the writer is trying to convince the reader of the logic of his or her argument.

The parts of an article or paper submitted to a magazine are similar to those of a report. Mickey's article, for example, contains four main parts:

Summary A synopsis that tells very briefly what the article is about. It should summarize the three major sections that follow. Like the summary of a report, it should catch and hold the reader's attention.

Introduction Circumstances that led up to the event, discovery, or concept that Mickey will describe. It should contain all the facts readers need to understand the discussion that follows.

The pyramid approach is equally valid for a technical article

From: Mickey Wendell <mwendell@hlwinman.com>
To: John Wood <jwood@hlwinman.com>
Date: June 24, 2004
Subject: Approval for Proposed Technical Paper

John,

I am asking for company approval to write an article on concrete additives for publication in a technical journal. Specifically, I want to describe our experiments with Aluminum KL, and the salt-corrosion resistance it caused with the concrete samples we tested for the Alaska transmission tower project. I believe that our findings will interest many municipal engineers in the northern US and Canada. Like us, they have been trying for years to combat pavement erosion caused by applying salt to melt snow.

Your proposal could also include an outline of the proposed paper

I was thinking of submitting the article to the editor of "Municipal Engineering" but I'm open to suggestions if you can think of other magazines or journals.

I need your response by July 8 so I can begin writing the article. The deadline for submissions is September 3.

Mickey

Figure 8-5 An email asking for approval to publish an article or paper.

Discussion How Mickey went about the project, what he found out, and what inferences he drew from his findings. The topic can be described chronologically (for a series of events that led up to a result), by subject (for descriptive analyses of experiments, processes, equipments, or methods), or by concept (for the development of an idea from concept to fruition). The methods are very similar to those used for writing the discussion of a formal report (see Chapter 6).

Conclusion A summing-up, in which Mickey draws conclusions from and discusses the implications of his major findings. Although normally he will not make recommendations, he may suggest what he feels needs to be done in the future, or outline work that he or others have already started if there is a subsequent stage to the project.

If, however, Mickey had decided to submit his paper to a peer-reviewed scientific journal, then its parts and his writing style would have had to comply with the requirements described in the section on scientific papers (pages 221 to 226).

Illustrations are a useful way to convey ideas quickly, to draw attention to an article, and to break up heavy blocks of type. They must be instantly clear and usefully *supplement* the narrative. They should never be inserted simply to save writing time, neither should they convey exactly the same message as the written words. For examples of effective illustrations, turn to any major publication in your technical field and study how its authors have used charts, graphs, sketches, and photographs as part of the story. For further suggestions on different types of illustrative material, see Chapter 9.

Don't feel rebuffed if an editor cuts some of your words!

Mickey should not be surprised when the editor who handles his manuscript makes some changes. An editor who feels the material is too long, too detailed, or wrongly emphasized for the journal's readers, will revise it to bring it up to the expected standards. Mickey may feel that the alterations have ruined his carefully chosen phrases, but readers will not even be aware that changes have been made. They will simply recognize a well-written paper, for which Mickey, rather than the editor, will reap the compliments.

ASSIGNMENTS

Project 8.1: Performing a Task Analysis

Think of a product you use often. It might be a microwave oven, a fax machine, a vacuum cleaner, or anything else you are familiar with. Perform a task analysis for the product and then group and label the topics. Remember to use "ing" words to indicate the user's perspective. List the tasks, then group and label them.

Project 8.2: Writing Instructions for a DVD Player or a Microwave Oven

Two days ago a retired couple who live near to you bought a new DVD player. They asked you to show them how to connect it to the TV and integrate it with their sound system. You show them the instruction manual and "walk" them through the instructions.

Part 1

Today they telephone you and say they couldn't make it work. "Can you just come do it for us?" Mr. Smithson asks. You realize that if you do not write it down for them, they will never be able to figure out all the remote controls for the various components (TV, VCR, DVD, and stereo).

Finish with a usability test: watch a neighbor use your instructions...

You borrow the manual and write them a step-by-step instruction, using good design and simple explanations. Make sure you explain how the system is connected in case they or someone else needs to move the system. Include directions on playing a DVD.

(Note: Use a DVD player you are familiar with to write this instruction.)

Part 2

Mr. and Mrs. Smithson were so pleased with your instructions for their new DVD player they ask you to write an instruction for their microwave oven.

"Something I can glue to the side of the oven," Mrs. Smithson explains. "We have no trouble operating the oven, but in the spring each year, when we change to daylight saving time, I have to reset the clock. And it's the same again in the fall. The instruction manual that came with the oven just confuses me."

...then revise them!

Write an instruction for resetting the clock on a microwave oven that you are familiar with.

Part 3

Design the instructions for a DVD player or a microwave oven using *only* pictures.

Project 8.3: User's Manual for a Meeting Timer

Mechanical engineer Darwin Haraptiniuk places a drawing on your desk (see Figure 8-6) and says, "You're a good writer. Write a user's manual to go along with this."

You ask what "this" is.

"It's a meeting timer I designed several years ago. Now I'm having 10 prototypes built to test in various companies locally. If their response is positive, I'm going to redesign it and go into mass production. I want the user's manual to ensure that whoever tests the timer uses it properly."

"How does it work?" you ask.

"Like a gasoline pump. You work out the average hourly rate for people attending a particular meeting and set it in one window, and the number of people sitting around the table goes in the other window."

"How?"

"By rotating the knobs on the left. Then you just press the buttons, one at the start of the meeting, and one at the end."

"But what's it for?" you ask.

A recipe for a short instruction

"To help speed up business meetings," Darwin says. "You put it on the table so everyone can see it. When they can *see* how long a meeting has lasted and what it has cost so far, I figure they'll concentrate on getting things done quickly."

"All right," you say. "I'll get right on it."

"One more thing," Darwin adds. "You'd better warn users not to change the number of people attending the meeting or the average rate while the machine is running. That tends to damage the mechanism."

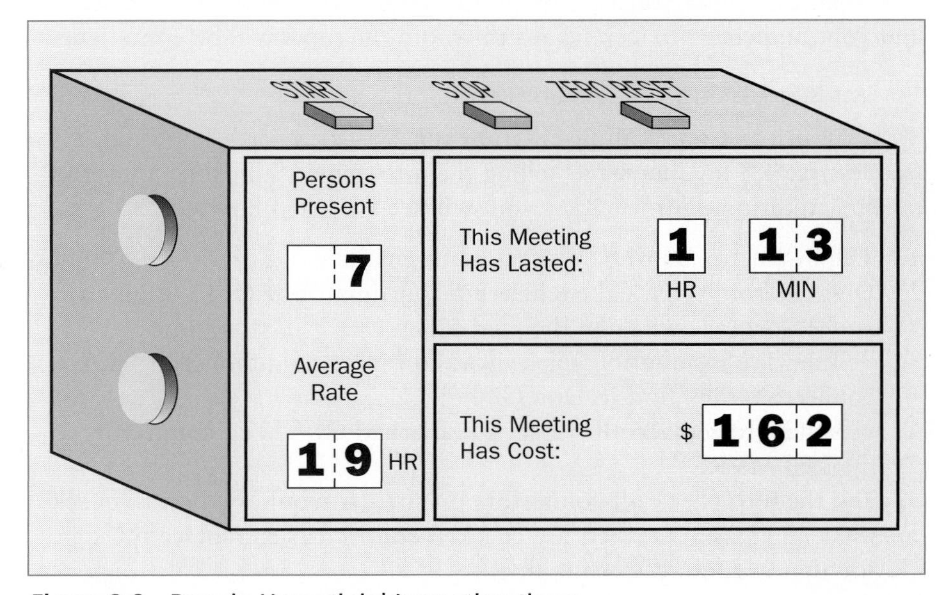

Figure 8-6 Darwin Haraptiniuk's meeting timer.

"What do you do if you have a problem, then?"

"Switch the machine off, make the change, then switch it on again."

"And if that doesn't work?"

"Call me: 774-1685."

"Okay," you say. And then you ask, "What do I call your machine?"

"Ah! Good point. What about the 'Darwin Meeting Timer Model No. 1'?"

"Fine!"

Now write the user's manual.

Project 8.4: Researching a New Manufacturing Material or Process

You are to research information on a topic allied to your technology and then prepare it for both written and oral presentation. The topic may be a new manufacturing material, method, or process. The written and spoken presentations must

A recipe for thorough research and a detailed description

1. introduce the topic,
2. state why it is worth evaluating,
3. describe the material, method, or process,
4. discuss its uniqueness and usefulness, and
5. show how it can be applied in your particular field.

To obtain data for your topic, you will have to research current literature and probably talk to industrial users, manufacturers, and suppliers. Typical examples of topics are: a new oil that can be used at very low temperatures, a method for supporting the deck of a bridge during concrete-pouring by building up a base on compacted fill, a new paint for use on concrete surfaces, a new materials-handling system, and a recently developed computer software program. Assume that both your readers and your audience are technicians to whom the topic will be entirely new.

Project 8.5: Testing Cable Connectors

Write an instruction to all installation supervisors at sites 1 through 17 (see Project 4.5 in Chapter 4) telling them to inspect all cable connectors on site. Additional information you will need is listed below.

1. The instruction is to be written as an email.
2. Don Gibbon, electrical engineering coordinator at H. L. Winman and Associates, will sign the memo.
3. Tell the site installation supervisors to report the number of GLA connectors they find to Don Gibbon.
4. Tell them to replace all connectors marked GLA with connectors marked MVK.
5. Tell them to check all connectors on site. (It would be best to check those in stock first, then use checked connectors to replace those in use that are found to be faulty.)

6. Tell them to send faulty GLA connectors to the contractor with a note to hold them for analysis under project 92A7.
7. Tell site supervisors to complete their tests within seven days.

Project 8.6: Installing a Mini-Minder

You are employed in the construction department of Midstate Telephone System, and you have been asked to write an installation instruction sheet to be shipped with the "Mini-Minder" (an electronic intrusion detection device developed by Midstate's engineering department). The Mini-Minder is a small detection unit that is mounted above windows, doors, or any other entries, where it automatically detects movement and transmits an alarm signal to a central unit. The central unit is concealed within the building and, when activated, sounds an audible alarm and sends a message to police headquarters.

The Mini-Minder is shown in Figure 8-7. It is fixed to the wall above the door or window by removing the backplate and screwing the plate to the wall. The unit is then snapped onto the backplate (removing the unit from the backplate also sounds the alarm). The unit is battery operated.

All materials and hardware are supplied with the Mini-Minder but the installer will need an electric drill with a $^{3}/_{16}$ in. masonry drill bit, a Robertson No. 2 screwdriver, and a sharp pencil to do the job. The sequence in which the installation should occur is shown by the circled numbers in Figure 8-7.

Project 8.7: Installing Encoder EC7

You work for Midstate Telephone System and you have been asked to write instructions for installing an EC7 encoder at all Midstate's microwave transmission sites. The instructions are to accompany the encoder, which is the box illustrated in Figure 8-8. The encoder removes unwanted signals and improves transmission performance by 8% to 10%.

In your instructions, tell the site technicians to find a suitable location for the encoder at the bottom right of the control panel (see Figure 8-8), drill two holes for mounting the encoder, and connect the encoder with four wires (the connections are shown on the diagram). All materials (such as mounting hardware and wire) are supplied with each encoder but the technician will need a soldering iron, some Ersin 60/40 resin core solder, a drill with a $^{3}/_{16}$ in. bit, and a Robertson No. 2 screwdriver to carry out the work. The circled numbers on the diagram show the sequence in which the installation is to be carried out.

You will have to extract pertinent points from the illustration

When the installation is complete, remind the technician to turn on the power to the control panel.

A reminder: write in the imperative mood

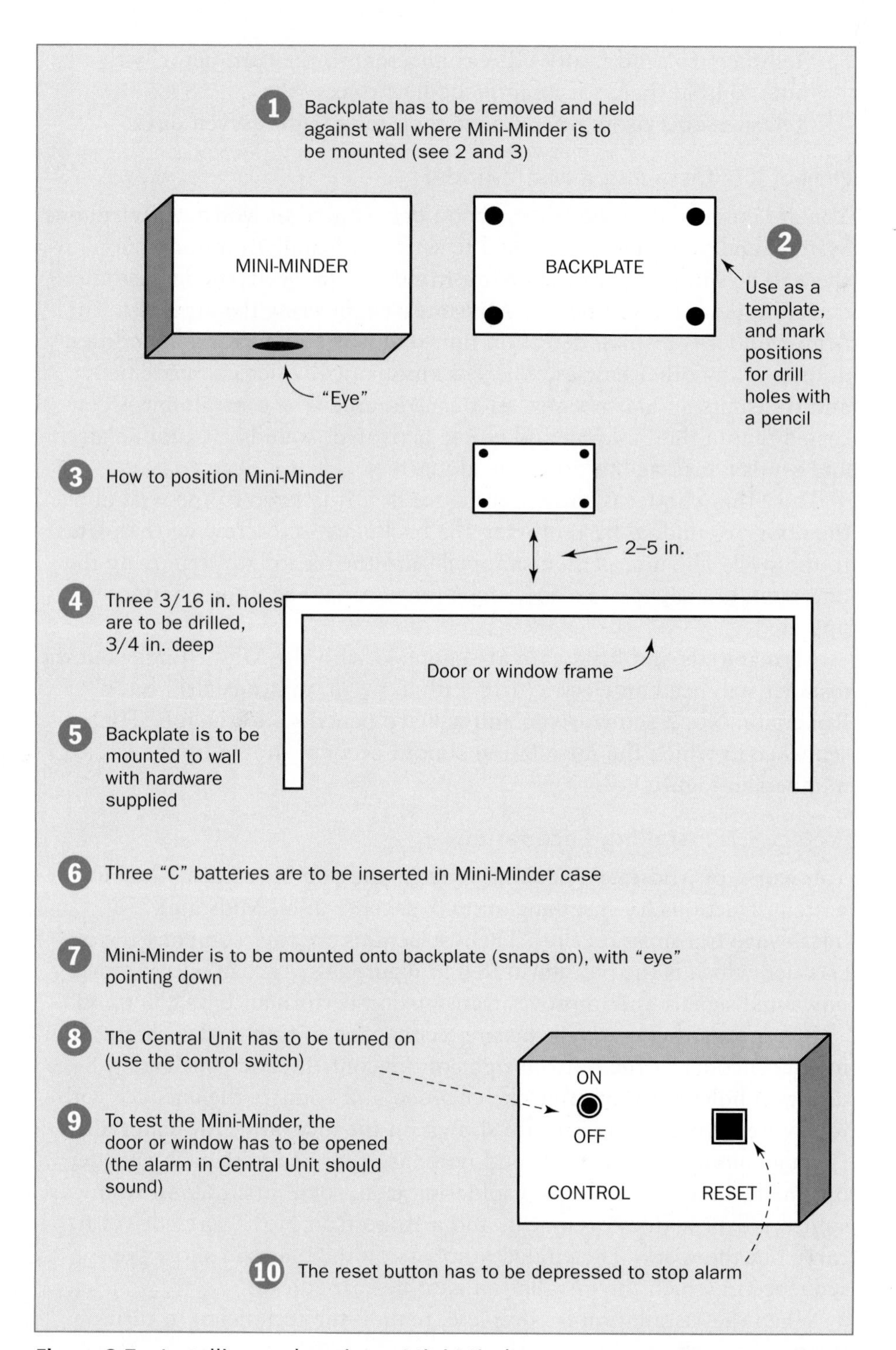

Figure 8-7 Installing and testing a Mini-Minder.

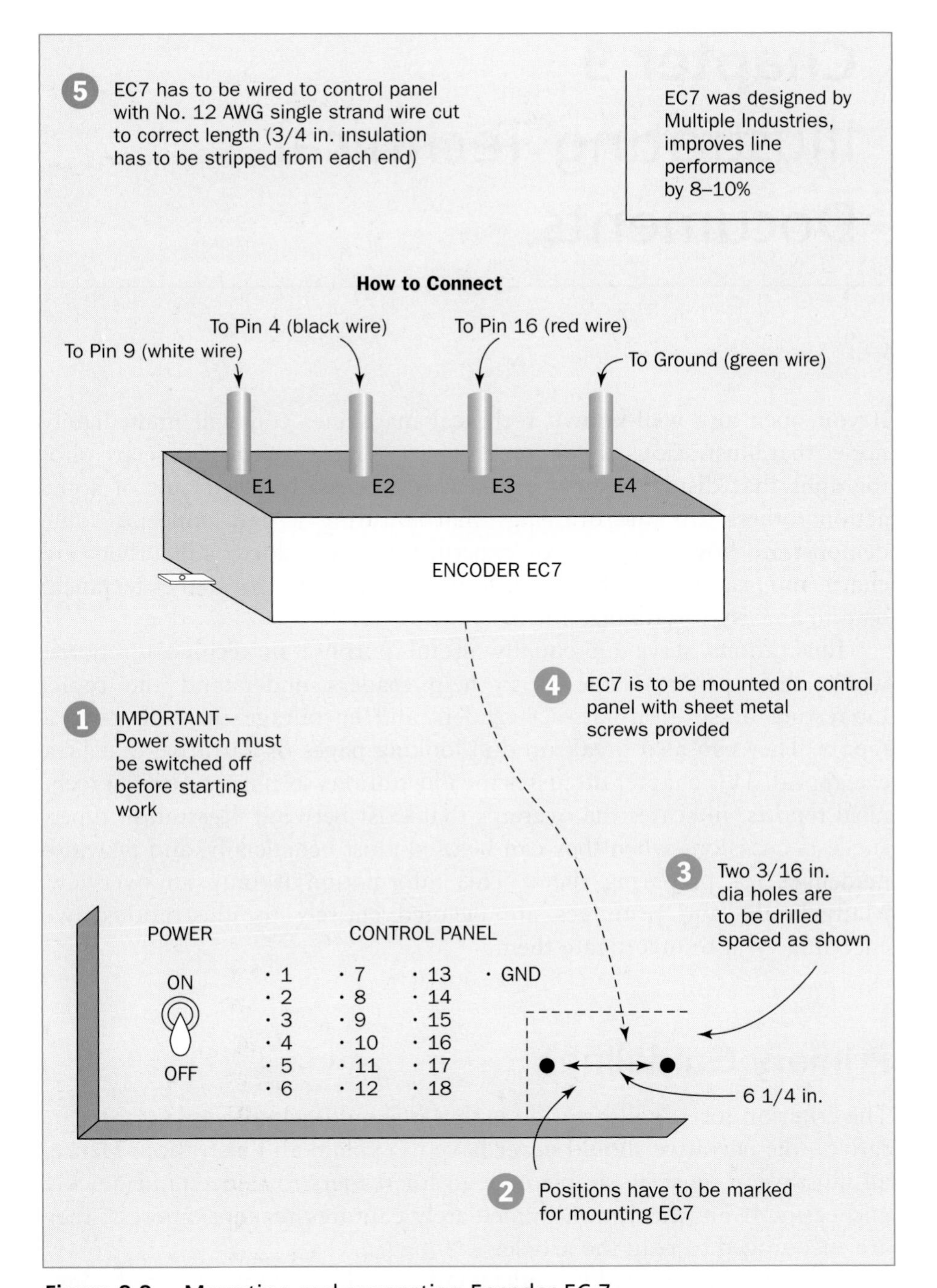

Figure 8-8. Mounting and connecting Encoder EC-7.

Consider writing in the active voice

Chapter 9
Illustrating Technical Documents

If you open any well-known technical magazine, you will immediately notice that illustrations are an integral part of most articles. Some are photographs that display a new product, a process, or the result of some action; others are line drawings that illustrate a new concept; some demonstrate how a test or an experiment was tackled; still others are charts and graphs that show the progress of a project or depict technical data in an easy-to-visualize form.

Illustrations serve an equally useful purpose in technical reports, where their primary role is to help readers understand the topic. Interesting illustrations attract readers and encourage them to read a report. They can also break up dull-looking pages of narrative that lack eye appeal. This chapter discusses the illustrations seen most often in technical reports, indicates the overlaps that exist between illustration types, suggests occasions when they can be used most beneficially, and provides guidelines for preparing them. This information is only an overview. Many books and resources are devoted entirely to illustrations. We encourage you to investigate them.

Primary Guidelines

The criterion for any illustration is that it should help to explain the narrative—the narrative should *never* have to explain an illustration. Hence, an illustration must be simple enough for readers to understand quickly and easily. If an illustration immediately captures readers' interest, they are encouraged to read the article.

To help readers readily understand the illustrations you insert into your reports, follow these seven guidelines:

1. Before selecting or designing an illustration, first consider the audience for whom you are writing, and what you want your readers to learn from each illustration.
2. Keep every illustration simple and uncluttered.
3. Let each illustration depict only one main point.

4. Position each illustration as near as possible to the narrative it supports (see "Positioning the Illustrations," at the end of this chapter).
5. Label each illustration clearly with a figure or table number and a title (with the figure number and title *beneath* a figure or chart, and the table number and title *above* a table).
6. Add a caption (that is, comments or remarks) beneath a figure title, to draw attention to significant aspects of the illustration.
7. Refer to every illustration at least once in the report narrative.

Good illustrations attract readers

Computer-Designed Graphs and Charts

Computer software has taken much of the drudgery out of preparing graphs and charts

There are a number of simple-to-use yet powerful computer software programs that will help you create illustrations and graphics. Software programs allow a report writer to enter the quantities of each function and then choose the type of graph or chart. It is possible to view the same information in several formats; for example, as a line graph, a bar chart, or a histogram. The following sections describe the different types of graphs and charts available. Understanding the benefits of each type will help you decide which is best for your information and audience.

Computer-designed graphics still need to follow the above guidelines. In addition, you need to be aware of how the computer truncates and positions the labels, because the results may make a chart incomprehensible. Many software programs create very attractive two- and three-dimensional illustrations, often in multiple colors on a computer screen, which may look fine in a glossy magazine or daily newspaper but look out of place in a business report.

Graphs

Graphs are a simple way to show a change in one function in relation to a change in another. Time is a function used frequently in such comparisons. The other function may be temperature, erosion, wear, speed, strength, or any factors that vary over time.

Table 9-1 shows readings that are part of a study engineering technologist John Greene is undertaking into the cooling rate of different components manufactured by Macro Engineering Inc. The information will also be used by the production department to establish how long manometer cases must cool before assemblers can start working on them with bare hands (the maximum bare-hand temperature has been established by management/union negotiation to be 38°C).

A quick inspection of this table shows that the temperature case drops continuously, is within 8.3° of the ambient temperature after 10 minutes (*ambient* means "surrounding environment"), and is down to the bare-hand temperature after 7 minutes. A closer examination identifies that the

Table 9-1 Cooling rate, manometer case MM-7.

Time Elapsed *(min:sec)*	Temperature *(°C)*	Time Elapsed *(min:sec)*	Temperature *(°C)*
:30	152.9	5:30	43.9
1:00	123.4	6:00	41.1
1:30	106.7	6:30	38.9
2:00	91.2	7:00	37.2
2:30	77.8	7:30	35.6
3:00	69.5	8:00	34.5
3:30	61.7	8:30	33.4
4:00	55.6	9:00	32.5
4:30	51.5	9:30	31.7
5:00	47.3	10:00	31.1

Ambient temperature 22.8°C   Oven temperature 180°C

A table provides specific details

temperature drops rapidly at first, then progressively more slowly as time passes.

Single Curve

A graph translates the details into an easily understood form

John Greene can make the data he has recorded in Table 9-1 easier to understand if he converts it into the graph in Figure 9-1. Now it is immediately evident that the temperature drops rapidly at first, then slows down until the rate of change is almost negligible. He chose a single curve graph because he had only one measurement over one function (time).

Multiple Curves

In Table 9-2 on page 242, John compares the temperature readings he has recorded for the cooling manometer case with measurements he has taken under similar conditions for a cover plate and a panel board. This time, however, he simplifies the table slightly by showing the temperatures at one-minute intervals.

What can we assess from this table? The most obvious conclusion is that in 10 minutes the cover plate has cooled down less than the manometer case, and even less than the panel board. We can also see that the initial rate at which the components cooled varied considerably: the panel board, very quickly; the manometer case, fairly quickly; the cover plate, seemingly quite slowly. But it is difficult to assess whether there were any *changes in the rate of cooling* as time progressed.

Again, this data can be shown more effectively in a graph, which John Greene has plotted in Figure 9-2. Because he has more than one measure-

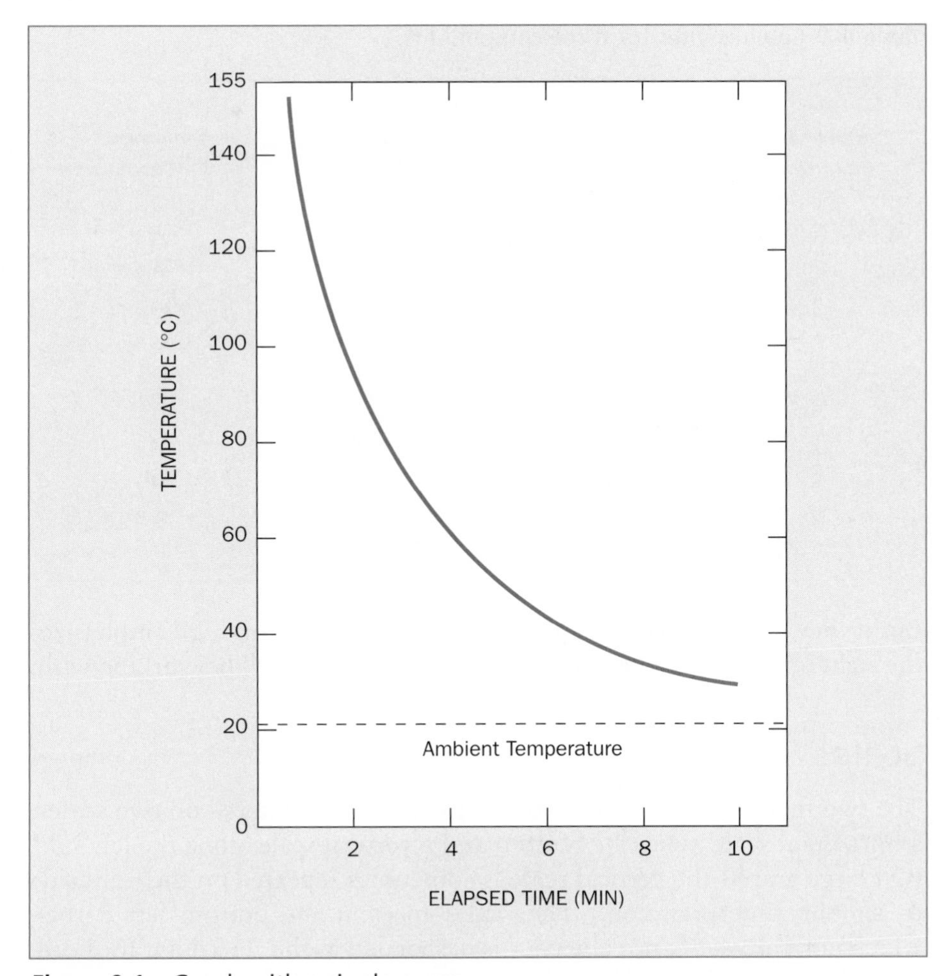

Figure 9-1 Graph with a single curve.

ment over one function (time), John uses a graph with more than one curve. The rapid initial drop in temperature is evident from the initial steepness of the three curves, with each curve flattening out to a slower rate of cooling after 2 to 4 minutes. The difference in cooling rates for the three components is much more obvious from the curves than in the table. (The lines in Figures 9-1 and 9-2 are commonly referred to as curves, even though in some cases they may be straight lines or a series of short straight lines joining points plotted on the graph.)

Technical Illustration
www.arcm.com/illustra.html
This site contains a discussion of the purpose of technical illustrations and has a link to descriptions and samples of product renderings, exploded diagrams, and cutaway views.

Eventually John will have to present this data in a report. If he plans to present only a general description of temperature trends, his narrative can be accompanied by graphs like these. But if he also wants to discuss exact temperatures at specific times for each material, then the narrative and graphs will have to be supported by figures similar to those in Table 9-2.

Constructing a graph usually offers no problems to technical people because they recognize a graph as a logical way to convey statistical data.

Table 9-2 Cooling rates for three components.

Time Elapsed *(minutes)*	Temperature (°C) Cover Plate	Panel Board	Manometer Case
0:30	154.5	145.1	152.9
1	136.7	97.3	123.4
2	112.3	67.8	91.2
3	95.1	51.7	69.5
4	82.3	42.3	55.6
5	71.7	36.1	47.3
6	63.9	32.2	41.1
7	55.6	29.5	37.2
8	49.5	27.2	34.5
9	43.4	26.1	32.8
10	38.9	25.0	31.1

A multi-factor table can be even more difficult to interpret

But if they are to construct a graph that both tells a story *and* emphasizes the right information, they must know the tools they will be working with.

Scales

Both functions may be variable, but only one depends on the other

The two functions to be compared in the graph are entered on two scales: a horizontal scale along the bottom and a vertical scale along the left side. (On large graphs the vertical scale is sometimes repeated on the right side to simplify interpretation.) The scales meet at the bottom left corner, which normally—but not always—is designated as the zero point for both.

Identify which function depends on the other

The two functions are commonly known as the dependent and independent variables, so named because a change in the dependent variable *depends* on a change in the independent variable. For example, if we want to show how the fuel consumption of a car increases with speed, we will enter speed as the independent variable along the bottom scale, and fuel consumption as the dependent variable along the left side, as in Figure 9-3. (Fuel consumption *depends* on speed; speed does not depend on fuel consumption.) The same applies to John Greene's temperature measurement graphs: temperature is the dependent variable because it depends on *the time that has elapsed* since the components came out of the oven (the independent variable).

When you construct a graph, the first step is to identify which function should form the horizontal scale and which the vertical scale. Table 9-3 lists some typical situations that show that the same function (e.g., temperature) can be an independent variable in one situation and a dependent variable in another. Selection of the independent variable depends on which function can be more readily identified as influencing the other function in the comparison.

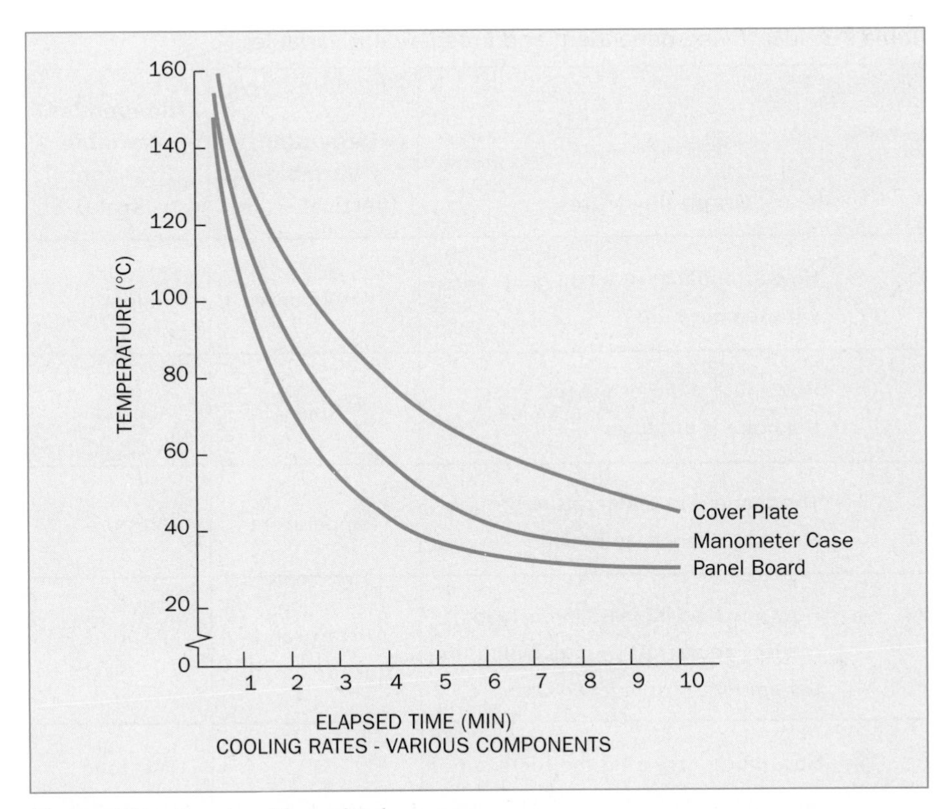

Figure 9-2 Graph with multiple curves.

A graph can make the same information much more accessible

Select scale intervals that will give a *balanced* appearance

The second factor to consider is scale interval. Poorly selected scale intervals, particularly scale intervals that are not balanced between the two variables, can defeat the purpose of a graph by distorting the story it conveys. Suppose John Greene had made the vertical scale interval of his time vs temperature graph in Figure 9-1 much more compact, but had

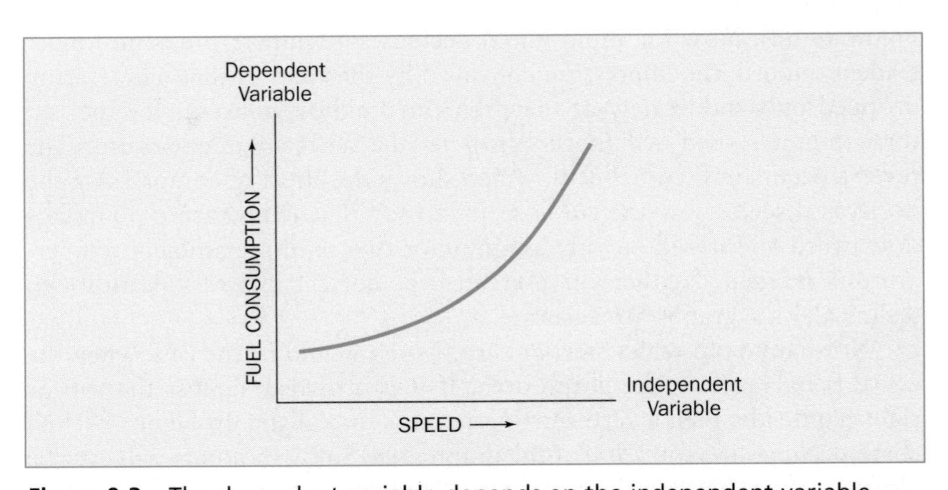

Figure 9-3 The dependent variable depends on the independent variable.

Table 9-3 Identifying dependent and independent variables.

Graph Illustrates	Dependent Variable (vertical scale)	Independent Variable (horizontal scale)
1. How attendance at a ball game varies with temperature	Attendance	Temperature
2. How much a motor's speed affects the noise it produces	Noise	Speed
3. The changes in temperature brought about by changes in pressure	Temperature	Pressure
4. How much an increase in payload reduces an aircraft's range by limiting the amount of fuel it can carry	Aircraft range (or fuel load)	Payload
5. How much increasing the fuel load of an aircraft to achieve greater range reduces its effective payload	Payload	Fuel load (or aircraft range)
Note: A function can be either dependent or independent, depending on its role in the comparison (see temperature in examples 1 and 3, and both functions in examples 4 and 5).		

retained the same spacing for the horizontal scale. The result is shown in Figure 9-4(a). Now the rapid initial decrease in temperature is no longer evident; indeed, the impression conveyed by the curve is that temperature dropped only moderately at first, remained almost constant for the last three minutes, and will never drop to the ambient temperature. The reverse occurs in Figure 9-4(b), which shows the effect of compressing the horizontal scale: now the curve seems to say that temperature plummets downward and it will be only a minute or two until the ambient temperature is reached. Neither curve creates the correct impression, although technically the graphs are accurate.

Adjust a scale's starting point to center the curve(s)

Normally both scales start at zero, which would be the case when the curve is balanced in the graph area. If it is crowded against the top or right-hand side, then a zero starting point is unrealistic. In Figure 9-1 the curve occupies the top 75% of the graph area. Since no points will ever be plotted below the ambient temperature (which will hover around 23°C),

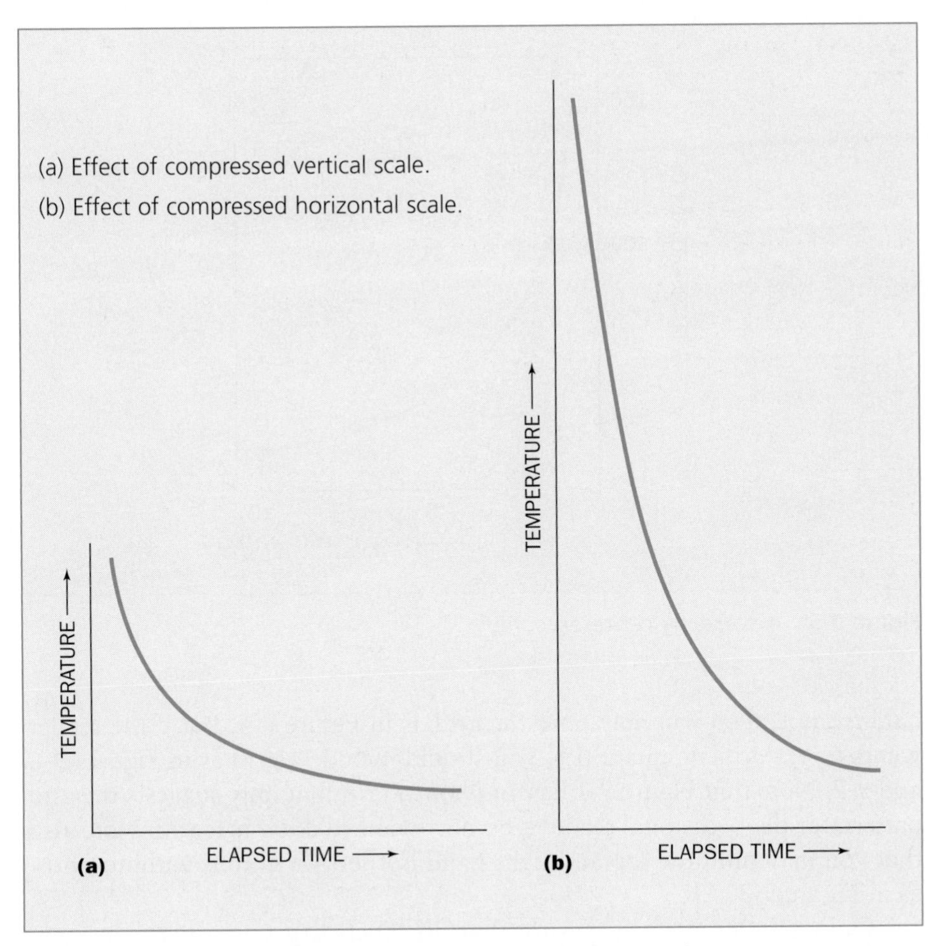

Figure 9-4 Poorly chosen scales can create an inaccurate image.

the bottom portion of the vertical scale is unnecessary. This can be corrected by starting the vertical scale at a higher value (say 20°, as in Figure 9-5), or by breaking the scale to indicate that some scale values have been omitted (Figures 9-2 and 9-6).

Emphasize the most significant curve(s)

Multiple-curve graphs should have no more than four curves otherwise they will be difficult to interpret, particularly if the curves cross one another. You can help a reader identify the most important curve by making it heavier than the others (Figure 9-6), and can differentiate among curves that cross by creating different weight lines (Figure 9-7). Avoid using colored lines because the average copier reproduces all the lines in one color (usually black).

Simplicity

Simplicity is important in graph construction. If a graph illustrates only trends or comparisons, and the reader is not expected to extract specific

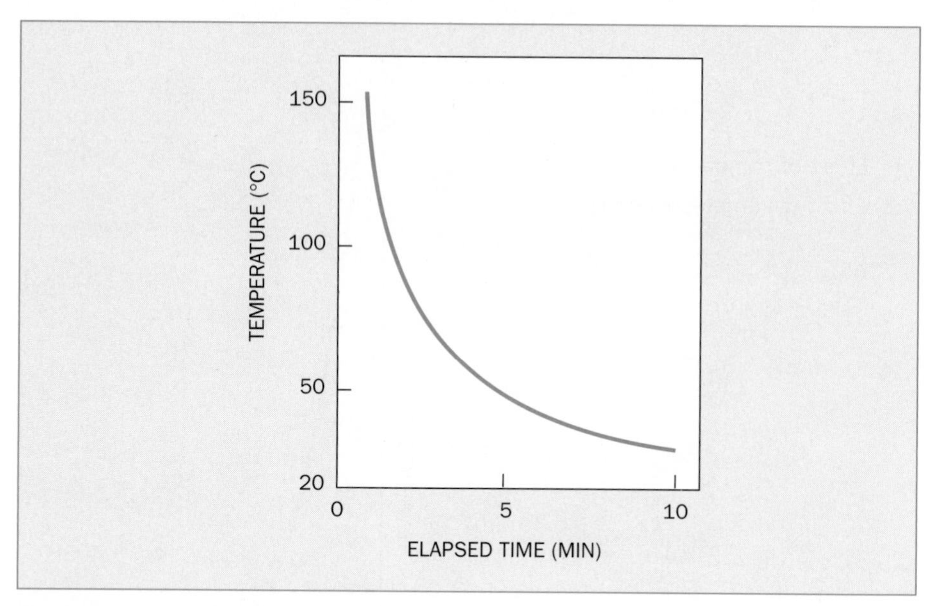

Figure 9-5 A correctly centered curve.

data from it, then you may omit the grid as in Figure 9-5. But if the reader wants to extrapolate quantities, you should include a grid as in Figures 9-6 and 9-7. Note that Figure 9-1 has an *implied* grid that only suggests the grid pattern for the occasional reader who may want to draw in a grid. Note also that you may omit the top and right-hand borders on graphs without grids, as in Figure 9-3.

Omit plot points and ensure that all labels are *horizontal* (the only label that may be entered vertically is the label for the vertical scale function). Insert labels for the curves at the end of the curve whenever possible (Figure 9-2) or, alternatively, above or below the curve (Figures 9-6 and 9-7). Never write a label along the slope of the curve.

Charts

Most charts show trends or compare only general quantities. They include bar charts, histograms, surface charts, and pie charts.

Bar Charts

You can use bar charts to compare functions that do not necessarily vary continuously. In the graph in Figure 9-1, John Greene plotted a curve to show how temperature decreased continuously with time. He could do this because both functions were varying continuously (time was passing and temperature was decreasing.) For the production department, however, he has to prepare a report on how long it takes various components

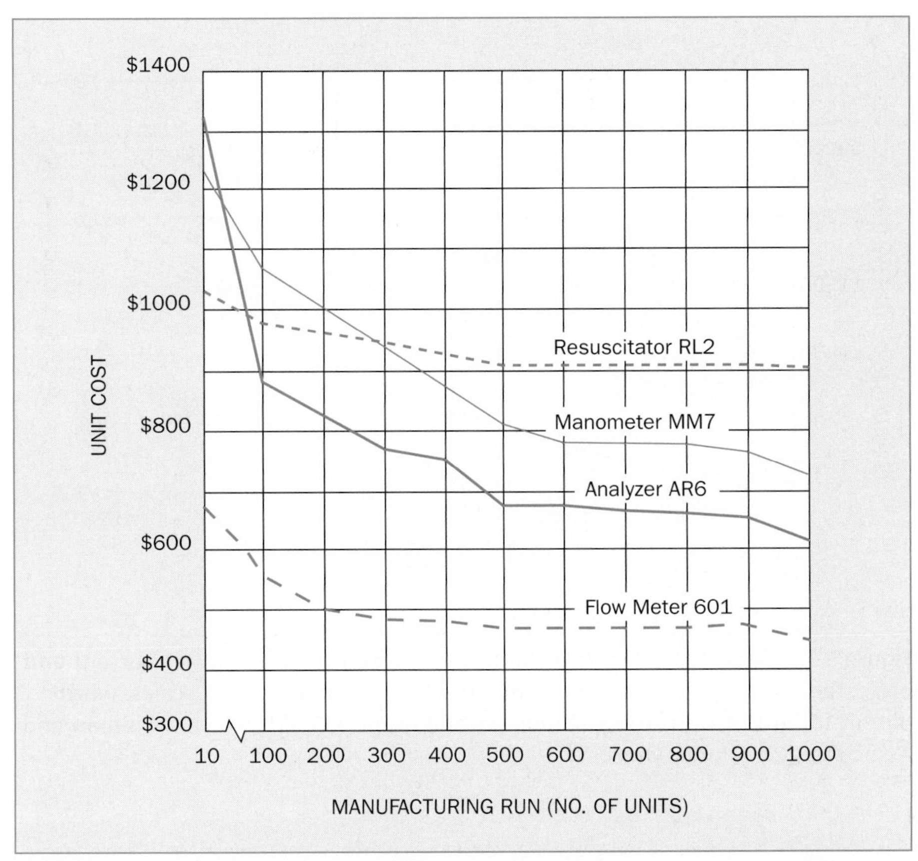

Figure 9-6 Bolder lines draw attention to the most important curve (those showing maximum benefit from quantity manufacturing). The grid permits readers to draw reasonably accurate data from the graph. (Courtesy Macro Engineering Inc., Phoenix, Arizona.)

Avoid confusion: Limit the number of curves

coming from the oven to cool to a safe temperature for bare-hand work. He prepares a bar chart to depict this because he knows the report will be read by both management and union representatives, and some of the readers may need easy-to-interpret data. He also has only one continuous variable to plot: elapsed time. The other variable is noncontinuous because it represents the various components he has tested. In this case elapsed time is the dependent variable, and the components are the independent variable. The bar chart John constructs is shown in Figure 9-8.

Scales for a bar chart can be made up of such diverse functions as time, age groups, heat resistance, employment categories, percentages of population, types of soil, and quantities (of products manufactured, components sold, software programs used, and so on). Charts can be arranged with either vertical or horizontal bars depending on the type of information they portray; when time is one of the variables, it is usually plotted

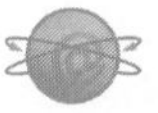

Illustrating Tech Documents
www.incrediblecharts.com/technical/chart_types.htm
This website shows the different charts that can be created for technical purposes.

Avoid writing along the slope of the curve

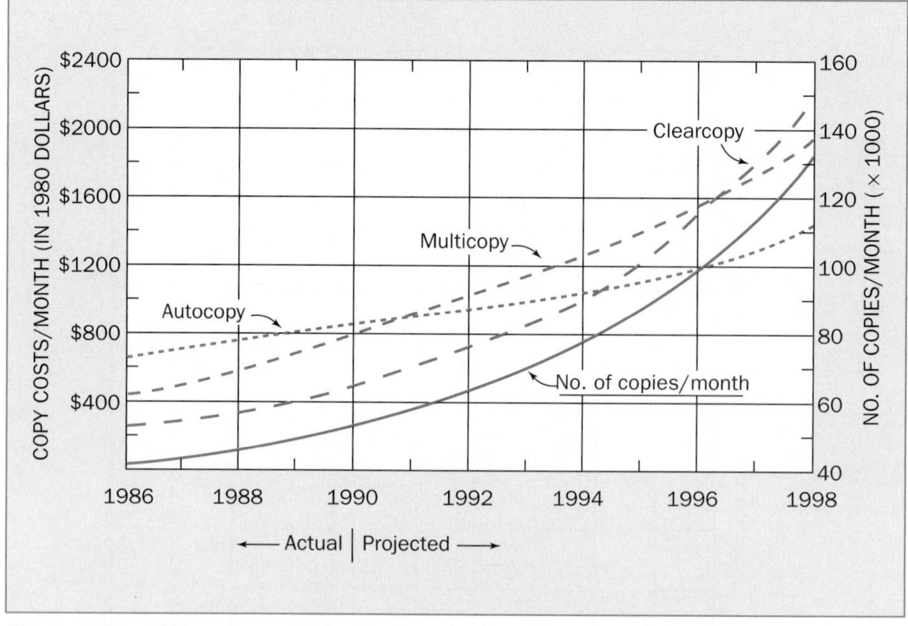

Figure 9-7 Different symbols distinguish between curves showing current and projected copying costs for three copiers. Note the two vertical scales, which permit three functions to be shown on one graph. (Courtesy H. L. Winman and Associates, Cleveland, Ohio.)

along the horizontal axis, as in Figure 9-8. The bars are normally separated by spaces the same width as each of the bars.

Some computer software provides only a narrow space between the bars

In a complex bar chart, the bars may be shaded to indicate comparisons within each factor being considered. The vertical bar chart in Figure 9-9 uses two shades to describe two factors on the one chart, individual bars can also be shaded to show proportional content, as has been done in Figure 9-10, in which case a legend must be inserted beside or below the graph to show readers what each shading represents. Alternatively, each segment may be labeled as in Figure 9-11. A computer-generated 3-D vertical bar graph showing the same information as presented in Figure 9-11 is shown in Figure 9-12.

Horizontal bar charts can be used in an unconventional way by arranging the bars on either side of a zero line. This can be done, for example, to compare negative and positive quantities, satisfactory and defective products, or passed and failed students. The chart in Figure 9-13 divides products returned for repair into two groups: those that are covered by warranty, and those that are not. Each bar represents 100% of the total number of items repaired in a particular product age group and is positioned at about the zero line depending on the percentage of warranty and nonwarranty repairs.

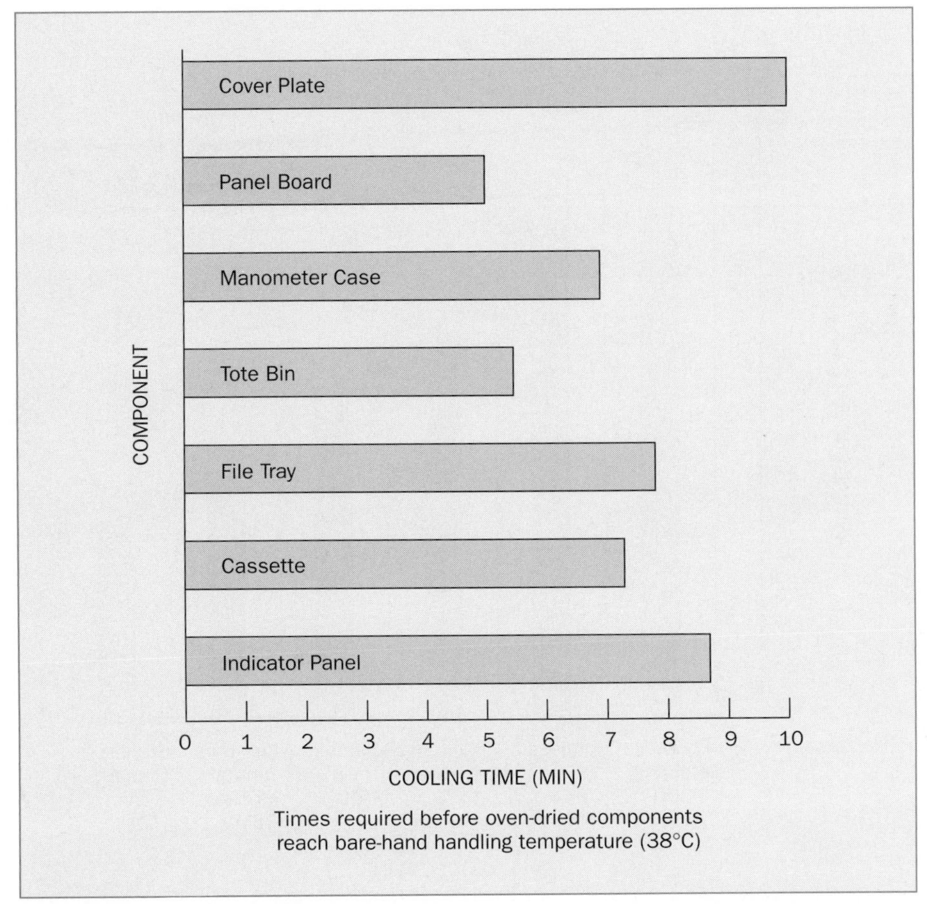

Figure 9-8 Horizontal bar chart with one continuous variable (cooling time).

Use horizontal bar charts to depict elapsed time

Histograms

A histogram looks like a bar chart, but functionally it is similar to a graph because it deals with two continuous variables (functions that can be shown on a scale to be increasing or decreasing). It is usually plotted like a bar chart because it does not have enough data on which to plot a continuous curve (see Figure 9-14). The chief visible difference between a histogram and a bar chart is that there are no spaces between the bars of a histogram.

Although seen rarely, histograms are useful when there is only limited data

Surface Charts

A surface chart (Figure 9-15) is like a graph, in that it has two continuous variables that form the scales against which the curves are plotted. But, unlike a graph, individual curves cannot be read directly from the scales.

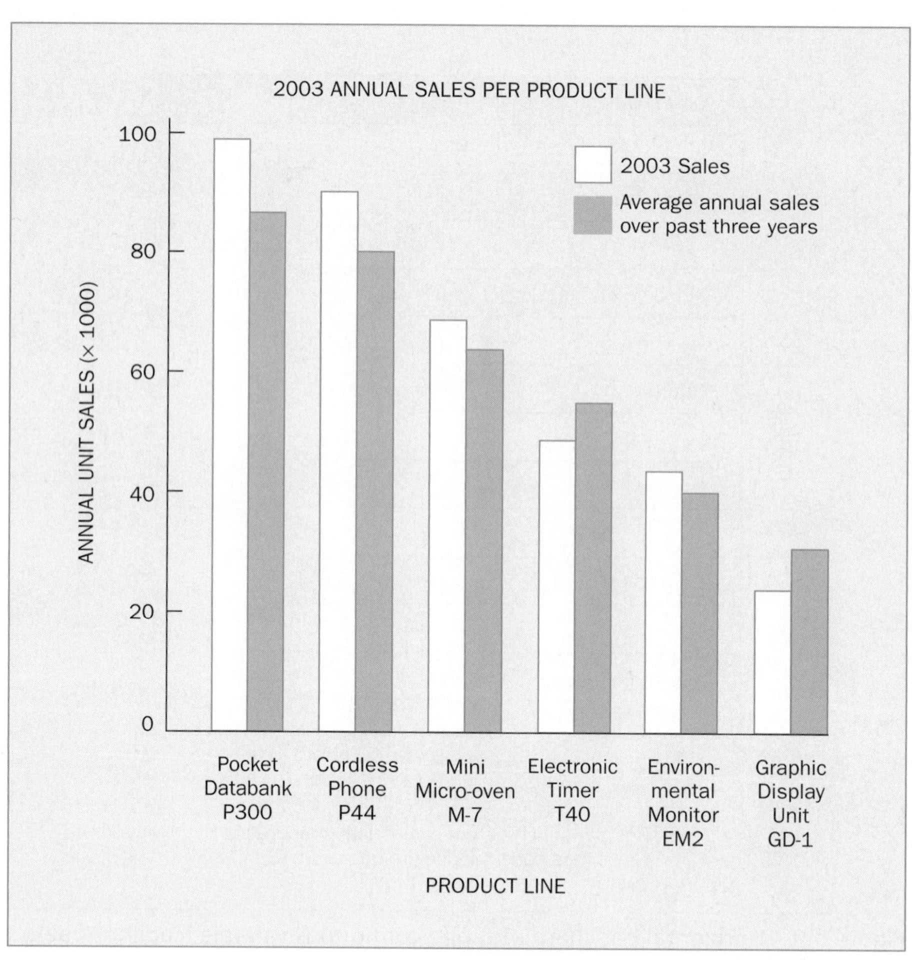

Figure 9-9 A vertical bar chart that lets readers compare current statistics with past statistics, and so determine trends.

Hand-drawn bar charts offer only limited scope for creative presentations

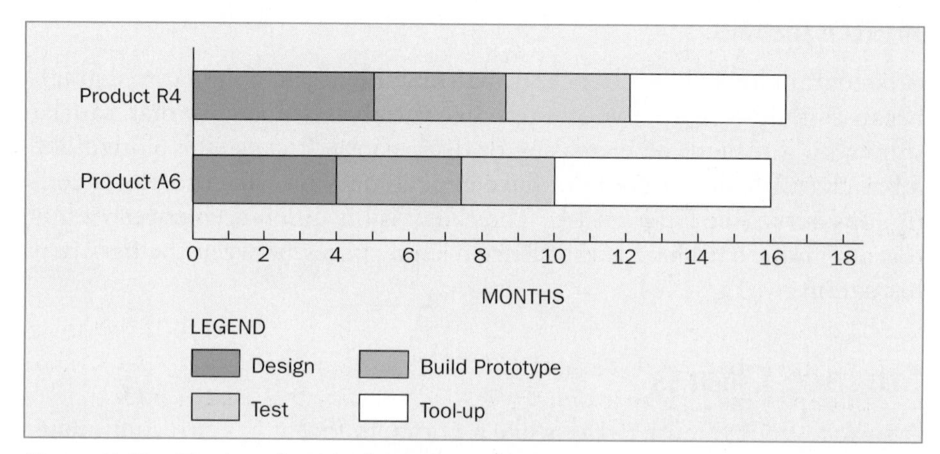

Figure 9-10 The bars in this chart show development times for proposed new products. The legend is included with the chart.

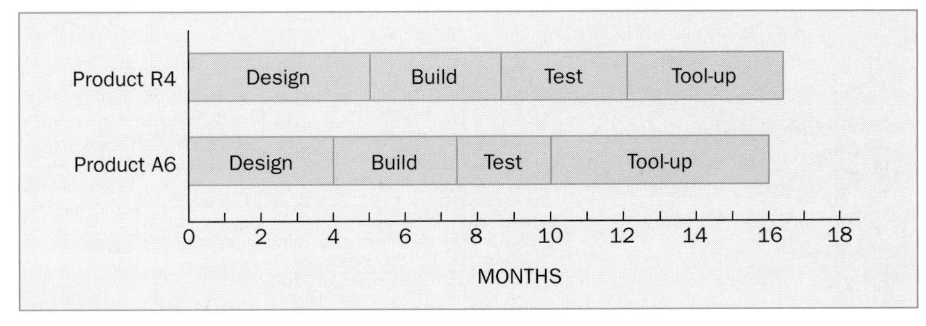

Figure 9-11 A segmented bar chart with internal labeling.

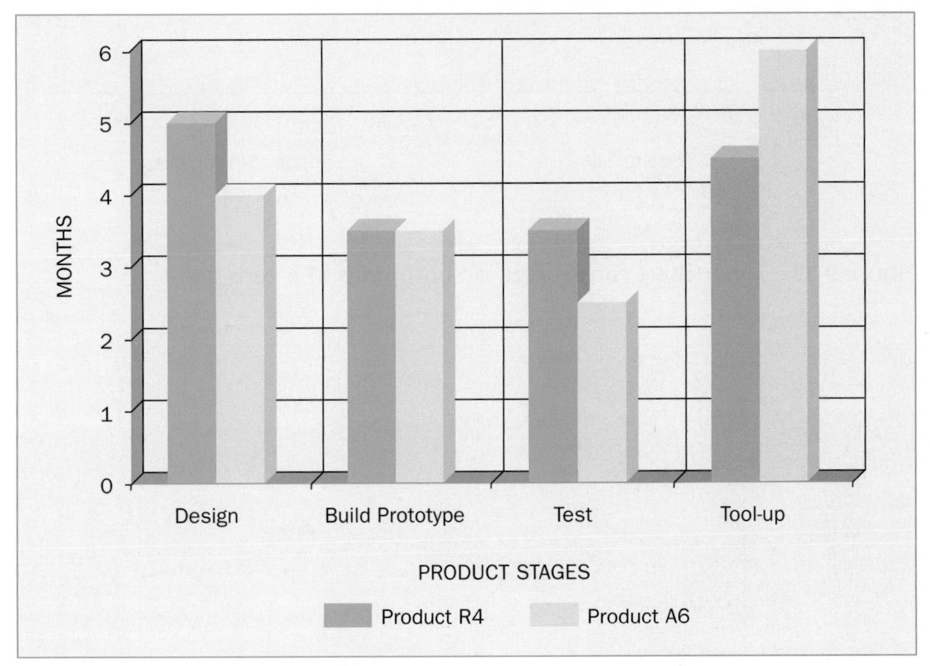

Figure 9-12 A computer-generated 3-D vertical bar graph showing the same information as presented in Figure 9-11.

Computer graphics automatically create imaginative presentations

The uppermost curve on a surface chart shows the *total* of the data being presented. This curve is achieved as follows:

1. The curve containing the most important or largest quantity of data is drawn in first, in the normal way. This is the Thermal curve in Figure 9-15.
2. The next curve is drawn in above the first curve, using the first curve as a base (i.e. "zero") and adding the second set of data to it. For example, the energy resources shown as being variable in 2000 are:

Thermal Power:	33,000 MW
Hydro Power:	14,500 MW

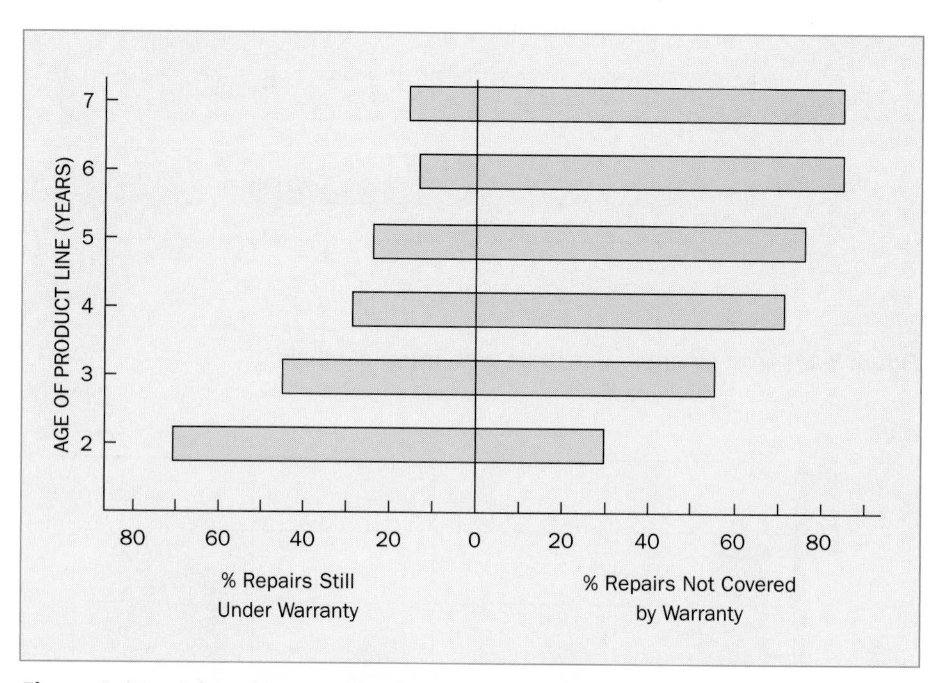

Figure 9-13 A bar chart constructed on both sides of a zero point.

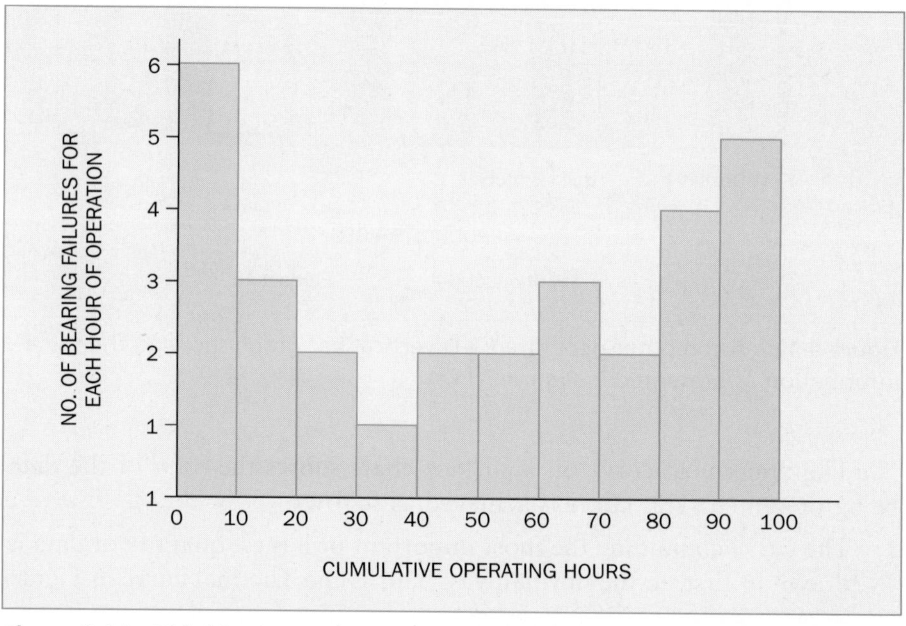

Figure 9-14 This histogram shows the number of bearing failures for every 10 hours of operation. Considerably more data would have been required to construct a curve.

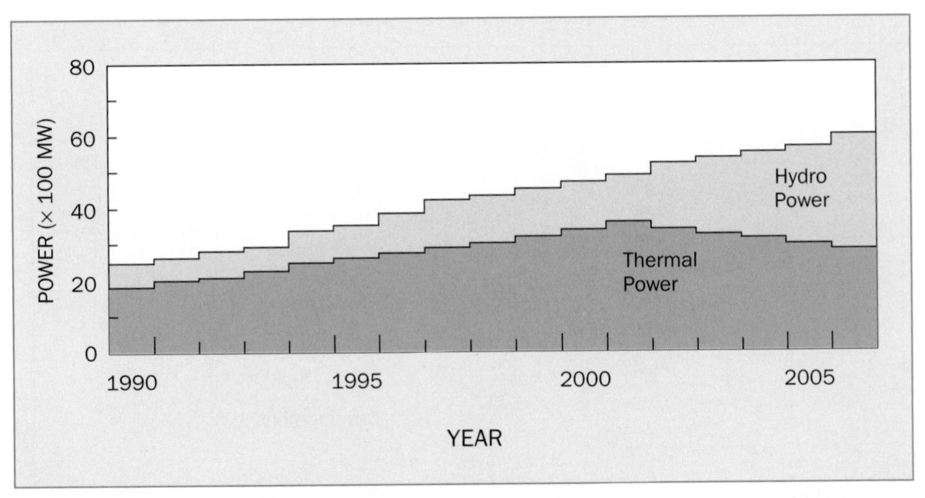

Figure 9-15 This surface chart adds thermal data to hydro data to show projected energy resources of a power utility.

In Figure 9-15, the lower curve for 2000 is plotted at 33,000 MW. The next curve is 14,500 MW, which is added to the first set of data so that the second curve indicates a *total* of 47,500 MW. (If there were a third set of data, it would be added in the same way.) Normally, the lowest set of data has the darkest shade.

Difficult to draw by hand, yet useful for depicting a cumulative effect

Pie Charts

A pie chart is a pictorial device for showing approximate divisions of a whole unit. The pie chart in Figure 9-16 depicts the percentage of work done by Macro Engineering Inc. in eight major product or service categories. Figure 9-17 is a computer-generated pie chart showing the stages of product development.

Probably the most widely recognized and readily understood

If a pie chart has several wedges that would be difficult to draw and hard to read, you may combine some of them into a larger single edge and give it a general heading, such as "miscellaneous expenses," "other uses," or "minor effects." All the wedges must add up to a whole unit, such as 100%, $1.00, or 1 (unity).

Diagrams

Diagrams include any illustration that helps the reader understand the report narrative yet does not fall within the category of graph, chart, or table. It can range from a schematic drawing of a complex circuit to a simple plan of an intersection. There is, however, one restriction: any illustration included in the narrative part of a report must be clear enough to be

A diagram must be readily understood; it should rarely need to be explained

Illustration Samples from NASA's Glenn's Graphics Group
http://grcpublishing.grc.nasa.gov/graphics/samillus.cfm
Several examples of detailed illustrations from NASA's Glenn Graphics Groups are available at this site. A contact address is included for readers who want more information about the preparation of technical illustrations.

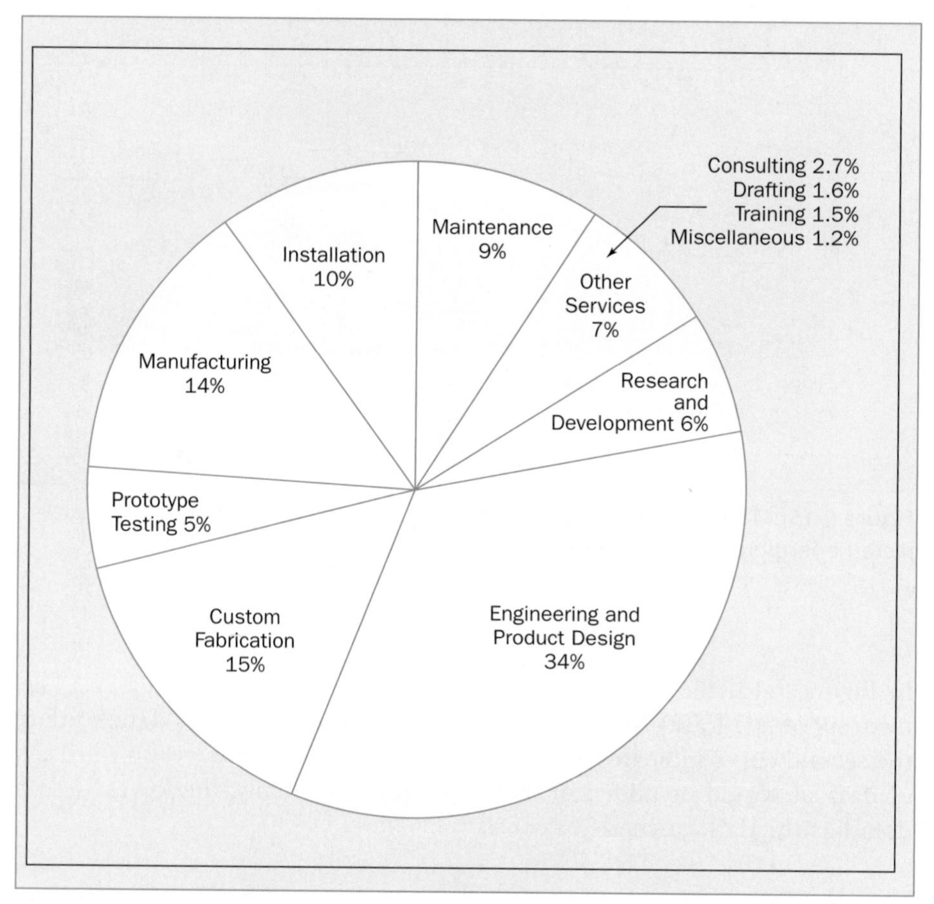

Figure 9-16 Pie chart shows Macro Engineering Inc.'s products and services. "Slices" add up to 100%.

understood easily. This means that complex drawings should be placed in an appendix and treated as supporting data.

Diagrams should be simple, easy to follow, and contribute to the narrative. (See Figure 8-3 on page 219 for two good examples.) They can comprise organization charts, flow diagrams (Figure 10-1), site plans (Figure 9-18), and sketches.

Photographs

A photograph helps a reader visualize shape, appearance, complexity, or size. The criterion when selecting a photograph is that it be clear and contain no extraneous information that might distract the reader's attention. For example, if you are trying to show damage to a building's foundation, the photograph should be a close-up of the area, showing cracks in the cement. You should remove any garbage cans, bicycles, car parts or other items that have nothing to do with what you are trying to demonstrate.

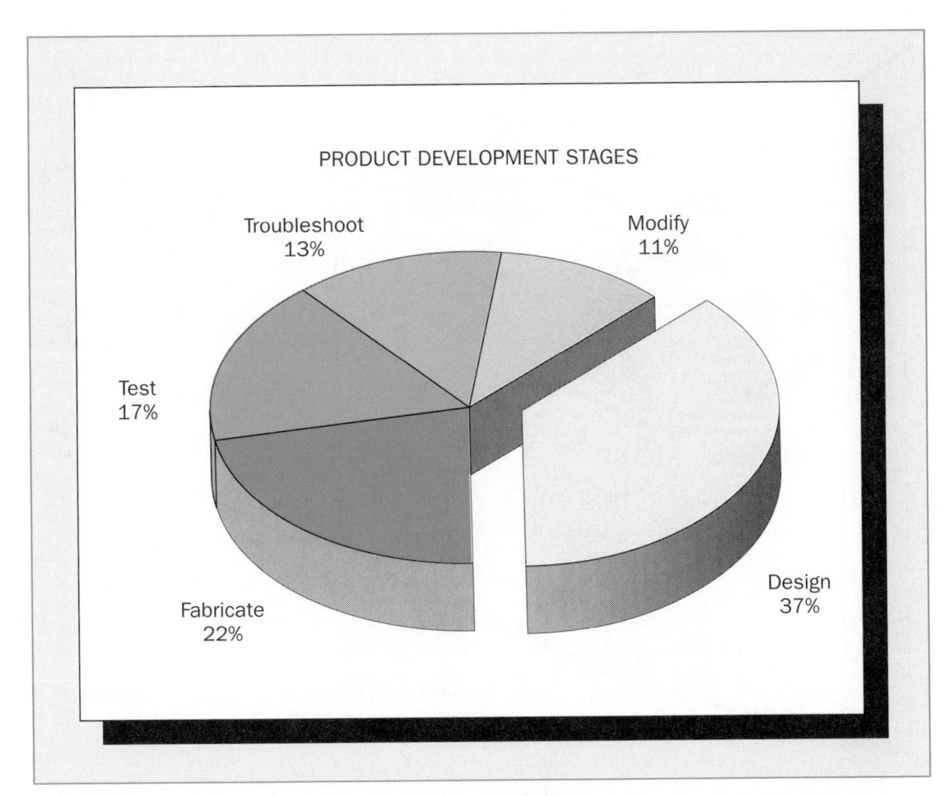

Figure 9-17 A computer-generated exploded-view pie chart, with one segment emphasized by being pulled partly away.

The pie chart particularly adapts to 3-D presentation

With digital equipment, it's becoming easier to use photographs

Digital equipment (scanners and cameras) makes taking photographs easy even for novice photographers. There are some things you need to consider, though. The minimum acceptable quality for an image created by a scanner or camera is 150 pixels. (A pixel is the tiny dot that forms digital images.) This ensures that the eye cannot see the individual dots. Your output device (normally a printer) should be set for a minimum of 300 dpi (dots per inch). A higher dpi will result in better image quality so if you are preparing a report for an external client on glossy paper, you should consider a higher dpi.

Pay attention to the contrast and lighting in a photograph. You can use image-editing software to manipulate the contrast (the difference in brightness between the light and dark areas of a picture) of an image. This is particularly important if you know the report will be photocopied, since photocopiers inherently add contrast in the copying process. You'll want the image to be of less contrast (a little darker). Always test it yourself by making several photocopies and paying attention to the image quality.

Often reports are distributed electronically and including images and photographs may increase the file to an unmanageable size. Save your photos as a JPG file (a compressed format for encoding graphic images) to

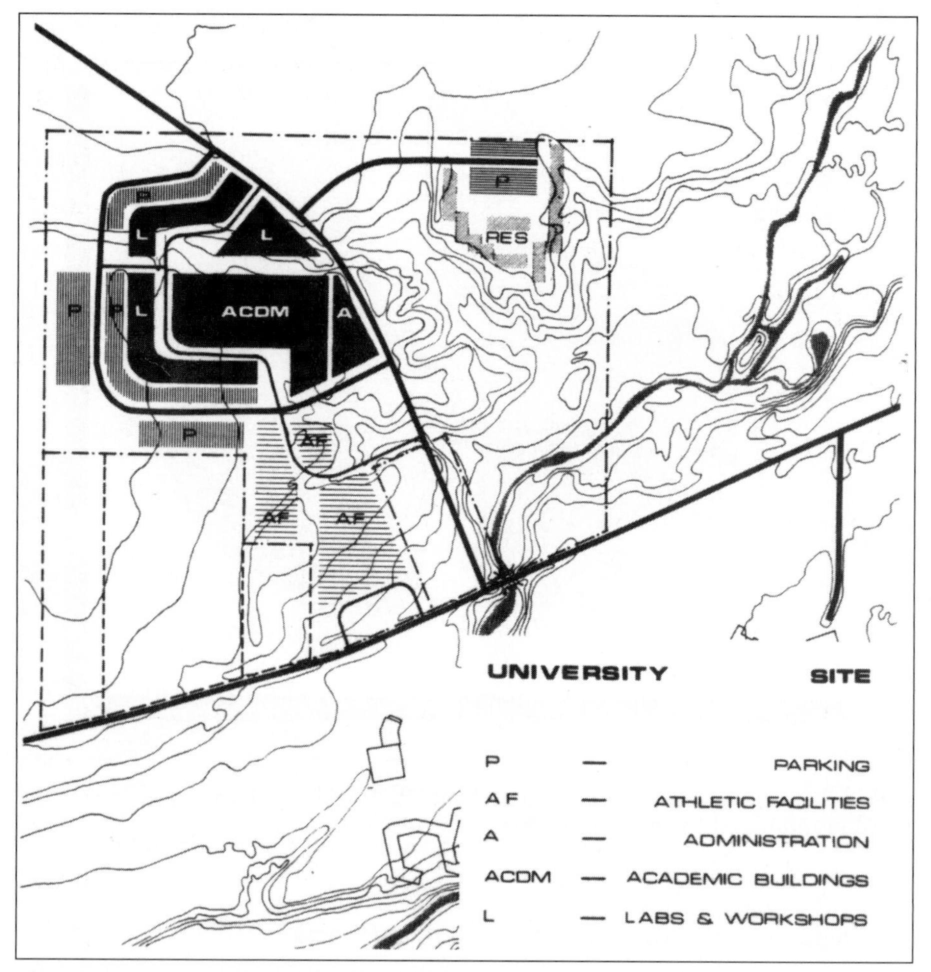

Figure 9-18 A site plan that illustrates where college facilities are to be located. (Courtesy Smith, Carter, Searle—W. L. Wardrop & Associates Ltd., Winnipeg, Man.)

retain the image quality. If, however, you are capturing computer screen images, you'll need to save them as GIF files (graphic interchange format) because the compression of the JPG format deteriorates the screen image.

Depending on the importance and legal liability of the content of your document, you might consider hiring a professional photographer who is trained in the finer points of imaging.

Tables

A table may be a collection of technical data, as in Tables 9-1 and 9-2, or a series of short narrative statements, as in Table 9-3. Whether you should insert a table into the report narrative or place it in an attachment or appendix depends on three factors:

1. If the table is short (i.e. less than half a page) and readers need to refer to it as they read the report, then include it as part of the report narrative (i.e. in the discussion), preferably on the same page as the text that refers to it.
2. If readers can understand the discussion without referring to the table as they read the report, but may want to consult the table later, then place the table in an attachment or appendix.
3. If readers need to refer to a table but the data you have will occupy a full page or more, then
   - summarize the table's key points into a short table to be inserted into the report narrative, and
   - place the full table in an attachment or appendix.

Use word-processing software to create tables, for words as well as numbers

There are four additional guidelines that apply to tables, particularly if the tables contain columns of numerical data:

1. Keep a table simple by limiting it only to data the readers will *really* need, and create as few columns as possible.
2. Insert units of measurement, such as decibels, volts, kilograms, or seconds, at the head of each column rather than after each column entry (see the "min:sec" and "°C" entries at the top of the columns in Table 9-1).
3. Insert a table number and an appropriate title above the table.
4. Draw readers' attention to a table by referring to it in the report narrative and commenting on a specific inference to be drawn from the table. For example:

> The voltage fluctuations were recorded at 10-minute intervals and entered in column 3 of Table 7, which shows that fluctuations were most marked between 8:15 and 11:20 a.m.

Positioning the Illustrations

Whenever possible place each illustration on the same page as or facing the narrative it supports. A reader who has to keep flipping pages back and forth between narrative and illustrations will soon tire, and your reasons for including the illustrations will be defeated.

Consider illustrations as an integral part of a document, not just an "add on"

When reports are printed on only one side of the paper, full-page illustrations can be difficult to position. The only feasible way to place them conveniently near the narrative is to print them on the back of the preceding page, facing the words they support. But this in turn may pose a printing problem. A more logical solution is to limit the size of illustrations so that they can be placed beside, above, or below the words.

When an illustration is too large to fit on a normal page, or is going to be referred to frequently, consider printing it on a foldout sheet and

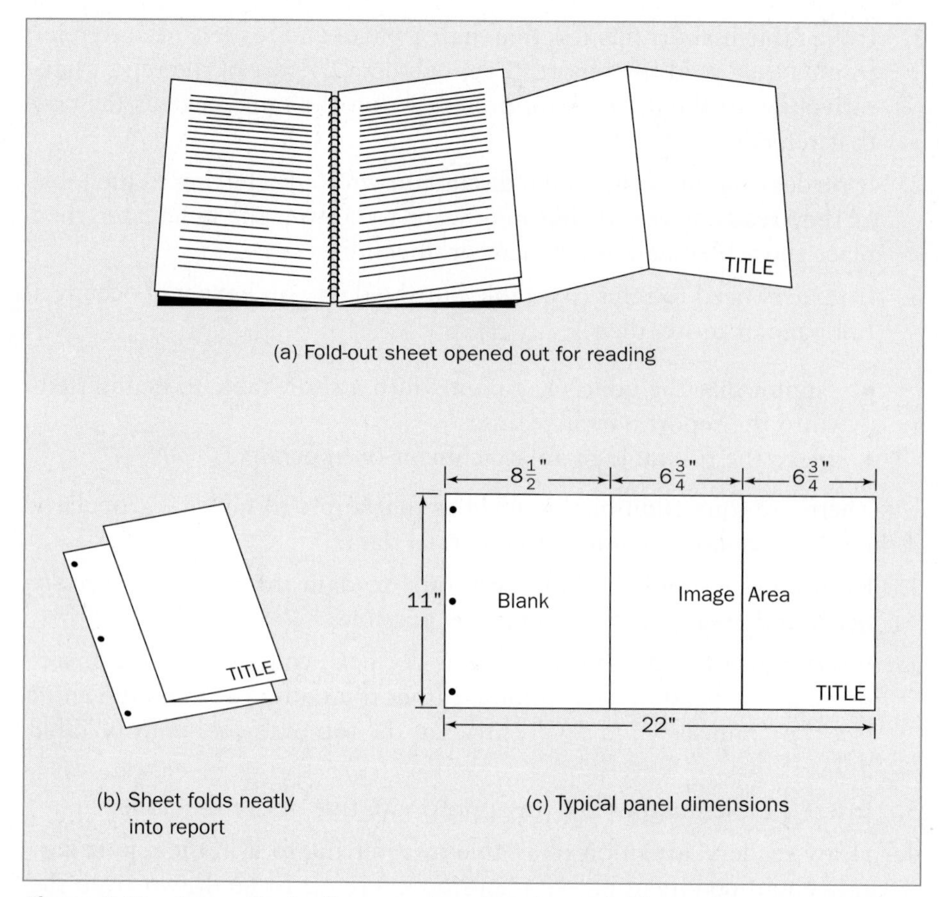

Figure 9-19 Large illustrations can be placed on a fold-out sheet at rear of report.

inserting it at the back of the report (see Figure 9-19). If the illustration is printed only on the extension panels of the foldout, the page can be left opened out for continual reference while the report is being read. This technique is particularly suitable for circuit diagrams and flow charts.

Position horizontal full-page illustrations sideways on a page so that they are viewed from the right (see Figure 9-20). This holds true whether they are placed on a left- or right-hand page.

If your reader wants to refer to any figures or tables from your report, we suggest you add a List of Figures and Tables right after your Table of Contents. List the figure or table number, the title or caption, and the page number.

Working with an Illustrator

Although you may prepare your own illustrations, in a large technical organization you may work with a company graphics person or illustra-

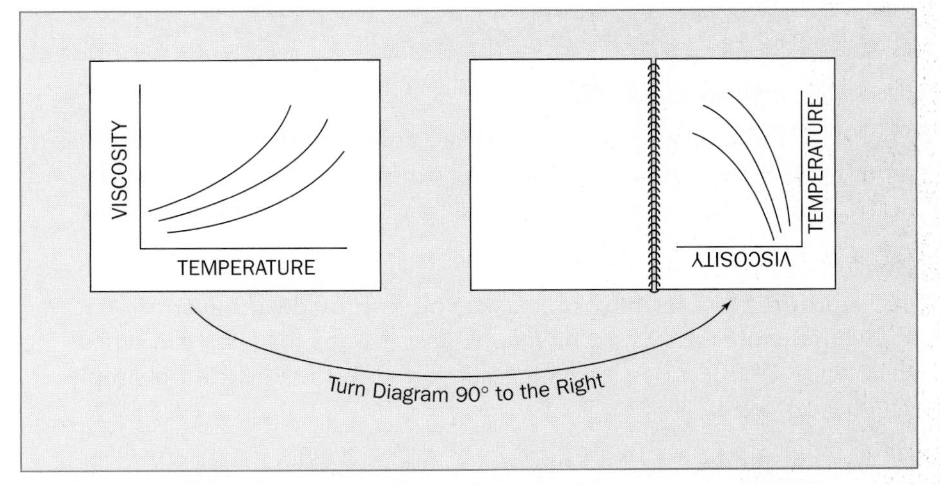

Figure 9-20 Page-size horizontal drawings should be positioned so they are viewed from the right.

You align horizontal illustrations this way, even if some words appear to be upside down

tor who will prepare your drawings, graphs, and charts according to your requirements. Good communication between you and the illustrator is essential if the drawings you want are to appear in the form you visualized when you wrote your report.

An illustrator needs to have much more than a bare, roughed-out sketch to work from. You will have to describe your project and its outcome in detail, so that the illustrator will know

- the background to and purpose of the report or oral presentation,
- who the readers or listeners will be, what their technical knowledge is, and how they will use the information contained in the report or presentation,
- what each illustration is to portray, and what particular aspects are to be emphasized,
- what size each illustration is to be (vertical and horizontal dimensions) and, for a talk, the size of the expected audience and their likely distance from the screen,
- how much the illustrations' size will be reduced when they are printed or converted into slides or transparencies (a drawing that is to be reduced must have lines that are not too fine), and
- when you need the illustrations.

If you're lucky, you will work with an illustrator; more often, you will create your own visuals

You can help an illustrator even more by providing a sketch of each proposed illustration and a copy of the words the illustration is to support. Better still, talk to the illustrator *before* you write your report or make your speaking notes, describe what illustrations you plan to use, and ask for suggestions for their preparation.

ASSIGNMENTS

Project 9.1: Who Buys "Planit"?

You work for a very successful software company that publishes a monthly user magazine. The company's most successful software has been "Planit," a program for organizing a user's business operations.

Part 1

You have to illustrate a consumer survey

The editor of the user magazine asks you to provide an illustration showing the breakdown of buyers by user groups for last year, when there were 14,236 sales, and suggests you keep the illustration simple. The buyers were:

Hospitals	1588
Small businesses	1011
Public utilities	2165
Consultants	233
Manufacturers	3176
Architects	217
Writers/editors	116
Land surveyors	245
Sales representatives	866
City/town planners	1155
Engineers	1732
Miscellaneous	433
Radio and television stations	1299

Part 2

When you give your illustration to the editor, the response is: "I like that. But I'd also like another one comparing last year's buyers with those of four years ago, which was the first year we marketed Planit."

The sales of Planit four years ago were:

Hospitals	174
Small businesses	1393
Public utilities	1132
Consultants	174
Manufacturers	2351
Sales representatives	958
City/town planners	1306
Engineers	784
Miscellaneous	261
Radio and television stations	174

Project 9.2: Comparing Electricity Costs

Your company markets heat pumps, and wants to demonstrate to electricity users that a heat pump can significantly reduce their electricity bills. Prepare an illustration based on the following actual monthly bills for four different dwellings last year. The residences are identical five-room homes built at the same time and in the same block on Margusson Avenue. The only differences are that Nos. 216 and 234 do not have air-conditioning (marked "No AC" in the table), while Nos. 227 and 248 do (marked "+AC"), and Nos. 234 and 248 each have a heat pump.

Use these figures to create an easy-to-read illustration

Monthly Electricity Bills

	Actual Bills No Heat Pump		Amount Saved With Heat Pump	
	216 No AC	227 +AC	234 No AC	248 +AC
January	$ 48.07	$ 49.23	$ 12.06	$ 11.59
February	45.15	46.01	11.68	11.86
March	43.20	42.86	10.90	11.13
April	41.11	42.20	4.15	3.80
May	38.37	42.15	0.36	3.48
June	35.20	48.16	—	6.10
July	33.06	57.19	—	18.26
August	32.87	56.80	—	17.44
September	34.11	47.10	2.64	7.21
October	38.62	39.20	7.85	8.60
November	41.67	41.10	9.62	10.51
December	43.20	42.89	11.58	11.77

Chapter 10
Technically-Speak!

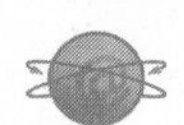

Crossing a Bridge of Shyness: Public Speaking for Communicators
www.eeicom.com/eye/shyness.html
Diane Ullius, the author of this article published in *The Editorial Eye*, teaches oral and written communication skills at Editorial Experts Inc. and at Georgetown University. She offers practical advice about getting over the fear of speaking in public.

This chapter covers two facets of public speaking, both concerned with the oral presentation of technical information. The first is the oral report, sometimes called the technical briefing, delivered to a client or one's colleagues. The second is the technical paper presented before a meeting of scientific or engineering-oriented people. Both depend on carefully honed public speaking skills for their effectiveness, although neither requires vast experience or knowledge in this field. The chapter also describes how to present information at and contribute effectively to office meetings.

The Technical Briefing

Your department head approaches your desk and says:

> "We've had a call from the RAFAC Corporation. They're sending in some representatives next Tuesday. I'd like you to give them a rundown on the project you're working on."

Every day visitors are being shown around industrial organizations, and every day engineers and technicians are being called upon to stand up and say a few words about their work. On paper, this sounds straightforward, but to those who have to make the presentation it can be a traumatic experience. Much of their nervousness can be reduced (it can seldom be entirely eliminated, as any experienced speaker will tell you) if they learn a few simple public speaking techniques.

Many people fear having to stand up and speak, even more than sky diving

Establish the Circumstances

Your first step is to establish the circumstances affecting your presentation. Go to your department head—or the person arranging the event—and ask four questions. The first two will identify your listeners:

1. **Who Will Be in My Audience?**
   If you are to focus your presentation properly, and use appropriate terminology for the people you will be addressing, you need to know

whether they are engineers and technologists knowledgeable in your area of expertise, technical managers with only a general appreciation of the subject, or laypersons with very little or no technical knowledge.

2. **What Will They Know Already?**
   If you are to avoid boring your listeners by repeating information they already know, or confusing them by omitting essential background details, you need to find out how much they know now about your subject, or will have been told before you address them.

It's just like writing a report: first, identify your audience

The second two questions deal with the briefing itself:

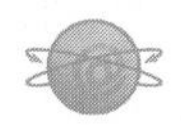

3. **How Long Do You Want Me to Talk?**
   Find out if you are to describe the project in detail or simply touch on the highlights. The answer will directly influence how deeply you cover the topic.

4. **Where Is the Presentation Taking Place?**
   Identify whether you are to make your presentation in your company's conference room or training room, or at a client's or some other premises. Within your own company you can easily identify what audiovisual facilities are available, and where your audience will be seated in relation to you. If you will be speaking at another location you should ask for a description of the facilities or, even better, be able to view them in advance.

Preparing Outstanding Presentations
www.cs.utexas.edu/users/ethics/Other/presentation1.html
This is a series of articles from Cheryl Reimold's "Tools of the Trade" column. Preparing Outstanding Presentations includes sections on understanding your audience; basic presentation structure; the introduction, body and summary of a presentation; effective visuals; and making visuals memorable.

Now you can start making your notes. Jot down the topics you intend to discuss, and arrange them in an interesting, logical order.

Find a Pattern

The best technical briefings follow an identifiable pattern, just as written formal reports do. You can establish a pattern for your briefing by mentally placing yourself in your listeners' shoes and asking yourself three questions.

A briefing is like a progress report...

1. **What Are You Trying to Do?**
   The answer will help you build your *Introduction*, as you would for a formal report. Offer your listeners some background information, which may comprise

   - how your company became involved in the project (with, perhaps, a comment on your own involvement, to add a personal touch),

- exactly what you are attempting to do (in more formal terms, your objectives), and
- the extent or depth of the project (i.e. its scope).

...describe what you have done; say what you're doing now; outline what you plan to do next

2. **What Have You Done So Far?**
   This is equivalent to the Discussion section of a formal report. Your answers should cover
   - how you set about tackling the project,
   - what you have accomplished to date (work done, objectives achieved, results obtained, and so on), and
   - preliminary conclusions you have reached as a result of the work done (if the work is complete, these will be the final conclusions).

3. **What Remains to Be Done?** (or **What Do You Plan to Do Next?)**
   This question is relevant only if the project is still in progress, in which case it is equivalent to the *Future Plans* section of a written progress report. Your answers to this question should cover
   - the scope of the planned future work,
   - results you hope to achieve, and
   - a time schedule for reaching specific targets and final completion.

   If the project is complete, this question is not relevant and is replaced by an alternative question: **What Are the Results of Your Project?** The answers are then combined with the final answer to question 2, and are thus equivalent to the *Conclusions* section of a written report.

Now you have a pattern for the main part of your presentation. But, as the flow diagram in Figure 10-1 shows, you still need to start with a quick synopsis of the project in easy-to-understand terms—the equivalent of a report *Summary*—and to end by repeating the key points and the main outcome (this becomes a *Terminal Summary).* Then invite your listeners to ask questions.

Prepare to Speak

Make Speaking Notes

Prepare your speaking notes on prompt cards no smaller than 4 × 6 inches. Write in large, bold letters that you can see at a glance, and use brief headings to develop the information in sufficient detail. A specimen prompt card is shown in Figure 10-2.

The amount of information you include will depend on the complexity of the subject, your familiarity with it, and your previous speaking experience. As a general rule, the notes should not be so detailed that you

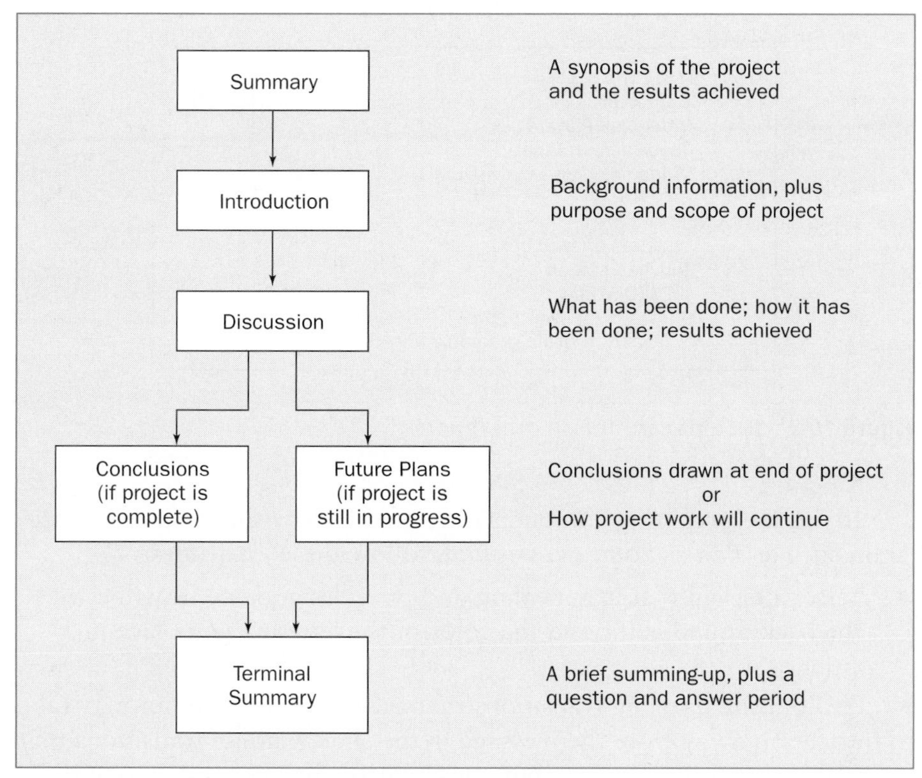

Figure 10-1 Flow diagram for a technical briefing.

Use the pyramid approach to structure your briefing

cannot extract pertinent points at a glance, nor so skimpy that you have to rely too much on your memory, which may cause you to stumble through your presentation.

Visual Aids in Presentations
www.adm.uwaterloo.ca/infotrac/visualaids.html
This site, from TRACE tip sheets at the University of Waterloo, contains a summary of the use of visual aids in presentations.

Prepare Visual Aids

Visual aids can help you give a clearer, more readily understood briefing. They may range from a series of steps listed as headings on a flip chart, through computer-generated slides for an LCD projector, to a working model that demonstrates a complex process. Here are some hints for creating effective visuals:

- Strive for simplicity: let each visual make just one point. A visual aid should support your spoken narrative; you should rarely have to explain it.
- Use large, bold letters that will be visible from the back of the room. Ideally, use upper and lower case letters rather than all capitals.
- Use color to accentuate key words or parts, but in moderation. Some colors, such as green and blue or red and orange, are difficult to tell apart from a distance.
- Place a short title above or below each slide.

Let each slide provide just one message

Write in bold letters that can be read from 18 to 22 inches distance

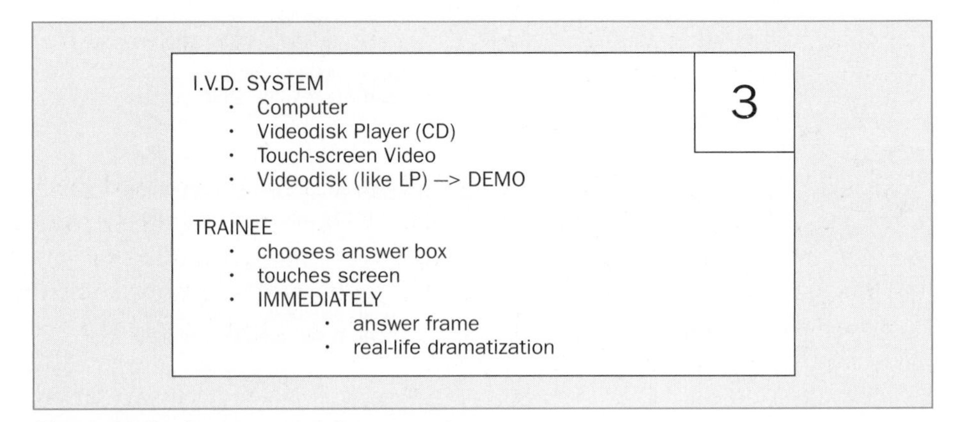

Figure 10-2 Prompt card for an oral report.

If you are preparing computer-generated slides using a software program such as *Power Point* or *Astound*, follow these guidelines:

- Select a design that is appealing and provides good contrast between the background color and the color of the lettering you place in front of it.
- Avoid using dazzling transitions that thrill the audience but may take attention away from the message in the slide. Choose transitions that make a smooth, effective, but conservative change between one slide and the next, and maintain continuity by using similar transitions throughout the presentation.
- Create a natural progression from one slide to the next. If you are presenting, say, four bulleted points on a slide, bring in each point with each successive slide. Also, when you bring in a new point, show it in dark, bold type, but put the previous points in softer, less bold type.
- Print copies of your slides onto 8 $^1/_2$ x 11 inch sheets, four or six slides to a page. You will need these as a prompt when presenting the slides.

Resist the temptation to use fancy transitions!

Chapter 9 contains additional suggestions for preparing visual aids such as graphs, charts, and diagrams.

Practice working with your visual aids, first on their own and then as part of the whole presentation. This gives you a chance to check whether you have keyed them in at the correct places, and whether the entries in your notes are sufficiently clear to permit you to adjust from speech to a visual aid and then back again without losing continuity.

Don't overlook the practical aspects of the briefing. If you have equipment to demonstrate, consider its layout in relation to the sequence of your presentation. Try to arrange the briefing so that you move progressively from one side of the display area to the other, instead of jumping back and forth. If the display is large and easy to see, let it remain unobtrusively at the back of the area. If it is small, consider moving it forward and talking from beside it.

Practice, Practice, Practice

Take a leaf from the experienced technical speaker's notebook and practice your briefing. Run through it several times, working entirely from your prompt cards, until you can speak without undue hesitation or stumbling over awkward words. If the cards are too hard to follow, or contain too much detail, amend them. Then ask a colleague to sit through your demonstration and give critical comments.

Modify your notes after each practice reading. Where necessary, insert more information; in other places, delete unneeded words. As you grow familiar with the notes you will find that your confidence increases and certain sentences and phrases spring readily to mind at the sight of a single word or topic heading; this will help you to maintain oral continuity.

Practice...practice... practice

Time yourself each time you rehearse your presentation. Aim to speak for slightly less time than allowed; for example, plan to speak for 17 or 18 minutes for a talk scheduled to last 20 minutes. This will give you time to include some previously unanticipated remarks, should you want to do so at the last minute.

Prepare Handout Notes

During the planning stages you need to decide whether to provide printed handouts for your listeners. If so, you will also have to decide whether they should be copies of your LCD slides (or overhead transparencies), or a specially prepared narrative-style record of the main topics you will cover.

There is a trend today for speakers to provide only copies of their slides, which are much simpler to make. However, we recommend you prepare a summary of your presentation which, although it may take longer, will provide your listeners with a useful resource when, in future months, they want to refresh their memories about the topic you presented.

Now Make Your Presentation

Control Your Nervousness

There are very few people who are not at least a little nervous when the time comes to stand up and speak before an audience. Some nervous tension is perfectly normal and can even help a speaker give a better performance. You will find that, once you start speaking, your nervousness will gradually decrease. A lot depends on the quality of your speaking notes: if you have done a thorough job preparing them, know they are reliable, and have practiced using them, you will find that the familiar phrases and sentences form easily. Then you will begin to relax and so speak with even greater confidence, which will help you relax even more.

Tell Your Story Three Times

Figure 10-1 shows an opening summary, a central discussion, and a closing or terminal summary. We call this the "Tell—Tell—Tell" method of presentation: you tell your story three times:

Start by telling your audience where you plan to take them...

Tell 1: Tell your readers what they most need to hear: the key points. Then outline very briefly the main topics you will cover.

Tell 2: Now provide all the details, in the same order you mentioned them in *Tell 1*.

...end by telling them where they have been

Tell 3: Sum up by very briefly repeating the key points, and possibly offering a recommendation.

Capture Audience Attention

Grab your listeners' attention by offering an *interesting* start: tell your audience immediately where you are going to take them, and why you are going to take them there. Here are two examples:

A dull start "Today, I want to tell you about the effects of poor quality control when manufacturing electrical products. This happened in the fall of 2002, at our plant in Dayton, Ohio. The problem began when a supplier failed to monitor production quality adequately..."

Get off to a flying start!

An exciting start "Have you, in your company, ever built a better mousetrap? A product that significantly outclasses the competition? Well, we did, in the spring of 2002. It was an immediate success and we had to start a second production line to keep up with the demand. But five months later we had a disaster on our hands: warranty returns were reaching an unprecedented 30%!! The reason: poor quality control at one of our suppliers' plants..."

Sharpen Your Platform Manner

Knowing some elementary platform techniques can help improve your performance. There are ten:

1. Arrive early and check that the computer, LCD projector, and microphone work. Simultaneously, identify the light switches you will need to control, if you have to dim the lights when presenting video programs or LCD slides.
2. Appear businesslike and cheerful.
3. Speak from notes. You will lose contact with your audience if you read from a prepared speech.

Be an energetic, enthusiastic speaker

4. Let your enthusiasm *show*. If you present your information vigorously, your audience will see that you really enjoy talking about your subject and will listen more attentively.
5. Look at your audience. Try to speak to individuals in turn, rather than the group as a whole. Pick out someone in one part of the room and talk

to that person for a few moments, then turn to someone in another part of the room. Let each listener feel he or she is being addressed personally.

6. Use humor sparingly, and only if it fits naturally into your presentation. Never insert a joke to "warm up" an audience. If you do use humor, make sure your audience is laughing *with* you, not *at* you.
7. Speak at a moderate speed. We recommend 120 to 140 words per minute.
8. Speak up. If possible try speaking without a microphone, since this gives you much greater freedom of movement and tonal flexibility. If the room is large and you have to use one, try to obtain a lavaliere (traveling) microphone that clips onto your clothing and has a long cord. Better still, ask for a radio microphone.
9. Pause occasionally to study your speaker's notes. Never be afraid to stop speaking for a few moments to consolidate your position and establish that you have covered every major topic.
10. Avoid distracting habits that divert audience attention. For example, avoid pacing back and forth or balancing precariously on the edge of the platform (the audience will be far more interested in seeing whether you fall off than in following your topic). Also avoid nervous behaviour, such as jingling keys or coins in your pocket (put them in a back pocket, out of reach), playing with objects on the speaker's table (remove them before you start speaking), or cracking your knuckles.

Watch your body language; make it work for you, not against you

At what point should you distribute your handout notes? There are three approaches:

1. If your handouts are simply copies of your slides, then hand them out right at the start of your presentation. Listeners can then make notes on them while you speak.
2. If you have charts or diagrams you want your listeners to refer to as you speak, hand them out at the appropriate moment during your talk.
3. If your handouts are a detailed narrative-style description of the points you will be making, then hand them out toward the end or immediately after your presentation. (If you hand them out at the start, your listeners will tend to leaf through them to identify "where you are" in your talk, which means they create a disturbance for their neighbors and irritate you because you will have lost their attention.) Tell your listeners at the start that they will be receiving detailed handouts, so they will know they do not need to take notes.

Reach Out to Your Audience

Although good pre-platform preparation and knowledge of platform techniques can give you confidence, they are not sufficient in themselves to break down the initial barrier between speaker and audience. Successful speakers develop a well-rounded personality they use continuously and unconsciously to establish a sound speaker-audience relationship. When you speak to an audience only once, and then only briefly, probably the most important attributes to develop are enthusiasm and sincerity: enthusiasm about your topic, and sincerity in wanting to help your audience learn about it.

The time and effort you invest in preparing for a briefing will depend on your confidence as a speaker and your familiarity with the subject. The more confident you are, the less time you will need to prepare. No one expects you to give a fully professional briefing at your first attempt, but your listeners (and your employer) will appreciate your efforts when they see that you have prepared your talk carefully and are presenting it in an interesting way.

Speaking at a conference can create even greater stress

The Technical Paper

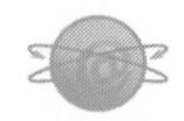

Giving a Scientific Talk: A Guide for Botanists
www.botany.uwc.ac.za/sciwriting/talks.htm
In this article, Derek Keats and Alan Millar argue that oral presentations are an inefficient means of giving information because little of it is retained. Such presentations, however, are effective for stimulating discussion, constructive criticism, and interest among colleagues and students. The authors describe the primary features of a good talk and provide detailed suggestions for creating audiovisual aids suitable for scientific presentations.

Chapter 8 discussed the steps Mickey Wendell would have to take to publish a magazine article or technical paper. (He is a senior lab technician who has discovered that an additive called Aluminum KL mixed with cement in the right proportions produces a concrete with high salt resistance.) This chapter assumes that the papers committee of the Combined Conference on Concrete liked Mickey's abstract and summary, and the chairperson of the committee has notified him that his paper has been selected for presentation at the forthcoming conference. Mickey has four months to prepare for it.

Presenting a paper before a society meeting is more demanding than delivering the same information at a technical briefing. The occasion is more formal, the audience is usually much larger, and the speaker is working in unfamiliar surroundings. Yet the guidelines for preparing and presenting a technical briefing still apply.

Never read your paper to the audience: the result will be a dull, monotonous delivery that can turn even a superior technical paper into a dreary, uninteresting recital. The key is to start preparing early, to make good speaker's notes, and to practice speaking from them.

The spoken version of a technical paper does not have to cover every point encompassed by the written version. In the 15 to 20 minutes allotted to speakers at many society meetings, there is only time to present the highlights—to capture listeners' interest, so they will want to read the published version.

Before presenting his paper, Mickey Wendell needs to prepare three versions of his speaking notes. The first will consist mainly of brief topic headings derived from the written paper. He should jot these headings onto a sheet of paper and then study them with four questions in mind:

1. Which points will prove of most interest to the audience?
2. Which are the most important points?
3. How many can I discuss in the limited time available?
4. In what order should I present them?

When Mickey wrote his paper, he was preparing information for a reader. Now he is preparing the same information for a listener and the rules that guided him before may not apply. The logical and orderly arrangement of material prepared for publication is not necessarily the order listeners will find interesting or easy to understand.

Focus attention on the paper's most interesting features

Mickey may assume that the audience at a society meeting is technically knowledgeable, has some background information in the subject area, and is interested in the topic. Most of his audience will likely be civil engineers and technologists, with a sprinkling of sales, construction, and management people. He must keep this in mind as he examines his list of headings, identifies which points he intends to talk about, and writes his speaking notes in the order he feels will most suit his listeners (see Figure 10-2).

(Mickey should avoid writing his speaking notes in the margin of his printed paper. The sequence may seem illogical to his audience and, if he is nervous, he may be tempted to read rather than speak extemporaneously.)

Next, Mickey needs to practice speaking from his notes, just as he would for a technical briefing. It will not be enough to scan or read the notes to himself and assume he is becoming familiar with them. He must speak from them out loud, as though he is presenting the paper to an audience.

Then, when he has practiced enough and feels his preliminary notes are satisfactory, Mickey can prepare his final speaking notes. These he should keystroke with a minimum 14 pt typeface, or handletter them (in black ink) in clearly legible letters.

Prepare well in advance; then relax, knowing you are ready

Mickey should plan to have his final notes ready at least three days before leaving for the conference. To a certain extent, the headings in the first set of notes have helped to trigger familiar phrases and sentences. Now he has to familiarize himself with the new pages. Ideally, during these last practice sessions he will hold a full dress rehearsal, presenting his paper before some of his colleagues, with someone qualified to comment on his platform techniques among them. If this is not possible, he should at least try speaking the paper alone, standing at a rostrum or desk to simulate actual conditions. This "dry run" will also give him a final opportunity to check how long it takes to present his paper.

Taking Part in Meetings

We all occasionally attend meetings. In industry, you may be asked to sit on a committee set up for a multitude of reasons, from resolving technical problems that are tying up production to organizing the company's annual picnic. The effectiveness of such meetings is entirely controlled by those taking part. Meetings attended by people *aware of their role* as participants can move quickly and achieve good results; those attended by individuals who seize the opportunity to air personal complaints can be deadly dull and cripple action.

Meetings can be either structured or unstructured, depending on their purpose. A structured meeting follows a predetermined pattern: its chairperson prepares an agenda that defines the purpose and objectives of the meeting and the topics to be covered. The meeting then proceeds logically from point to point. An unstructured meeting uses a conceptual approach to derive new ideas. Only its purpose is defined, since its participants are expected to introduce suggestions and comments that may generate new concepts (this approach is frequently called "brainstorming"). We will discuss the structured meeting here, because you are much more likely to encounter it in industry.

The chairperson may run the meeting, but the participants control its progress

A meeting is composed of a chairperson and two or more meeting participants, one of whom often is appointed to be secretary for that particular meeting. (The secretary makes notes of what happens during the meeting and, after the meeting, writes the "minutes," or meeting record.) Each person's role is discussed here.

The Chairperson's Role

Good chairpersons are difficult to find. A good chairperson controls the direction of a meeting with a firm hand, yet leaves ample room for the participants to feel they are making the major contribution to the meeting. The chairperson must be a good organizer, an effective administrator, and a diplomat (to smooth ruffled feathers if opinions differ widely). Much of the success of a meeting will result from the chairperson's preparation before the meeting starts and ability to maintain control as it proceeds.

Prepare an Agenda

Approximately two days before the meeting the chairperson should prepare an agenda of topics to be discussed and circulate it to all meeting participants. The agenda should identify three things:

- The date, time, place, and purpose of the meeting.
- The topics that will be discussed (numbered, and in the sequence they will be addressed), divided into two groups:

600 Deepdale Drive, Phoenix, AZ 85007

To: Members, E-Learning Research Committee

The monthly meeting of the E-Learning Research Committee will be held in conference room B at 3 p.m. on Friday, September 19, 2003. The agenda will be:

Decision Items:

1. Accelerated completion of the e-learning program for the DPS-2A ultra-narrowbeam system installation. (R. Taylor)
2. Purchase of off-the-shelf e-learning software. (W. Frayne)
3. Proposals for papers: 2004 International E-Learning Conference and Symposium, Atlanta, GA (D. Thomashewski)

Discussion Items:

4. Report on investigation into significance of streaming e-learners according to their ability. (C. Bundt)
5. Proposal for beta-testing completed e-learning modules. (R. Mohammed)
6. Plans for Annual Research Division banquet. (C. Tripp; J. Kosty)
7. Other business: (Please send topics to me by 10 a.m. Thursday, September 18.)

This month's meeting secretary: J. Kosty.

Daniel K. Thomashewski

Daniel K. Thomashewski
Chairperson, ELR Committee
September 13, 2003

***Name* the person who will speak about each particular topic**

Figure 10-3 An agenda for a meeting.

Place the most important points at the top of the agenda

1. Decision Items
2. Discussion Items

(This arrangement ensures that participants deal with the most important items early in the meeting.)

- The person who is delegated to record the meeting's minutes.

A typical agenda is shown in Figure 10-3.

Run the Meeting

The chairperson's first responsibility is to start the meeting on time: a person with a reputation for being slow in getting meetings started will encourage latecomers. The second responsibility is to keep the meeting as short as possible without seeming to "railroad" decisions. The third responsibility is to maintain control.

The meeting should be run roughly according to the rules of parliamentary procedure. (Since most in-plant meetings are relatively informal, full parliamentary procedure would be too cumbersome.) The chairperson should introduce each topic on the agenda in turn, invite the person specializing in the topic to present a report, then open the topic for discussion. The discussion offers the greatest challenge for the chairperson, who must

An effective chairperson encourages reluctant contributors, dampens over-enthusiastic ones

- permit a good debate to generate among the members, yet steer a member who digresses back to the main topic,
- sense when a discussion on a subject has gone on for long enough, then be ready to break in and ask for a decision, and
- know when strong opinions are likely to block resolution of a knotty problem, and assign one person or a small subcommittee to investigate further and present the results at the next meeting.

Sum Up

Effective "summing up" helps the secretary identify what should be said in the minutes

Before proceeding from one topic to the next, the chairperson should summarize the outcome of the discussion on the first topic. The outcome may be a general conclusion, a consensus of members' opinions, a decision, or a statement of action defining who is to do what, and when. In this way all members will be aware of the outcome, and the secretary will know what to enter into the minutes.

The chairperson should also sum up at the end of the meeting, this time reviewing major issues that were discussed and the main results. At the same time, the chairperson should point the way forward by mentioning any important actions to be taken and the time, date, and place of the next meeting (presuming there is to be one).

The best way to learn to be a good chairperson is to watch others undertake the role. Study those who seem to get a lot of business done without appearing to intrude too much in the decision making. Learn what you should not do from those whose meetings seem to wander from topic to topic before a decision is made, have many "contributors" all speaking at the same time, and last far too long.

The Participants' Role

You can contribute most to a meeting by arriving prepared, stating clearly your facts, ideas, and opinions when called on, and keeping quiet the remainder of the time. If you observe these three basic rules you will do much to speed up affairs.

Come Prepared

If a meeting is scheduled to start at 3:00 p.m. do not wait until 2:30 to gather the information you need. Arriving with a sheaf of papers in your hand and shuffling through them for the first 15 minutes creates a disturbance and causes you to miss much of what is being said. Start gathering information as soon as the agenda arrives, sort your information to identify the specific items you need, then jot down topic headings.

Plan your presentation like a technical briefing

Preparation becomes even more important if you have been researching data on a particular topic and will be expected to present your findings at the meeting. Start by dividing your information into two compartments:

1. Facts your listeners *must* have if they are to fully understand the case you are making and reach a decision (if, for example, you are requesting their approval to take a specific course of action). These become "Need to Know" facts.
2. Details your listeners only *may* be interested in, and do not necessarily need to understand your case or reach a decision. These are "Nice to Know" facts.

Plan to present only the **Need to Know** facts in your prepared presentation, but have the **Nice to Know** facts ready in case one of the listeners asks questions about them.

The second step is to examine the Need to Know facts to identify the two or three pieces of information your listeners will *most* need to hear. This will become your Main Message, or Summary Statement.

The three compartments form a pyramid-style speaking plan, as illustrated in Figure 10-4.

Be Brief

In your opening remarks summarize what your listeners most need to hear from you (your Main Message), and then follow immediately with facts and details from the Need to Know compartment. If you have statistical data to offer, either project transparencies or computer-generated images onto a screen or print copies and distribute them when you begin to speak.

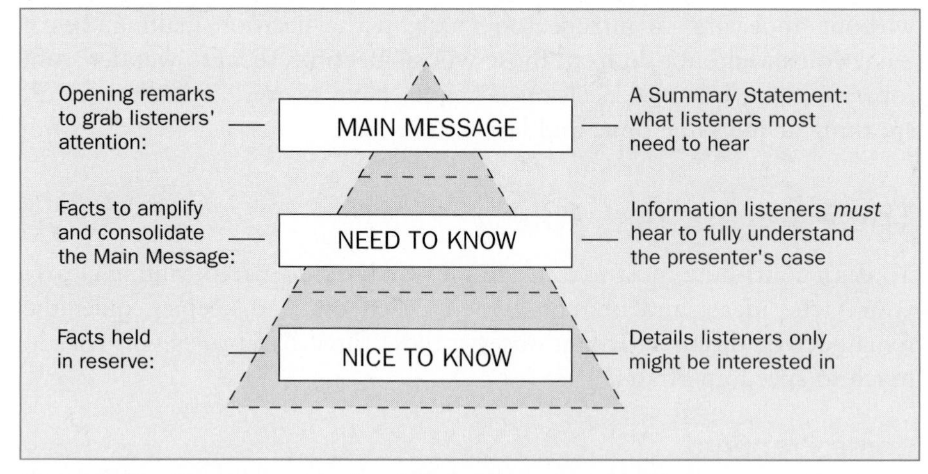

Figure 10-4 Plan for presenting information at a meeting.

Structure your information like a written report or proposal

(If you have a lot of information to distribute, print copies ahead of the meeting and ask the chairperson to distribute them with the agenda, so that everyone can examine your data before coming to the meeting.)

Keep your cool; defend your point logically and rationally

At the end of your presentation, invite questions from the meeting participants. For your answers, draw on the data you have prepared for the Nice to Know compartment. If some listeners seem to resist your ideas, try to avoid becoming defensive. Say that you can see their point of view, and then explain why your approach is sound or offers a better alternative than the one they may be suggesting. In particular, avoid getting into a personal confrontation with one or more of the meeting participants. And when the chairperson calls on the members to indicate whether they accept your ideas or will approve your proposal, if they respond negatively be ready to accept their decision gracefully.

Contribute Constructively

Many parts of a meeting require you to be only an interested observer. At these times keep quiet unless you have a relevant question, additional information to contribute, or an educated opinion to offer. Discussion of a topic should be a one-to-one conversation between you and the chairperson, or sometimes between you and the topic specialist. It should never become a free-for-all with each person arguing a point with his or her neighbor.

The Secretary's Role

Sometimes an administrative assistant is brought in to record the minutes of a meeting, but more often the chairperson appoints one of the participants to take minutes. If you happen to be selected, you should know how to go about it.

600 Deepdale Drive, Phoenix, AZ 85007

E-Learning Research Committee

Minutes of Meeting

Friday, September 19, 2003, 3:00 p.m.

In Attendance:	C. Bundt	R. Mohammed
	W. Feldman	R. Taylor
	W. Frayne	D. Thomashewski (Chair)
	J. Kosty (Secretary)	C. Tripp
Regrets:	D. Wilton	

Minutes	*Action*
Decision Items:	
1. The e-learning program for the DPS-2A ultra-narrowbeam system is incomplete. Changes had to be made to the storyboard to reflect learning difficulties experienced during usability testing of module 4. The program is now scheduled for completion on October 16.	***R. Taylor***
2. The committee approved W. Frayne's proposal to purchase three copies of the Telesat software Version 4.3 at a total cost of $2200. The cost will be drawn equally from the E-Learning Research and the Operating budgets.	***W. Frayne*** ***R. Mohammed***
3. Three papers are to be submitted to the program committee for next year's International E-Learning Conference and Symposium. Deadline for proposals is November 30.	***C. Bundt*** ***W. Feldman*** ***J. Tripp***
Discussion Items:	
4. C. Bundt completed Phase 1 of the investigation into streaming e-learners on September 12. She estimates Phase 2 will be complete on October 2.	
6. The Research Division banquet will be held in collaboration with the annual Awards Dinner on March 13, 2004. The banquet committee will establish a joint plan with the Awards Committee.	***C. Tripp*** ***J. Kosty***

J Kosty

J. Kosty, Secretary

Minutes should reflect key outcomes, not describe what everybody said

Figure 10-5 Minutes of a meeting. Note the "Action" column, which draws individuals' attention to their post-meeting responsibilities. (The dotted line indicates a break between two pages and some omitted information.)

Recording minutes does not mean writing down everything that is said. Minutes should be brief, so there is only room to mention the highlights of each topic discussed. Items that must be recorded are

- main conclusions reached,
- decisions made (with, if necessary, the name[s] of the person[s] who made them, or the results of a vote),
- actions agreed upon, and who is to take the action and by when,
- what is to be done next, who is to do it, and by when, and
- the exact wording of any policy statements derived during the meeting.

The best way to get this information quickly is to write the agenda topics on a lined sheet of paper, spacing them about two inches apart vertically. In these spaces jot down the highlights in note form, leaving room to write in more information from memory immediately after the meeting.

Write the minutes while the events are still fresh in your mind

The completed minutes should be distributed to everyone present, preferably within 24 hours. They should be a permanent record on which the chairperson can base the agenda for the next meeting, and participants can use as a reminder of what they are supposed to do. The format shown in Figure 10-5 also provides an "action" column to draw participants' attention to their particular responsibilities.

ASSIGNMENTS

Speaking situations you are likely to encounter in industry will develop from projects on which you are working. Hence, assignments for this chapter are assumed to grow naturally out of the writing assignments presented in other chapters.

Project 10.1: Speaking Situations Evolving from Other Projects

You have to attend a meeting to present the results of a study or investigation you have carried out, at which you will brief managers or a client on your findings and recommendations. In each of the following instances, which are drawn from projects in Chapters 3, 4, 5, 6, and 7, list in point form the information you would convey

- as your "Main Message," and
- as your "Need to Know" details.

In some projects—mainly in Chapters 3 and 4—you will be able to draw on the details provided in the assignment instructions without first doing the report writing project itself. In others—primarily in Chapters 5 and 6—it will help if you have first completed the study and written the report.

Be ready to present this information orally.

1. Project 3.1
Two government representatives involved with Environmental Studies Contract WM-23357 are to visit H. L. Winman and Associates next week. Your manager asks you to return to the office to attend a meeting with them to explain that the two-week delay in completing the lake level measurement program has been caused by faulty manufacture of the water stage manometer's spring and drive assembly, not by poor operation or slow work.

2. Project 4.1
When you return to your office after checking the damaged crates with Noella Redovich, there is a request from manager Vern Rogers among your email messages: "Please come in as soon as it's convenient and describe your findings."

3. Project 4.3
You drive back to the office to pick up a van to replace the one that was damaged during your drive to the Mooswa River construction site. Your manager—M. B. Corrigan—telephones and asks you to see him before you return to Mooswa River. "Please describe the incident to me," he says when you enter his office. "I need the details for when I talk to Meadows Electronics (where you rented the video camera), and the insurance company."

4. Projects 4.4 and 7.2
The project manager from Terrapin Control Systems flies in to your city a few days after the power outage that caused you to discard the electronic switches in Batches 87H and 84C. Your manager—J. H. Grayson—asks you to join him and the Terrapin project manager for a 2 p.m. meeting at which you are to describe the cause of the problem and what steps you will be taking to prevent a recurrence. "It's essential we assure him there will be no further work stoppages," Grayson says. "We've been told privately that Terrapin Control Systems is issuing an even bigger contract to us, so we must be sure they understand we have taken positive action to prevent further delays."

5. Project 4.5
Maintenance crew supervisors from microwave sites 1 through 8 are attending a meeting at H. L. Winman and Associates. Because you have just returned from microwave site 14, where you have been investigating connector problems, Andy Rittman asks you to come in to the meeting and brief the supervisors on your findings.

6. Project 5.1
Robert Delorme telephones you and asks, "Have you finished the Quillicom landfill study?" You tell him you have, but you have not yet

written the report. "Then I want you to go to the Quillicom Town Council meeting at 7:30 p.m. tonight. The councillors want to hear what recommendations you will be making."

7. Project 5.2
Paullette Machon (vice president of operations at Baldur Agri-Chemicals—BAC) telephones to say she will be coming to your office tomorrow and bringing BAC's manager of human resources with her. Rather than wait for your report to reach them, they want you to brief them on your findings into BAC's power house problems, and then they want to discuss the implications with you.

8. Project 6.1
Highways engineer Carlos Alvarez asks you to brief his highway engineers on the results of your highway paint study.

9. Project 6.2, Part 3
When you deliver your sound level study to Trudy Parsenon, you tell her there is a problem at Mirabel Realty and it's going to cost a fair sum to remedy it. She telephones you two days later. "Three people from head office will be here next week on a routine visit. I'd appreciate it if you could come in one afternoon to describe your findings to them. You'll do a better job than me, because you know the study better than I do."

10. Project 7.1
You have written the proposal for purchasing handheld computers and you handed it to Wilson Harcourt three days ago. Today he emails you with this message: "I want you to present your idea to purchase handheld computers in person at next week's Capital Budget meeting. The committee may want to know specifically which model you are recommending and why it is your preferred choice."

11. Project 7.3
When you deliver your proposal to Student Council president Sharon Gilchrist, she says: "It's a good idea. I'd like you to come to the next meeting and present the proposal in person."

Project 10.2: Informing Technicians of a New Product

Make an in-depth presentation to unknowledgeable listeners

You have researched information on a new manufacturing material, method, or process, as described in Project 8.4, and have prepared a written description. Now you have to inform other technicians about the product at a lunchtime briefing organized by your department head.

Part 1
On a sheet of paper write brief notes in point form identifying what you will say under each of the following topic headings:

- Summary Statement
- Purpose (of product, material, or process)
- Details (what it does and why it is unique)
- Conclusions

Part 2
Make your presentation.

Chapter 11
Communicating with Prospective Employers

As head of the administration and personnel department of H. L. Winman and Associates, Tanys Young is responsible for hiring new staff. Recently she advertised for an engineer to coordinate a new project, and received 148 applications from across the country. Since it would have been impractical to interview all the applicants, she narrowed the field down to the nine applicants she felt had the best qualifications, basing her selection on the information contained in the 148 application letters and resumes she received.

One of the applicants was Eugene Koenig of Indianapolis, Indiana, who believed—quite rightly—that he was probably the most qualified person for the job. But his name was not one of the nine on Tanys's short list, so he was not interviewed. Tanys has since met the nine selected applicants and offered the job to the most promising person. She will never know that Eugene would have been a better person to hire. And Eugene will never know why he was not considered.

What went wrong? The fault was entirely Eugene's, whose letter and resume failed to persuade Tanys that she should talk to him.

In today's highly competitive employment market, job seekers have to tailor each resume and application letter so that together they capture the interest of a particular employer. (To mail or email copies of an identical resume and similar letter to every employer is a wasted effort.) Job seekers must carefully orchestrate the whole employment-seeking process, from preparing their resumes to presenting themselves personally at an interview.

This chapter describes the various stages of the job-seeking process, and the careful steps that you, as an applicant, must take if you want every employer you contact to consider you seriously as a potential employee. The process starts with preparing a personal data record, then proceeds through preparing a resume, writing an application letter, completing an application form, attending interviews, and accepting or declining a job offer.

The Employment-Seeking Process

Figure 11-1 illustrates the five steps a *successful* applicant has to take before being employed. At each step the number of contenders for a particular job is reduced until only one person remains. Tanys Young received letters and resumes from 148 potential employees, but asked only 25 of them—those she felt most nearly met the company's requirements—to complete a company application form. Thus in one step she cut the field by one-sixth. Then, from the application forms and resumes, she selected nine people to interview.

The significant factor here is that Tanys narrowed the field down to only 6% of the original applicants *based solely on their written presentations*. Like Eugene Koenig, you cannot afford to be one of the over 94% who were eliminated from the interview stages because of inadequate written credentials.

The image you convey—on paper or online—has a major impact on a potential employer

The five steps of the job-seeking process are outlined below. At each step you have to present a confident, positive image of yourself if you are to proceed to the next step.

1. **Initial Contact.** Your first step as a job seeker is to approach a prospective employer and ask to be considered for employment, either by responding to an advertisement (this may be over the Internet) or by approaching the employer "cold" (in the hope that the employer either has or shortly will have an opening). You may make this initial contact by presenting yourself at the employer's door, by writing a letter or email message, or by telephoning. The personal visit and the letter or email are better, because they let you *place your resume in the employer's hands*.

A resume may be submitted electronically

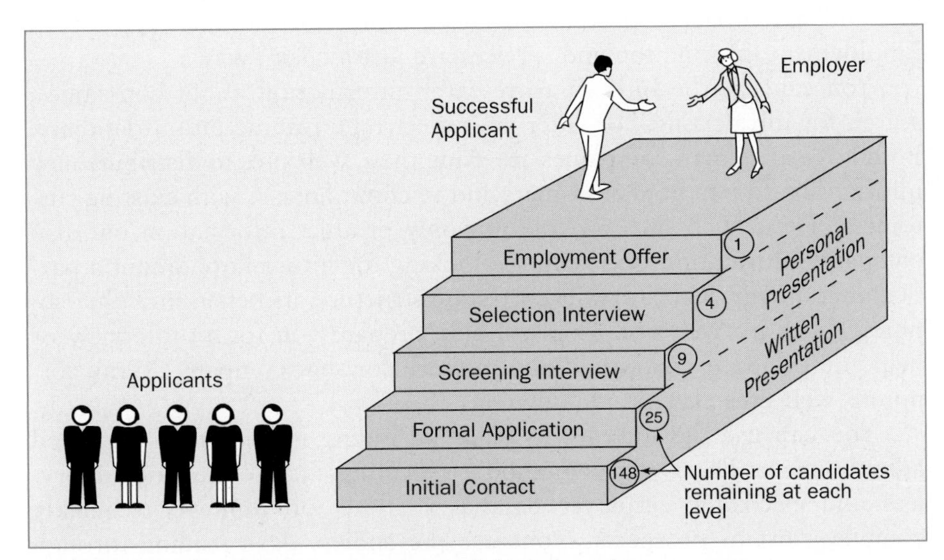

Figure 11-1 The five stages or steps of the job-seeking process.

2. **Formal Application.** Often the employer will ask you to complete a company application form so that all applicants are documented in the same way.
3. **Screening Interview.** The first interview you attend helps the employer identify which applicants have the strongest potential. In a large firm such as H. L. Winman and Associates, the screening interview may be conducted by only one person, usually the employment manager or an employment representative.
4. **Selection Interview.** The most promising candidates are asked to attend a second interview. This time the manager of the department where the successful applicant will work is also present, and is sometimes accompanied by technical specialists.
5. **Job Offer.** The employer makes a formal offer of employment to the successful applicant, often by telephone first and then by letter. The applicant responds, also by telephone and letter, to confirm his or her acceptance.

Not all employment-seeking processes develop exactly along these lines. Sometimes an applicant will obtain a company application form beforehand and submit it in step 1, with his or her letter and resume. At other times there may be only one job interview or, in some cases, there may be three or more. With advances in technology and the Internet, the initial contact may be made by the company after retrieving your electronic resume from a database.

Using the Internet in Your Job Search

The Internet has opened up a world of opportunities and possibilities to people seeking employment. It provides a completely new way for employers and potential employees to meet and learn about each other. Employment information and services are just a click away.

You research corporate information online...

You can use the Internet to research information about companies, search for job postings, or enter your electronic resume into a database. Both large and small companies are using their Web sites to distribute new information to potential customers and to communicate with existing customers. These Web sites provide not only product information but also company information, so if you are looking for information about a particular company, its history, its corporate structure, its beliefs and philosophies, check its Web site. This will help prepare you for an interview or help you decide if you would like to work for that company. Many corporate Web sites also list job openings.

...or announce your availability as a potential employee

You can use the Internet to place an online international classified advertisement. This is a new method for locating a job: online job ad services and electronic employer databases allow job hunters to quickly compile a list of prospects. Gone are the endless days reading through newspapers and directories!

Companies from around the world in all types of industries use the Internet to advertise employment opportunities. Search capabilities allow you to narrow your criteria to specific locations or job requirements. There are also a number of commercially run Internet services that specialize in matching people to positions, and nonprofit Bulletin Board networks that allow anyone to list or look at job openings.

Another way to use the Internet in your job search is to post your resume so potential employers can find you. It is not only a quick and effective way of transmitting your credentials, but also demonstrates your understanding of and ability with the new technology. When H. L. Winman and Associates were looking for a computer services manager, they turned to the Internet and found Susan Jenkins. Since one of the computer services manager's responsibilities would be to give the company a presence on the Web, looking for someone presenting themselves on the Web was a good place to start the hiring process.

Consequently, we recommend using an Internet resume service as the best way to have your electronic resume seen. These companies specialize in organizing, indexing, and distributing resumes (see the section "The Electronic Resume" later in this chapter for suggestions on preparing this type of resume). Independent database services are companies that match people with jobs.

Start by contacting an Internet resume specialist

Developing a Personal Data Record

There are three ways you can go about writing a resume: you can rely solely on your memory; you can dust off and update a previous resume; or you can create a new resume from a permanent personal data record (PDR). Using a PDR is best, because it provides a much broader information base for you to draw on.

A PDR becomes particularly useful in later years, when our ability to recall names, addresses, dates, and specific details of earlier employment diminishes. It can also be invaluable if, when calling initially on a potential employer, you are asked to complete an application form on the premises.

Start a databank: store your history online

If you do not already have a PDR, prepare one now. You may find it's a pain to start, but not difficult to update and keep current. You will need to record details (and update them approximately once a year) for four topic areas:

- Education
- Work Experience
- Extracurricular Activities
- References

Remember, this is not a resume. You will draw from this information when you begin to organize your resume. As you progress in your career you will need to remove older, unrelated experience and add the newer, most relevant ones.

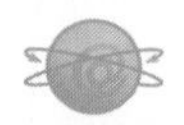

Writing in the Job Search
http://owl.english.purdue.edu/handouts/pw/#sub2
The Online Writing Lab at Purdue University has a section devoted to writing job search materials, including applications, resumes, cover letters, acceptance letters, references, and personal statements.

Education

List the schools, colleges, and universities you have attended or are attending. Start at junior high school and record the name of each school, the address and telephone number, and the dates you were there, and for high school, your graduation date and your area of specialization. For college and university, particularly note courses taken, special options, and the full name of the degree, diploma, or certificate you were awarded. Include grades or at least a grade point average (GPA) for each year. List any additional courses or seminars you completed and the date attended. Don't forget about courses taken while working for an organization.

Work Experience

For each job you have held in the past—and, if you are currently employed, the job you now hold—list

Be as detailed as possible: don't rely on your memory

- the full name, address, and telephone number of the company or organization, and the full name and title of each supervisor you worked for,
- the dates you started and finished employment and, if you held several positions within the company, the name of the position and the date you were appointed to it,
- your job title, or titles if you held several positions,
- your specific responsibilities and duties for each position, paying particular attention to the supervisory aspects and responsibilities of any job that you carried out without supervision,
- any special skills you learned on the job,
- special awards or words of praise you received, or results you achieved, and
- projects you were involved in, including the type of technology you learned or used.

Extracurricular Activities

List your activities in organizations that were not necessarily part of the job you have held or your education, but which show your participation and leadership qualities. These help identify you as a well-rounded, balanced person. For example:

- Membership in a club, society, or group, particularly noting your responsibilities as an active participant or committee member. (For example, member of sports committee or secretary of administrative committee.)
- Participation in community activities such as the Big Brother or Sister Organizations, YMCA or YWCA, YMHA, 4-H Club, Parent-Teacher Association (PTA), or local community club. Particularly describe any executive or administrative positions you held, with special responsibilities and dates.
- Involvement in a technical society on a local or national level, with particular mention of any conferences you attended or papers you presented or published.
- Participation on a sports team, with special mention of your role as a team leader or coach.
- Involvement in hobby activities such as stock car racing or rebuilding, a computer club, or dog breeding.
- Awards you have received for any activities you have been involved in.

Don't overlook your personal attributes and life experience

For each activity, include the dates of your involvement and the name, address, and telephone number of a person who can vouch for your participation. Make sure you indicate whether you were elected to the position or are doing the work voluntarily.

References

List the names of people you feel are best fitted to speak on your behalf. They fall into two groups: those who can vouch for your *capabilities* (as an employee, student, or committee member), and those who can speak for your *character*. Always contact these people first and ask if you may list them as a reference. Then, for each person write down

Be thorough: you cannot tell in advance who you may want to use as a reference

- full name, professional title (such as Plant Engineer), place of employment, and job position,
- employer's address and telephone number,
- home address, email address, and telephone number, and
- how long you have known them.

(If a reference has changed jobs, list details of both the previous and current employer.)

For each person you worked with in an extracurricular activity, also list

- the name of the organization you both were involved with, and the reference's position within that organization, and
- whether the person prefers to be called, or written to, at home or at work.

Preparing a Resume

A resume contains key information about yourself, carefully assembled and presented so that prospective employers will be impressed not only by your qualifications but also by your ability to present yourself effectively. (The correct spelling is "résumé," but common usage has made the accentless "resume" acceptable. In some countries it's called a *curriculum vitae*, or CV.)

Technical people tend to be conservative when they write their resumes, yet today's employment environment demands they be *competitive*. If a resume is to capture an employer's attention, it must display its writer's capabilities to full advantage. The resume is a sales tool and the writer is the product.

There are no specific rules about the "right" way to prepare a resume. A resume is a very personal document and should show your personality. However, there are generally accepted guidelines of what employers expect to see and the type of information they require when evaluating you as a prospective employee. This section discusses three different formats for a paper resume. Later in this chapter we discuss the various types of electronic resumes and present some guidelines for submitting resumes electronically. You can submit your resume either electronically or on paper. It really depends on what the employer or job search agency has asked for. Don't fall into the trap of creating just one version of your resume and using it both electronically and on paper. The designs are very different.

Resume Formats

Your resume style must fit your personality...

The three resumes presented on the following pages range from fairly conservative to clearly provocative. You will have to decide which you want to use, keeping four factors in mind: which will best represent you as an individual; which will best present your qualifications; which will most suit the position you are applying for; and which will most likely appeal to the particular employer. These examples are designed as paper resumes and do not follow the guidelines for electronic resumes.

For ease of reference, we will refer to the three styles as the traditional resume, the focused resume, and the functional resume. All three have one important feature in common: they open with a summary statement, often called an objective, that (1) describes the applicant's strongest qualifications from the *employer's* point of view, and (2) identifies that the writer is seeking work in a particular field. Ideally, there is a logical connection or development between these two pieces of information, and they are presented in a short paragraph of no more than two or three sentences. For example:

Objective
I have four years' experience supervising the installation and testing of wire and fiberoptic telephone communication systems, and a recent M.S. in electronics engineering with a major in fiberoptics. I am now seeking employment where I can apply my knowledge and experience in fiberoptics engineering.

An assertive statement such as this at the start of a resume draws the employer's attention rapidly to the applicant's primary experience and education, and to the employment direction the applicant wants to pursue. If the resume "hits the mark" successfully, the employer automatically reads further to learn more about the applicant.

To be of most value, the opening statement is focused to suit the needs of a particular employer, or sometimes a group of employers engaged in similar work. The implications for job applicants are far-reaching: now they have to invest much more time, care, and research into resume preparation to ensure their resumes are clearly directed toward a specific audience.

...and that of the particular employer you are contacting

Job Star Central
http://jobsmart.org/tools/resume/index.cfm
Job Star Central gives you all the information you need to write your resume, with sample cover letters, resume resources on the Web, and samples of different kinds of resumes.

The Traditional Resume

For decades the most widely recognized approach to resume writing has been to divide a job applicant's information into five parts, each preceded by an appropriate heading:

Objective
Education
Experience
Extracurricular Activities
References

The traditional resume is particularly suitable for recent university or college graduates who have limited work experience, or for students who expect to graduate shortly. Alison Witney is a biological sciences undergraduate who has held two previous jobs, totaling three years of full-time employment. Her resume is shown in Figure 11-2. Comments on the resume, plus guidelines you can use to write a resume of your own in the traditional format, are presented below and keyed to the circled numbers in the figure.

A short, concise, directed resume is welcomed by employers!

1. Job applicants with only limited work experience should try to keep their resumes down to one page.

2. Each line of Alison's name, address, and telephone number is centered to give the top of the resume a balanced appearance.

An appealing, uncrowded appearance, coupled with good words, will catch an employer's attention

1 2

Alison V. Witney
1670 Fulham Boulevard
Amiento, FL 32704
Tel: (305) 474-6318
email: avwitney@flonline.net

OBJECTIVE

To work in a position related to Animal Biology or Health Science, where I can use to good advantage both my Diploma in Biological Science and my experience as a veterinary assistant.

3

EDUCATION AND TRAINING

- Will graduate with a Diploma in Biological Science from Amiento Technical College, June, 2004 (GPA to date: 3.7).
- Graduate of Morton Stanley High School, Corisand, FL, 2000 (avg: 92.3%).

4 5

WORK EXPERIENCE

2002 to date **Animal Treatment Centre**, Amiento, FL. Veterinary assistant, responsible for reception, grooming, and exercising of animals, assisting veterinarian during operations, changing dressings, administering injections and anaesthetics, and performing administrative duties such as accounting and ordering of supplies. (One year full-time, two years part-time.)

6

2000 to 2002 **Remick Airlines**, Orlando, FL. Accounts clerk in air freight department; coordinating billings, preparing invoices, following up lost shipments, assisting clients, and writing monthly reports. For nine months assisted in payroll preparation.

1994 to date **Bar None Riding Stables**, Corisand, FL. Part-time employment teaching the care and handling of horses, and basic riding techniques, to young riders. Assisted in grooming, cleaning, feeding, and saddling-up.

7

ADDITIONAL INFORMATION

- Winner of two educational awards: Morton Stanley Science Scholarship (1999) and Amiento Technical College Biology Scholarship (2004).
- Member of YWCA since 1995, where I now teach swimming and lifesaving.
- Interests: horseback riding and jumping, swimming, and water skiing.

8

REFERENCES

The following people have agreed to supply references on my behalf:

Dr. Alex Gavin
Veterinary Surgeon
Animal Treatment Centre
2230 Wolverine Drive
Amiento, FL 32704
Tel: 474-1260
Fax: 474-1355
email: a.gavin@atlantic.vet.net

Mr. Charles Devereaux
Owner-Manager
Bar None Riding Stables
2881 Westshore Drive
Corisand, FL 32715
Tel: 632-2292
Fax: 631-3105
email: bar.none@galaxy.net

Figure 11-2 A traditional resume or biography of experience.

3. There is no need for Alison to list all the primary and secondary schools she attended; it is enough to state the name of her senior high school and the year she graduated. She should then list each college or university she attended, plus the type of course enrolled in, the diploma or degree received, and her year of graduation (or expected graduation). Alison has decided to include her grade point average (GPA) because it is high. This is optional, but if you do include it be sure to include it for all schools.

4. Experience is usually presented in reverse order, with the most recent work experience appearing first and earliest experience last. You should provide more details about recent experience (as Alison has done), and about earlier work that is similar to that of the position you are seeking, than for less-related work. Quote dates as whole years for long periods of employment, but as month and year for short periods; for example, Jun 2000–Feb 2001.

5. For each employer, state the name of the company or organization first, emphasize it with bold type, and then identify the city and state in which it is located. Then describe the position held (give the job title), and what the work involved. Particularly draw attention to the *responsibilities* of the job rather than merely listing the duties you performed. Use words that create strong images of your self-reliance, such as:

Let the words you use convey a positive impression

coordinated	organized
monitored	implemented
presented	supervised
planned	directed

(Note that Alison uses "responsible for," "administering," "coordinating," and "teaching.")

If you have held several short part-time jobs, describe them together and draw attention to the most important, like this: "Several after-school jobs, primarily as a stock clerk in a grocery store."

6. The two-column arrangement of dates and work experience is important because it gives a less crowded appearance to the page. If the job descriptions were carried to the left—under the dates—the job details would appear as heavy, less visually appealing blocks of information.

An employer wants to see "the whole picture"

7. Employers are particularly interested in an applicant's activities and interests outside normal work. They want to know if the person is more than a routine employee who arrives at 8 a.m., works until 4:30 p.m., then drives home, eats supper, and presumably watches television all evening. Information on your hobbies, interests, and participation in sports and community activities tells prospective employers that you recognize your role in society, are not too rigid or too narrow, and adapt well to your environment. Employers reason that such an applicant will make an interesting, active employee who will not only contribute to the company, but also take part in social and sports functions. Outside activities represent a balanced lifestyle and provide outlets for stress.

8. Try to draw your list of references from a cross section of people you have worked for, been taught by, or served with on committees, and ensure that their relevance is apparent (their connection to one of your previous jobs or activities must be clear). Before including them in your list, check that all are willing to act as references. We suggest you always provide references on your resume. Resumes that list "References Available on Request" make the reader take one more step. By providing them in advance you are making the selection committee's work easier and shorter.

Both Alison Witney and Colin Farrow (whose resume appears in Figure 11-3) are well aware of the important role a resume's appearance plays in a prospective employer's readiness to consider an applicant. Submit a carefully arranged and printed resume. (When printing your resume, if possible use a laser jet printer; inkjets can smudge.)

The Focused Resume

Focus your resume to match the employer's primary interest

Job applicants who have more extensive experience to describe do better if they focus an employer's attention on their particular strengths and aims. This means asking themselves what a prospective employer is *most likely to want to know* after reading their opening statement. (Probably it will be: "What have you done that specifically qualifies you to achieve the objective you have presented?") To answer, applicants must focus on their work experience rather than their education, particularly on work that is *relevant to the position they are seeking* (which means they must first research information about the company).

If their experience is sufficiently varied, then they can go one step further and divide the Work Experience section of their resume into two parts: (1) work related to the position they are seeking; and (2) work in unrelated areas. They must place all of this information *ahead* of the "Education" section, so that there is a natural flow from their Objective to their Related Experience. Thus, the parts of a focused resume are:

Objective (or Aim)
Related Experience
Other Experience
Education
Extracurricular Activities
References

Colin Farrow's two-page resume in Figure 11-3 adopts this sequence. The circled numbers beside the resume refer to the comments below.

1. Colin has sufficient information to warrant preparing a two-page resume, but he should not run over onto a third page. A third page can be used, however, if an applicant has published papers and articles or has obtained patents for new inventions, which can be listed on a separate sheet and identified as an attachment. A separate page can also be used to list references.

2. Colin's objective clearly shows his thrust toward structural engineering and his desire to obtain employment in that field.

3. The positions described within each Experience section should be listed in reverse order, the most recent experience being described first and the earliest experience described last. The most recent and most relevant experience should be described in considerably greater depth than early or unrelated experience (compare the descriptions of Colin's Northwestern Steel Constructors' experience with his Bowlands Stores' experience).

Divide your work experience into "directly related" and "less related" compartments

4. As in the traditional resume, each employer's name is listed first (in boldface type) and followed by the city and province. The person's position or job title is identified next, and then a description of what the job involved. If several positions have been held within the same firm, each is named and its duration stated so that the applicant's progress within the firm is clear. Each position should draw particular attention to the personal responsibilities and supervisory aspects of the job, rather than just listing specific duties. Verbs should be chosen carefully, so they make the position sound as comprehensive and self-directed as possible. If the paragraph grows too long, it can be broken into subparagraphs like these:

...appointed crew chief responsible for

- installing interconnection and distribution systems
- hiring, training, and supervising local labour
- ordering and monitoring delivery of parts and materials
- arranging and supervising subcontract work
- preparing progress and job completion reports.

Economize on space yet appeal to the eye

5 Single-spaced typing should be used as much as possible to keep the resume compact. At the same time there should be a reasonable amount of white space on each side and between major paragraphs to avoid a crowded effect. Although we normally recommend setting the right margin "ragged right," for Colin's resume a justified right margin does not seem too rigid. See the Electronic Resume Formats section on page 302 if you are submitting your resume electronically. The guidelines are different.

6 Education can be listed either in chronological or reverse sequence. If a resume is to be sent to another province, or if the applicant was educated in another province, he or she should identify the city and province of each educational institution attended.

7 Employers are *interested* in a job applicant's accomplishments and extracurricular activities, particularly those describing community involvement and awards or commendations. This part of a resume can be preceded by a heading such as "Extracurricular Activities" instead of "Additional Information."

8 Both people Colin has chosen as references can be cross-referenced to his previous work experience. Telephone numbers and email addresses are important, because most employers prefer to talk to rather than receive a letter from a reference.

The Functional Resume

Of the three resumes discussed here, the functional resume goes furthest in *marketing* a job applicant's attributes. For some employers its approach may seem too forthright—too blatantly "pushy"; for others, particularly employers seeking someone for a technical sales position, its approach helps demonstrate that the applicant has strong capabilities.

The functional resume is not for everyone, yet for certain people and jobs its direct approach is ideal

The functional resume is the only one to offer *opinions*: its objective identifies in general terms what the applicant believes he or she can do to improve the quality of the employer's product or service, and then follows immediately with the applicant's key qualifications—the capabilities the

Colin R. Farrow, P.E.
408 Medwin Street
St. Cloud, MN 56301
Tel: (612) 548-1612
email: c.farrow@mnonline.com

❶

OBJECTIVE

After four years comprehensive experience as an engineering technologist installing and testing transmission line towers in northern Minnesota, I returned to university where I obtained a B.S. in Structural Engineering. Now I am seeking employment where I can use my experience and education to research and test tower anchors and grouts in permafrost areas.

❷ Immediately announce your strengths and show how they can be used by the employer

RELATED WORK EXPERIENCE

June 2001 to October 2003 — **Fairborne and Warren Associates,** Consulting Engineers, St. Cloud, MN. Project engineer managing construction of microwave transmission towers and associated structures between Brainerd and Little Falls, MN, for General Telephone and Electric. Wrote specifications, coordinated and monitored contractors' work, prepared progress reports, and maintained liaison with client.

❸

June 1994 to August 1998 — **Northwestern Steel Constructors Inc.,** Lincoln, NE. Crew chief, supervising team installing high-voltage transmission line towers between Weekaskasing Falls, NE and Bismark, ND. After 30 months was assigned to assist project engineers of Ebby, Little and Company, testing concretes and grouts installed in discontinuous permafrost (10 months). For final year, appointed installation coordinator, responsible for scheduling and supervising installation crews. Resigned to attend university.

❹

OTHER WORK EXPERIENCE

January 1989 to February 1993 — **United States Air Force**, Construction and Maintenance Directorate. For first two years, member of crew installing communication systems (buildings and towers) at USAF bases between Bangor, ME, and Evansville, IN. For final two years, antenna installation and maintenance technician at USAF Lackland, San Antonio, TX. Attained rank of corporal.

❺

September 1986 to December 1988 — **Bowlands Stores**. Stock clerk in Store No. 26, Duluth, MN. (One year part-time while at high school, 1 1/4 years full-time.)

/2...

Figure 11-3 A focused resume for a job applicant with a varied background.

Colin R. Farrow – page 2

EDUCATION

- B.S. in Structural Engineering, University of Minnesota, 2001.
- Diploma in Civil Engineering Technology, Technical Vocational Institute, Minneapolis, MN, June 1994.
- Graduate (Grade 12), Henderson High School, Henderson, MN, 1987.

If applying to an educational institution, consider placing the Education section ahead of Work Experience

ADDITIONAL ACTIVITIES/INFORMATION

- Member, Association of Professional Engineers of Minnesota (APEM).
- Member, Certified Technicians and Technologists Association of America (CTTAA).
- Awarded Orton R. Smith Scholarship for proficiency in applied mathematics, Technical Vocational Institute, 1993.
- Courses attended in United States Air Force:
  * Construction Techniques, 1989.
  * Supervisory Skills Development, 1991.
  * First Aid and Safety Methods (various courses), 1990 to 1992.
- Junior Leader, Henderson YMCA, 1985 to 1988, teaching swimming and aquatic activities to boys and girls age 9 to 15. Awarded Red Cross Bronze Medallion, 1986. Lifeguard at Grand Beach, Minnesota, summers of 1986 and 1987.

REFERENCES

The following have agreed to provide information regarding my qualifications and work capabilities:

Martin G. Warren, M.S.	Philip G. Karlowsky
Projects Coordinator	Contracts Manager
Fairborne and Warren Associates	Northwestern Steel Constructors Inc.
360 Rosser Avenue	3335 Notre Dame Avenue
St. Cloud, MN 56302	Lincoln, NE 68528
Tel: (612) 544-1687	Tel: (612) 632-1450
Fax: (612) 544-1628	Fax: (612) 632-2177
email: mgw13@aol.com	email: p.karlowsk@norsteel.com

We recommend including two references, rather than writing "References available on request"

applicant believes best demonstrate that he or she is qualified to do what the objective proclaims.

To prove that the applicant's opinions are valid, the third section establishes—with clear facts and figures—what he or she has done for previous employers or organizations. This results in a revised arrangement of the resume's parts:

Objective
Qualifications
Major Achievements
Employment Experience
Education
Awards/Other Activities
References

The intent of this arrangement is to target the resume not just for a particular employer but also for a particular position. It is especially useful under two circumstances: for job applicants who have experience in marketing and want to be employed in technical sales; and for applicants who have a lean educational background but have proven and demonstrable practical experience that can be of value to a specific employer.

The resume in Figure 11-4 shows how Reid Qually uses the functional method to capture the attention of the marketing manager of a company engaged in selling cellular telephone services. The circled numbers beside his resume are keyed to the following comments.

Use subtle marketing techniques to promote yourself

1. Reid has positioned his name in the top right corner of the page because, in a pile of resumes, his name will stand out just where the person's hand is placed to flip through the pages. The line underneath his name helps draw the reader's eye to it. Reid has also saved a few lines by putting his contact address centered, all on one line. If you are gong to submit your resume electronically or if you know it will be scanned, the techniques are different. See the Electronic Resumes Formats section on page 302.

2. Reid has written his Objective with a specific employer in mind. He has heard that King Cell—a relatively new West Coast player in the cellular telecommunications field—is planning to expand and hopes to become a major provider of cellular telephone services across the country. By echoing the company's philosophy, he is almost certain to catch management's attention.

3. Reid is aware that, as soon as the personnel manager at King Cell has read his Objective, he or she is likely to think: "You have told

Reid G. Qually **1**

7 - 2617 East 38th Avenue • Seattle, WA • 98105 • Tel: 206-263-4250 • email: qually@interex.net

Reid sets the scene with a commanding, professional approach

Objective **2**

To use my proven skills in marketing to increase market share for a West Coast company providing cross-country cellular telephone services and selling cellular telephone systems.

Qualifications

I have proven capability to **3**

- Identify special-interest client groups and develop innovative marketing strategies for them.
- Create results-oriented proposals and focus them to meet specific client needs.
- Follow through with clients, both before and after a sale.
- Supervise and coordinate the efforts of small groups.
- Establish strong interpersonal relations with clients, management, and sales staff.

Major Achievements

For previous employers and organizations I have **4**

- Devised an innovative lease/purchase marketing plan for first-time customers, resulting in a 34% increase in lease agreements and a 23% increase in follow-on sales over a 12-month period (for Morton Sales and Leasing, in 2001).
- Increased sales and leases of facsimile machines by 31%, and answering machines by 26%, over a nine-month period (for Advent Communications Limited, in 2003–2004).
- Received a company-wide "Salesperson of the Year" award (from Provo Department Stores, in 1998).
- Advised and coordinated Electronic/Computer Technology students who won a nationwide IEEE "Carillon Communication Award" (for Pacific Rim Community College, 2003).

Achievements must be stated strongly and supported with specific details

/2...

Figure 11-4 A functional resume identifies in detail what an applicant feels he or she can do for a particular employer.

Resume of **Reid G. Qually** – page 2

Employment Experience

5

June 2003 to the present	**Advent Communications Inc.**, Seattle, WA. Assistant Marketing Manager, responsible for coordinating four representatives selling facsimile transmission (fax) and telephone answering equipment to commercial customers.
November 1999 to June 2002	**Morton Sales and Leasing**, Seattle, WA. Sales representative marketing fax machines and cordless telephones to business accounts and private customers.
July 1996 to October 1999	**Provo Department Stores**, Store No. 17, Portland, OR. Sales representative in Home Electronics Department. Responsible for over-the-counter sales of stereos, videocassette recorders, and portable radios.

Education

6

June 2003	Certificate in Commercial and Industrial Sales, Pacific Rim Community College, Seattle, WA (placed 2nd in course with GPA of 3.84).
1998 to 2002	Various courses in theoretical and applied electronics, at Pacific Rim Community College, Extension Division (partial credit toward electronics technician certificate).
June 1996	Graduated from Rosemount High School, Rosemount, WA.

Awards and Other Activities

7

October 2002 and November 2003	Coordinator, IEEE "Papers Night," Pacific Rim Community College, at which students of Electronics and Computer Technology presented term projects.
2000 to present	Associate Member, Institute of Electrical and Electronics Engineers Inc. (IEEE).
1999 to present	Member, Northwest Sales and Advertising Association; currently vice-president.

Information in a functional resume must be easy to find

References

8

Two people will provide immediate references; other names are available.

James B. Morton
President, Morton Sales & Leasing
330 Pruden Avenue
Seattle, WA 98107
Tel: (206) 475-3166
Fax: (206) 475-2807
email: j.morton@bconline.com

Dr. Fergus Radji
(Chairman, Seattle Section, IEEE)
Pacific West HV Power Consultants
1920 – 784 Thurlow Street
Seattle, WA 98102
Tel: (206) 488-1066
Fax: (206) 489-2722
email: radji@pacwest.net

me what you want to do. Now tell me *why* you think you can do it." So he immediately offers five reasons, each demonstrating that he can handle the job. Note particularly that

- each is short, so that the reader assimilates the information quickly,
- each starts with a strong "action" verb (i.e. *identify, create, establish*), which creates a strong, definite image, and
- each is an opinion (although not recommended for other types of resumes, opinions can be used here because Reid will follow immediately with *evidence* to support his assertions).

Opinions must be supported by solid evidence

4 Reid's evidence provides *facts*, which demonstrate he has already established a solid track record. Reid keeps each piece of evidence short and offers definitive details (i.e. percentages, names, and dates), which adds credibility to his statements.

5 Reid can keep details of his work experience short because he has already identified his major accomplishments. For each employer he provides

- start and finish dates (by month),
- employer's name (emphasized, in bold or italic letters),
- employer's location (city and province), and
- his job title and major responsibilities.

To maintain continuity, he lists his employment experience in reverse sequence.

6 Reid has only limited formal education, so he draws attention to his high grade point average (GPA) on returning to school after a long absence.

7 In a functional resume, the "other activities" section provides additional information to support statements in the Qualifications and Major Achievements sections.

An unusual yet conservative appearance can help "sell" you as a strong, imaginative applicant...

8 Reid has asked several people to act as references but lists only two, partly because they are best able to speak about his qualifications, and partly to keep his resume down to two pages.

Reid's use of bullets on page 1 and a two-column format with dates on the left of page 2 provide variety in his resume's overall layout yet continuity within each page. The bulleted items on page 1 can be read

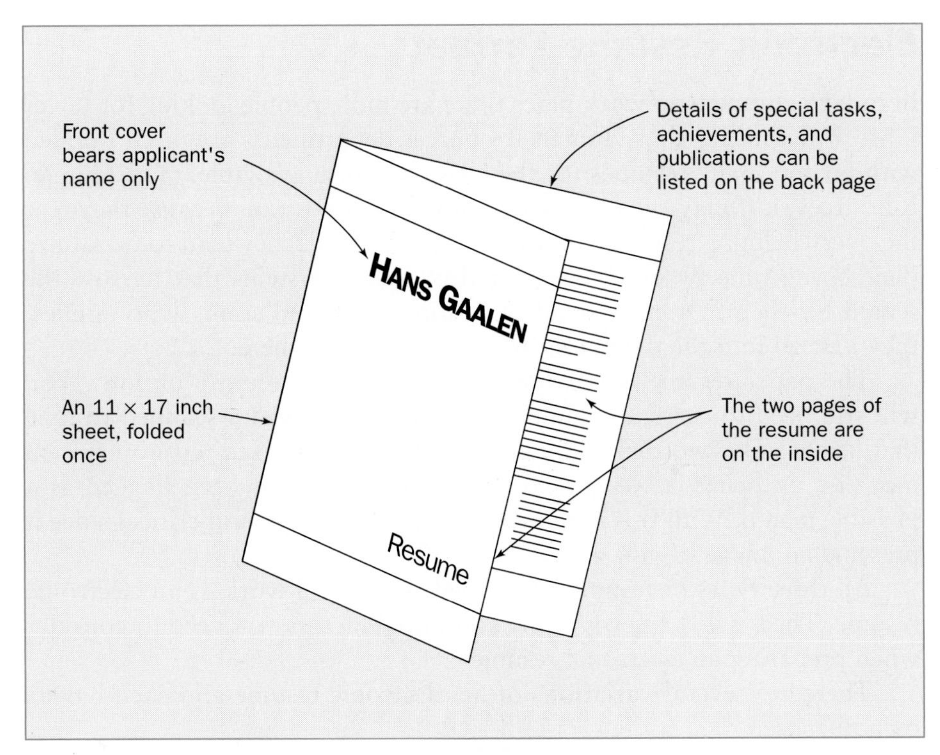

Figure 11-5 An imaginatively prepared resume.

easily—Reid wants his readers to learn quickly about him—while the facts on page 2 can be examined in more detail.

The functional resume is an effective way for a job applicant like Reid to present himself to a particular employer, but it *must* be done well if it is to create the right impact. Ideally, an applicant should use it only if he or she is confident that the employer will not be "turned off" by its non-traditional approach.

... but these techniques cannot be applied if you send your resume electronically

Never be afraid to use a display technique for your resume that will enhance its professional quality and make it stand out among other resumes. (We do not mean you should make your resume "flashy," because an overdone appearance can evoke a negative reaction from a reader.) An engineer with technical editing experience recently prepared a two-page resume that he had printed side-by-side on 11 x 17 inch (280 × 420 mm) paper, and then folded the sheet so that the resume was inside. On the outside front he printed only his name and the single word "Resume" (see Figure 11-5). On the back he created a table in which he listed the major projects he had worked on and, for each, itemized his degree of involvement. When employment managers placed his resume among other resumes submitted for a particular job opening, its professional appearance captured their interest and resulted in his being called in for more interviews than he had anticipated.

Electronic Resume Formats

If you submit electronically be aware of the different requirements

In today's competitive work place there are more people looking for fewer jobs, which means that Human Resources departments are often flooded with far too many resumes for the jobs they have available. (The Human Resources staff may even hesitate to advertise a position because they fear the overwhelming response they might get!) Instead they tend to maximize their hiring time by using automated computer systems that narrow the search for them. Often, instead of posting a job and seeing who applies, they first go into the databases and see who is out there.

A smart job seeker uses every avenue to "record a hit"

The paper resume is *not* obsolete: you just have more options. You will still need to carry a resume to the interview. Even a scanned resume that has been keyworded and indexed for a computerized retrieval system may end up being viewed or downloaded, once it has been flagged as a possible match. With this in mind, your word choice is still critical since it presents an image of you.

All three types of resumes described earlier will work as an electronic resume. There are, however, some additional factors you need to consider when preparing an electronic resume.

There are several variations of an electronic resume and each has its own purpose:

Translating Resumes for the Internet
www.nytimes.com/library/jobmarket/0107sabra.html
This "Careers" article from *The New York Times*, gives practical advice about how to design an electronic resume.

If you are sending your resume...	We suggest this format...
As part of an email message	ASCII plain text
To post to a database	ASCII plain text with keyword summaries
As an attachment to an email message	A word processed document in ASCII rich text format (RTF)
To be scanned into a database	Follow the guidelines listed below

It never hurts to follow-up with a paper copy of your resume.

Plain Text Resumes

Although plain ASCII text is not very appealing to look at, it is the safest way to transfer electronic information to guarantee your document converts properly. Special formatting, boldface, italic, indents, and bullets, for example, are not used. Figure 11-6 shows the resume Susan Jenkins emailed to an agency knowing it would be posted to their database.

You can use any word processing program to create your ASCII plain text document. After you have typed your resume (using formatting with bullets, bold, etc), follow these steps:

1. Select all of the text and change it to a non-proportional font, such as Courier 12. This will give you 65 characters per line, which will accommodate most email programs.

```
SUSAN R. JENKINS
517-210 Oliva Crescent
Batavia, NY 14020
Tel: (585) 438-0761    email: s.Jenkins@interact.net

KEYWORD SUMMARY
Engineering firm. Computer services. Computer systems design. Computer maintenance. Computer
specialist. Computer integration. Computer engineering. BS Computer Science. RIT. Manager.
Web site design. Consulting. UNIX. C++, C, SQL. HP-UX.

OBJECTIVES
To obtain a position as a computer services consultant/coordinator for a major engineering
firm, so I may use my expertise in computer system maintenance and design.

QUALIFICATIONS SUMMARY
Five years of experience in designing, installing, and troubleshooting computing systems; a
proven track record in identifying problems and developing innovative solutions.

TECHNICAL SKILLS
* PROGRAMMING: C, C++, Visual BASIC, SQL, OSF/Motif, UNIX Shell Script, and JAVA scripting.
* OPERATING SYSTEMS: UNIX, MS Windows, MS DOS, MS Windows NT, Solaris, and HP-UX.
* NETWORKING: TCP/IP, OSI, Microsoft LAN Manager, and Novell Netware.
* APPLICATIONS: Microsoft Office, Microsoft Access, Microsoft Visual C++, Microsoft Project,
  Microsoft Publisher, Lotus 123, Lotus Freelance, and others.

PROFESSIONAL EXPERIENCE
Information Technologist
Superior Manufacturing Systems, Phelps, NY.  May 2003 to Present
* Responsible for upgrading software, configuring new systems and managing computer accounts
  and server space for a research and development lab of 138 employees.

Independent Consultant
Jenkins Communication Services, Batavia, NY.  April 2001 to Present
* Part-time business designing and developing Web sites for small businesses and organizations.

Computer Specialist
Woolland Computer Services, Batavia, NY.  June 1998 to 2001
* One year full time, after high school graduation; two years part-time while attending
university. Duties included
  * direct sales of computers and software
  * onsite servicing of computers and training of users

EDUCATION
* BS in Computer Science, Rochester Institute of Technology, Rochester, NY. 2003 GPA 3.8
  Senior Project: Developed a hypertext information system for athletic department.
* East Elms High School, Chili, NY 1998: 84.6%

EXTRACURRICULAR ACTIVITIES
Westferry Ski Club, Pittsford, NY  1998-present
* Received ski instructor certification in 1989, served as club secretary 2002 - present

Theatre for Youth, Batavia, NY  1996-2002
* Actor-in-training for one year, then as electrical/computer technician for five years,
  responsible for designing and implementing computer generated dramatic effects.

SPECIAL AWARDS
* Recipient of Miller Foundation scholarship in Computer Science, Rochester Institute of
  Technology, 2002
* Awarded Maitland Trophy for best overall performance, East Elms High School, 1997

REFERENCES
Margaret Ferbrache, Owner/Manager
Woolland Computer Services
313 Oak Street, Batavia, NY 14020
585-323-6647  email: ferbrache@woollcom.com

David Singh, Program Director
Theatre for Youth
PO Box 212, Batavia, NY 14020
585-717-6690  email: d.singh@players.net
```

Figure 11-6 Resume saved in ASCII plain text format.

2. Save your resume as a "text only" file with "line breaks." This will be an option listed in the document types of your Save Dialog Box. If the agency or person you are sending your resume to has requested you use "hard" carriage returns at the end of the paragraphs, save as "text only" without the line breaks. This instructs the software to break the lines whenever it needs and forces a break where you have entered one.
3. Use a text editor, such as Notepad, to open your new resume. This is what your recipient will see when you email your resume. The text editor will show you any characters that are not ASCII characters, such as bullets or bold. Replace all unsupported characters with an ASCII equivalent. (You can use any character that you can find on a standard keyboard as a replacement) For example, bullets appear as a question mark when opened in Notepad. They can be replaced with asterisks or hyphens since they are easier to understand than a question mark.
4. Now you can copy and paste this ASCII plain text resume into the body of an email message. Use the same technique to create a short cover letter and paste it into the email message above the resume text.

We suggest that before you send it to the agency or employer, you send yourself or a friend a copy of the message to make sure it converts properly.

Effective keywords are the key to getting noticed!

Keyword Summary Resumes

The resumes that get listed first are the ones that have matched the most keywords. So, when developing an electronic resume, you need to think like a Human Resources Manager and include as many keywords as possible.

A keyword is usually a noun, not an action verb. This is a change from how we recommend you write paper resumes, using strong action verbs like *managing, implementing, installing*. The Human Resources Manager would search for words that describe the qualities or skills needed for a particular position, words like *account manager, CAD skills, member IEEE*. The search often includes other company names, particular tools or technologies, schools, degrees, universities, years experience, and responsibilities.

H. L. Winman and Associates entered the keywords *Computer Specialist, three years experience, Web pages, Web site design* and *manager* when they were looking for someone to fill the new Computer Services Manager position. And that's how they found Susan. You can see her keyword summary in the resume shown in Figure 11-6.

Scanned Resumes

Sometimes an agency or employer will scan a paper resume and convert it to an electronic format. It may than be posted to a database or entered into a resume tracking system. With this in mind, it is imperative that you include keywords as described earlier in this chapter. If you know your resume will be scanned by a service there are certain guidelines you should follow to make sure your paper resume scans well.

Always send an original resume printed from a laser jet. Ink jet printers can smear the text and a dot-matrix is too outdated. Photocopies or faxed copies do not scan well. The best paper to use is light-colored, standard-sized 8 $^1/_2$ x 11 in. printed on one side only. Earlier we suggested being creative and using a brochure or pamphlet design for your resume; however, this is not conducive to scanning. The same is true for the font selection. Pick a simple, easy to read sans-serif font such as Arial. Avoid complex formatting like graphics, shading, italics, bold, parentheses and brackets or horizontal and vertical lines. Make sure there are no folds or staple marks in your original. Your resume may appear very bland and generic visually, but it will scan much cleaner and be more useful to your potential employer.

HTML Web Portfolios

The more global and technical our society becomes the more competitive the job market becomes. Many technical professionals are using the Internet to advertise their capabilities. You can do this too, by creating your own Web portfolio. It sounds more sophisticated and complex than it is. All you need is an HTML coded version of your resume and a URL to point to it.

Susan Jenkins used a basic menu structure with links to additional information. The first page of her Web portfolio is shown in Figure 11-7. Unlike ASCII plain text resumes or scanned resumes, using backgrounds, graphics, and special fonts and characters enhances HTML resumes. Be careful though: you still want your words to describe you and your capabilities. The danger is that you may get carried away designing your Web page and the only quality you will demonstrate is your programming skills. Remember, this is still your resume and it needs to sell you.

A major difference between electronic resumes posted to a database service and a Web portfolio is how people access them. If you create a Web portfolio it is your responsibility to attract people to it. When you post your resume to a database, people go there by themselves to search for it.

Most word processing packages today have tools to help you develop HTML pages. If you need additional help, search the Web for a tutorial or find a book on simple HTML coding.

An employer accessing this home page clicks on a particular bulleted heading to bring specific details onscreen

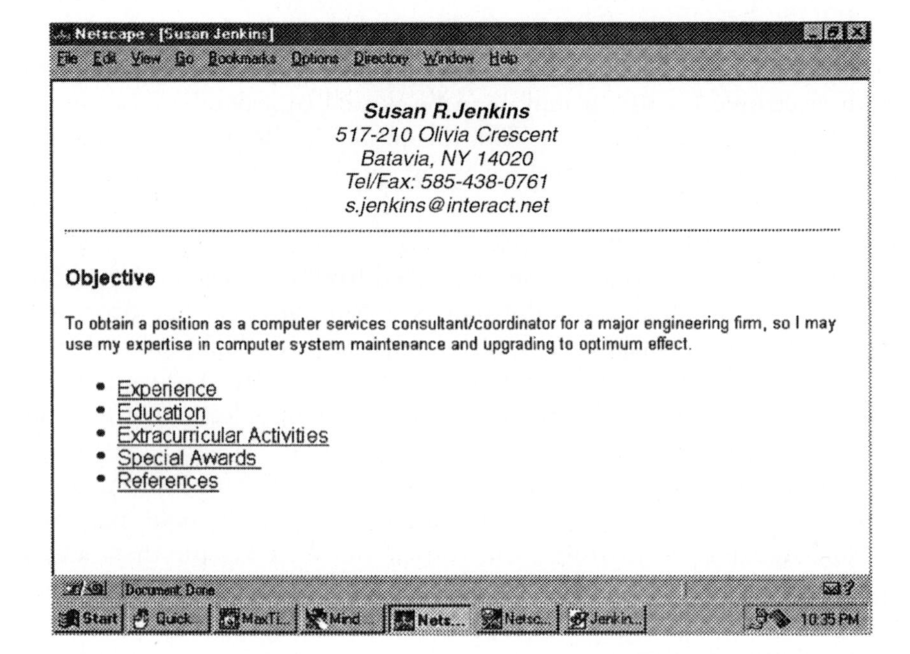

Figure 11-7 An HTML coded Web portfolio prepared by Susan Jenkins, which she used on her World Wide Web personal Home Page.

Writing a Letter of Application

Although some resumes may be delivered personally, the majority are mailed or submitted electronically, both with a covering letter. Because potential employers will probably read the letter first, it must do much more than simply introduce the resume. The letter needs to state your purpose for writing (that you are applying for a job) and demonstrate that you have some very useful qualifications that the reader should take the time to consider. It should never simply repeat what your resume says.

Write using a firm yet personable style

An assertive, interesting, and well-planned application letter can prompt employers to place your letter and resume with those whose authors they want to interview. Conversely, a dull, unemphatic letter may cause the same employers to drop your application on a pile of "not nows" because its approach and style seem to imply you are a dull, unemphatic person.

A letter of application should adopt the pyramid method of writing: it should open with a brief summary that defines the purpose of the letter, continue with strong, positive details to support the opening statement, and close with a brief remark that identifies what action the writer wants the reader to take. These three parts are illustrated in Figure 11-8.

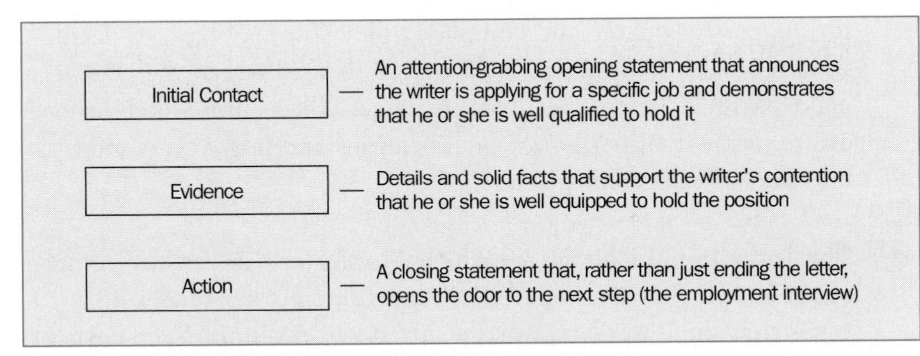

Figure 11-8 Writing plan for a job application letter.

Keep your letter down to one page and no more than four paragraphs

There are two types of application letter. Those written in response to an advertisement for a job that is known to be open, or at the employer's specific invitation, are known as solicited letters. Those written without an advertisement or invitation, on the chance that the employer might be interested in your background and experience even though no job is known to be open, are referred to as unsolicited letters. The overall approach and shape of both letters are similar, but generally the unsolicited letter is more difficult to write.

The Solicited Application Letter

The main advantage in responding to an advertisement, or applying for a position that you know to be open, is that you can focus your application letter on facts that specifically meet the employer's requirements. This has been done by Alison Witney in Figure 11-9, which responds to an advertisement in a Florida local newspaper.

The following comments and guidelines are keyed to the circled numbers beside Alison's letter.

1. You can create your own personal letterhead showing your contact information. Include your name, address, telephone number and email address. Some people have the information all on one line like Reid Qually did with his resume (Figure 11-4) and others center the information, with each item on a separate line.

Never write "Dear Sir or Madam" or "To Whom It May Concern"

2. Whenever possible, personalize an application letter by addressing it by name to the personnel manager or the person named in the advertisement. This gives you an edge over applicants who address theirs impersonally to the "Personnel Manager" or "Chief Engineer." If the job advertisement does not give the person's name, invest in a telephone call to the advertiser and ask the receptionist

for the person's name and complete title. (You may have to decide whether to send your letter and resume to someone in the personnel department or to a technical manager who is more likely to be aware of the quality of your qualifications and how you could fit into his or her organization.)

3 This is the **Initial Contact,** in which Alison summarizes key points about herself that she believes will most interest her reader and states that she is applying for the advertised position. Note particularly that she creates a purposeful image by stating confidently "I am applying...." This is much better than writing "I wish to apply...," "I would like to apply...," or "I am interested in applying...," all of which create weak, wishy-washy images because they imply only interest rather than purposefully applying for a job. An equally confident opening is "Please accept my application for...."

Draw on key information in your resume to support your opening statement

4 The **Evidence** section starts here. It should offer facts drawn from the resume and expand on the statements made in the first paragraph. Avoid broad generalizations such as "I have 13 years experience in a metrology laboratory," replacing them with shorter descriptions that describe your exact role and responsibilities, and stress the supervisory aspects of each position. The name of a person for whom you worked on a particular project can be usefully inserted here because it adds credibility to the role and responsibilities you are describing.

5 The **Evidence** section covers the key points an employer is likely to be interested in and draws the reader's attention to the attached resume. If the paragraph grows too long, divide it into two shorter paragraphs (as Colin Farrow has done in Figure 11-10).

6 This paragraph is Alison Witney's **Action Statement,** in which she effectively opens the door to an interview by drawing attention to her imminent visit to the advertiser's premises. She avoids using dull, routine remarks such as "I look forward to hearing from you at your earliest convenience" or "I would appreciate an interview in the near future," both of which tend to close rather than open the door to the next step.

7 Contemporary usage suggests that most business letters should end with a single-word complimentary close such as "Regards" or

Alison V. Witney
1670 Fulham Boulevard
Amiento, FL 37204
Tel: (305) 474-6318
email: avwitney@flonline.net

1

March 23, 2004

Dr. Eugene Coppola
Animal Science Experimental Institute
Mount Ashburn University
Three Hills, AL 35107

2

Dear Dr. Cartwright:

I am applying for the position of Research Technician (Animal Sciences) advertised in the March 18, 2004, issue of the *Amiento County Herald*. I have been involved with animals and their care and treatment for many years, and shortly will receive my Diploma in Biological Science.

3

My interest in animals dates back 12 years, to when I first learned to care for, groom, and ride horses. I now teach horseback riding in my spare time. For the past three years my employer has been Dr. Alex Gavin, veterinary surgeon at the Amiento County Animal Treatment Center, where I assist in the medical treatment of small animals. It was my interest in horses, plus Dr. Gavin's influence, that led to my enrollment in the two-year Biological Sciences course at Amiento Technical College, from which I will graduate in early June. The attached biographical details provide further information on my education, employment background, and work experience.

4

5

I will be visiting your research station from April 21 to 23, as part of my college term research project. May I call on you then, while I am at Three Hills?

6

Regards,

7

Alison Witney

Alison V. Witney
enc

Figure 11-9 A solicited letter of application prepared by an undergraduate.

"Sincerely," rather than the more formal but less meaningful "Yours very truly."

The Unsolicited Application Letter

An unsolicited application letter has the same three main parts as a solicited letter and looks much the same to the reader. To the writer, however, there is a subtle but important difference: it cannot be focused to fit the requirements of a particular position an employer needs to fill. This means the job applicant has to take particular care to make the letter sound both positive and directed. Here are some guidelines to help you shape such a letter.

- Make a particular point of addressing your letter to the person, by name and title, who will most likely be interested in you. This may mean selecting a particular department or project head, who will immediately recognize the quality of your qualifications and how you would fit into the organization, rather than applying to the personnel manager. Never address an unsolicited letter to a general title such as "Manager, Human Resources," because, if the company does not use such a title and you have not used a personal name, it will likely be the mail clerk who decides who should receive your letter.

Start your research with the company's Web site

- Try to find out enough information about a firm so that you can visualize the type of work it does and how you and your qualifications would fit the company's needs. This will enable you to focus your letter on factors likely to be of most interest to the employer.
- Try to make your initial contact positive and interesting even though you are not applying for a particular position, as Colin Farrow has done in his unsolicited letter in Figure 11-10.

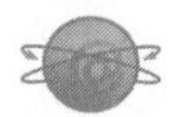

Entry Level Job Site
www.collegegrad.com
This site is designed for new college graduates. Tips on resumes, cover-letters, researching companies, and posting resumes.

Like Alison Witney, Colin has used the modified block format for his letter. It is longer than Alison's because he has more information to present, and to do so he has created two **Evidence** paragraphs.

Completing a Company Application Form

Filling in company application forms can become a boring and repetitive task, yet any carelessness on an applicant's part can draw a negative reaction from readers. Each company or organization usually uses its own specially designed form that, although it asks for generally the same basic information, may vary in detail. Consequently the suggestions below apply primarily to the *approach* you should take rather than suggest what you should write:

- When visiting prospective employers, always carry your personal data record with you so you can readily search for details such as dates, telephone numbers, and names of supervisors.

Colin R. Farrow, P.E.
408 Medwin Street
St. Cloud, MN 56301
Tel: (612) 548-1612
email: c.farrow@mnonline.com

December 15, 2003

The letter's appearance and sans-serif type create a crisp, professional impression

Vern A. Rogers, P.E.
Branch Manager
H. L. Winman and Associates
574 Lincoln Avenue
Minneapolis, MN 56565

Dear Mr. Rogers:

A

As a Structural Engineer who has specialized in constructing and maintaining transmission line towers and associated buildings for 10 years, and who has particular experience working in permafrost, I am applying for a position with H. L. Winman and Associates so I can use my expertise to good effect.

B

My experience evolves from three periods of employment. For four years I installed and maintained communications systems with the United States Air Force. Subsequently I became a crew chief and installation coordinator with the Northwestern Steel Construction Company, where for four years I was responsible for erecting and testing high-voltage transmission line towers between Weekaskasing Falls, NE, and Bismarck, ND. For the past two years I have been a project engineer supervising the construction and installation of microwave towers on General Telephone and Electric's Brainerd–Little Falls extension.

The enclosed resume describes my responsibilities in greater detail and my particular involvement in testing structures erected on discontinuous permafrost. I hold a Diploma in Civil Engineering Technology from The Technical Vocational Institute in Minneapolis, and a B.S. in Structural Engineering from the University of Minnesota. I am keen to return to the north and the challenge of building on unstable soil.

C

I would welcome the opportunity to meet you and learn more about your project at Winterton Lake. As I travel frequently between St. Cloud and Minneapolis, I will call you when I next expect to be in your city.

Sincerely,

Colin R. Farrow

Colin R. Farrow
enc

Figure 11-10 An unsolicited letter of application prepared by an experienced engineer. A is the initial contact, B is the evidence, and C is the action statement.

- Treat every application form as though it is the *first* one you are completing—write carefully, neatly, and legibly. Never let an untidy application form subconsciously prepare an employer to meet an untidy worker.

Take care: every word you write conveys an image of how you approach a task

- Complete *every* space on the form, entering N/A (not applicable), Not Known, or a short horizontal line in spaces that do not apply to you or for which you genuinely do not have information. This will prevent an employer from thinking you carelessly (or, worse, intentionally) omitted answering the question.
- Take care that your familiarity with your city and street names does not cause you to abbreviate or omit them. If you write "Mpls" for Minneapolis or omit the "St.," "Ave.," or "Crescent" from your street name (because you *know* it is a street, avenue, or crescent), you may create the impression that your approach to work is to take shortcuts whenever possible.
- Use words that describe the responsibility and supervisory aspects of each job you have held (as you would for a resume) rather than list only the duties you performed.
- Particularly describe extracurricular activities that show your involvement in the community, or activities in which you held a teaching or coaching role.

The most difficult part to write! You should prepare in advance for such a question

- Pay particular attention if there is a section on the form that asks you to comment on how your education and past experience have prepared you for the position. Think this through very carefully before you write so that what you say shows a natural progression from past experience to the job you are applying for. If you can, and if they fit naturally, add a few words to demonstrate how the position fits your overall career plan. This can be a particularly difficult section to write so do not be afraid to obtain an opinion of its effectiveness from another person.

Attending an Interview

This is the third step in the job application process and the first time you meet a prospective employer (or, more often, the employer's representative) face to face.

Prepare for the Interview

The key to a good interview is thorough preparation. If you have prepared yourself well, the interview will likely run smoothly and you will present yourself confidently.

As soon as you are invited to attend an interview—or, better still, before you are called—start researching facts about the company (or organization, if it is a government establishment). Presumably, you will have done some research before submitting your letter of application. Now you need to identify additional information, such as the number of people the company employs, specific fields in which it is involved, work for which it is particularly well known, its major products and services, important contracts it has received (news of which has been released to the media), locations of branch offices, and the company's involvement in community activities. (An ideal way to do this is to carry out a search on the Internet.) Such knowledge can be extremely useful during the interview, because it permits you to ask intelligent questions at appropriate places—questions that indicate to the interviewer that you have done your homework.

Attending an interview can create as much anxiety as speaking before an audience

You also need to prepare for difficult questions an interviewer may pose to test your readiness for the interview and the sincerity of your application. You may be asked:

- *Why do you want to join our organization?*
- *How do you think you can contribute to our company?*
- *Why do you want to leave your present employer?* (Asked only of persons who are already employed.)
- *Why did you leave such-and-such company on such-and-such date?* (Asked of persons whose resumes show no explanation for a previous employment termination.)
- *What do you expect to be doing in five years? Ten years?*
- *What salary do you expect?*

If you have not prepared for such questions, and so hesitate before answering, an interviewer may interpret your hesitation to mean that you find a question difficult to answer or that there are factors you would rather conceal. In either case, you may inadvertently give an entirely misleading impression of yourself.

Lack of preparation will show up in your body language and how you answer questions

An interviewer who asks what salary you expect is partly testing your preparation for the interview and partly assessing how accurately you value yourself. For an undergraduate at a university or college, the question is largely academic: undergraduates compare notes and quickly learn what starting salaries are being offered. But for a person who recently has been or is currently employed, the question is important and must be anticipated. Always know the salary you would like to receive and think you are worth. Avoid quoting a salary range, such as "between 42 and 44 thousand dollars," because it indicates unsureness. Quote a definite figure, such as $43,000, and you will sound much more confident. If you fear that the salary you want to quote may be too high, you can always add the qualification "...depending, of course, on the opportunities for advancement and the fringe benefits your company offers."

An interview is like a two-way street: traffic should flow in both directions

You should be ready to ask questions during the interview. The interviewer wants to acquire information about you, but you should also learn things about the company and the opportunities it can offer. Consider what questions you would like answered, jot them onto a small card, and store the card in a pocket or purse. Then when the interviewer asks, "Now, do you have any questions?" you can pull out the card.

Make the entries on your card brief and clearly legible, and keep the list short so you can scan it quickly. Remember, too, that the quality of your questions will demonstrate how carefully you have thought about the interview.

Create a Good Initial Impression

Remember that you are being evaluated from the moment you step into the interview room. Consequently,

- walk in briskly and cheerfully,
- shake hands firmly, because a limp handshake creates an image of a limp, indefinite applicant,
- repeat the person's name as you are introduced and look him or her directly in the eye, and
- sit when invited to do so, pushing yourself well back in the chair, making yourself comfortable, and avoiding folding your arms across your chest (which psychologically suggests you are resisting questions).

Participate Throughout the Interview

An experienced interviewer will try to ease your nervousness

An interview normally falls into three fairly easy-to-distinguish parts. The initial part is an exchange of pleasantries between yourself and the interviewer, who wants you to be at ease. To help you adjust to the interview environment, he or she may ask questions on topics you can answer confidently, such as a major news item or something from the hobbies and interests section in your resume. Normally, this initial part of the interview is short.

In comparison, the middle part of the interview is quite long. So that the interviewer can find out as much as possible about you, he or she will want to hear your opinions and have you demonstrate your knowledge of certain topics. Although the interviewer will want to control the direction the interview takes, you will be expected to develop your answers and to comment on each topic in sufficient depth to establish that you have real knowledge and experience, backed up by well-thought-out opinions.

The closing portion of the interview is also short. The interviewer will ask if you have any questions and will discuss details about the company and employment with it. By this stage the interviewer should have a good impression of you, and you should know whether you want to be employed by the company.

An effective interviewer will pose questions and subsequent prompts in such a way that you are carried easily from one discussion point to the next and are automatically encouraged to provide comprehensive answers. If, however, you face an inexperienced or inadequately prepared interviewer, the responsibility to develop your answers in greater depth than the questions seem to call for becomes yours.

Be ready to offer information, but don't monopolize the conversation!

For example, the interviewer may ask, "How long did you work in a mobile calibration lab?"

You might be tempted to reply "Three years," and then sit back and wait for the next question. You would do much better to reply: "For three years total. The first year and a half I was one of four technicians on the Minneapolis to Sioux City circuit. And then for the next year and a half I was the lab supervisor on the Fort Westin, to Manomonee route." An answer developed in this depth often provides the prompt (piece of information) from which the interviewer can frame the next question.

Sometimes you will face a single interviewer, while at other times you may face an interview board of two to five people. In a single-interviewer situation you will naturally direct your replies to the interviewer and should make a point of establishing eye contact from time to time. (To maintain continuous eye contact would be uncomfortable for both you and the interviewer.) In a multiple-interviewer situation

- direct most of your questions, and your responses to general questions, to the chairperson (but if an answer is long, occasionally look briefly at and talk momentarily to other board members),
- if a particular board member asks you a specific question, address your response to that person, and
- if a board member has been identified as a specialist in a particular discipline, direct questions to that board member if they especially apply to that field.

In certain interviews—often when applicants are being interviewed for a high-stress position—you may be presented with a stress question. A stress question is designed to place you in a predicament to which there may be two or even more answers or courses of action that could be taken. You are expected to think *briefly* about the situation presented to you and then to select what you believe is the best answer or course of action. Often you will be challenged and expected to defend the position you have taken.

Be prepared for questions that challenge your thinking or your ethics

The secret is not to let yourself be rattled and to defend your answer rationally and reasonably even though the questioner's challenging may seem harsh or unreasonable. Remember that the interviewer is probably more interested in seeing how you cope in the stress situation than in hearing you identify the correct answer.

Here are seven additional factors to consider:

- Use your voice to good effect; make sure everyone can hear you, speak at a moderate speed (think out your answers before speaking) and, where appropriate, let your enthusiasm *show*.
- Be ready to ask questions, but have a clear idea of what you want to ask before you pose them. The interviewer will recognize a good question and the clarity of thought behind it.
- If you do not know the answer to a question, say you don't know rather than try bluffing your way through it.
- If you do not understand the question, again don't bluff. Either say you do not understand or, if you think you know what the interviewer is driving at, rephrase the question and ask if you have interpreted it correctly. (Never imply that the interviewer posed the question poorly.)
- Use humor with great care. What to you may be extremely funny may not match the interviewer's sense of humor.
- Bring demonstration materials to the interview if you wish (such as a technical proposal or report you have authored, or a drawing of a complex circuit you designed) but be aware that you may not have an opportunity to display them. If the topic they support comes up during the interview, introduce them naturally into your response to a question. But remember that the interviewer does not have time to read your work, so the point you are trying to make must be readily identifiable. Never force demonstration materials on an interviewer.
- Do not smoke, even if the interviewer smokes and invites you to do so.

Above all, present an image of "the real you"

Finally, try to be yourself. Remember that interviewers want to see the kind of person you really are. If you relax and answer questions comfortably and purposefully, they will gain a good impression of you. If you try too hard to be the kind of person you think the interviewers want you to be, or to give the kind of answers you think they want rather than the answers you really believe in, they may detect it and judge you accordingly.

Accepting a Job Offer

The telephone rings and the personnel representative you met during your interview tells you that the company is offering you employment at a salary of $xxxx. You accept the offer! And then she asks when you can start work. (Employers recognize that if you are attending college there will be a waiting period until your course is finished and you have graduated; similarly, an employed professional has to resign from his or her present position, normally giving either two weeks' or one month's notice.) You quote a starting date to the personnel representative, which she agrees to. She then says she will confirm the offer in writing. She also asks you to write a letter confirming your acceptance of the position.

The two letters become, in effect, a contractual agreement: the employer offers you work under certain conditions, which you agree to. The letters can also prevent any misunderstandings from developing, which can occur if arrangements are made only by telephone. Consequently your acceptance letter should:

- Announce that you are accepting the offer of employment.
- Repeat any important details, such as the agreed salary and starting date.
- Thank the employer for considering you.

Accept a job offer *in writing*, like sealing a contract

The following acceptance letter conforms to this pattern:

Dear Ms. Tataryn:

I am confirming my telephone acceptance of your May 19 offer of employment as an engineering technologist in the controls department. I understand that I am to join the company on June 15 and that my salary will be $36,500 annually.

Thank you for considering me for this position. I very much look forward to working for Magnum Electronics.

Sincerely,

Sometimes an applicant may receive two offers of employment at the same time and will have to decline one. The letter declining employment should follow roughly the same pattern:

- Decline the offer.
- Briefly explain why.
- Thank the employer.

Here is an example:

Also decline a job offer in writing, with a smile on your face

Dear Mr. Genser:

I very much regret that I will be unable to accept your offer of employment. Since my interview with you I have been offered employment elsewhere and now must honor my commitment to the other company. Thank you for considering me for this position.

Regards,

Declining a job offer pleasantly and formally in a carefully worded letter like this is insurance for the future: one day you may want to work for that employer!

ASSIGNMENTS

Project 11.1: Preparing a Resume

You are to prepare a resume describing your background, education, work experience, extracurricular activities, and other interests. Do it in three parts.

Part 1
If you do not already have one, prepare a personal data record (PDR), using 5 × 8 inch file cards.

Part 2
Write down the following information:

- The name of a real employer for whom you would like to work at the end of your course.
- The type of position you would be qualified to hold with that particular employer.
- The type of resume that would be most effective to use.

Part 3

Create a resume

Prepare the resume (assume that you will be graduating from your course in two months). You can decide which format: paper or a particular electronic format.

Project 11.2: Applying for a Locally Advertised Position

From your campus student employment center or your local newspaper, identify a company currently advertising a position that you could apply for at the end of your course.

Part 1
Write a letter applying for the position (assume that you will be graduating in six weeks). Also assume that you are attaching a resume to your letter. If you are replying to a newspaper advertisement, attach a copy of the advertisement to your letter.

Part 2

Rehearse applying for a real position

Assume that the company you wrote to in Part 1 sends you an application form. Obtain a standard application form from your campus student employment center and complete it as though it is the advertiser's form.

Part 3

Now assume that the company has telephoned and asked you to attend an interview next Tuesday. On a sheet of paper write down five questions you would ask during the interview. After each question explain why the question is important and what answer you hope it will elicit from the interviewer.

Project 11.3: Unsolicited Application for Employment

This project assumes that you are seeking permanent employment at the end of a technical training program, but few job openings have been advertised in your field. Write to Macro Engineering Inc. in Phoenix, or to H. L. Winman and Associates (to the attention of one of the department heads in Cleveland, or local branch manager Vern Rogers), applying for employment. Use your knowledge of the company and your real background. If it is still early in your training program, you may update the time and assume that it is now two months before graduation.

Project 11.4: Replying to Other Advertisers

Apply for one of these positions

This project assumes that you are seeking permanent employment at the end of a technical or engineering-oriented educational program. You are to reply to any one of the following advertisements, using your present situation and actual background. If it is still early in your training program, you may update the time and assume that it is now two months before your graduation date. In each case enclose a resume with your application letter.

ROPER CORPORATION (NORTH CAROLINA DIVISION)

requires a

CHEMICAL TECHNICIAN

to join a project group conducting research and development into the organic polymers associated with the coating industry. The successful applicant will also assist in the development of control techniques for producing automated colour tinting. Apply in writing, stating salary expected, to:

Phyllis Cairns
Personnel Manager
P.O. Box 1728
Hillsborough, NC 27278

(Advertisement in Engineering Times*, April 17)*

INTER-MOUNTAIN PAPER COMPANY

Offers excellent opportunities for recent graduates to join an expanding manufacturing organization in the pulp and paper industry.

Engineers and Engineering Assistants

Positions are available for mechanical engineers and technologists to assist in the design, installation, and testing of prototype production equipment. Previous experience in a manufacturing plant would be helpful. Innovative ability will be a decided asset.

Electrical Engineering Technician

This person will assist the Plant Engineer in the maintenance of power distribution systems. Applicants should be graduates of a two-year course in Electrical Technology with good knowledge of automatic controls and machine application. Ability to read blueprints and working drawings is essential.

Electronics Engineer or Technician

Two positions are open for electronics specialists who will maintain and troubleshoot microprocessor-controlled production equipment.

Computer Specialist

This position will suit either a graduate of a Computer Engineering course or an Electronics Technician who has specialized in Computer Electronics. Duties will consist of installation, maintenance, and troubleshooting of mainframe and personal computers.

Environment Specialists

Persons selected will test air pollutants and water effluents from our paper mill and production plant, and assess their environmental impact. Applicants should be graduates of a recognized course in the environmental or biological sciences.

Salaries for the above positions will be commensurate with experience and qualifications. Excellent fringe benefit program available. Write in confidence to:

Manager of Industrial Relations
INTER-MOUNTAIN PAPER COMPANY
Montrose, OH 45287

(Advertisement in your local newspaper, March 10)

H. L. Winman and Associates
475 Lethbridge Trail
Cleveland, OH 44104

Civil Engineers and Technicians

This established firm of consulting engineers needs three Engineers or Technicians to assist in the construction supervision of several grade separation structures to be built over a two-year period. Graduation from a recognized Civil Engineering or Technology course is a prerequisite. Successful applicants will be hired as term employees. Opportunities are excellent for transfer to permanent employment before the project ends.

Apply in writing to:

A. Rittman, Head
Special Projects Department

(Advertisement on college notice board)

Project 11.5: Preparing Different Versions of Your Resume

Assume you have identified a large, multi-location company you want to work for after graduation and have decided to send your information to them even though they are not advertising any open positions. You have talked to the Human Resources Manger who has told you she needs an electronic resume to post in their internal database. She also requested you send her a paper version.

Part 1

Prepare a paper version of your resume and convert it to a format appropriate for posting to the company's resume database.

Part 2

Besides sending the paper and electronic resumes to the Human Resources Manager, you decide to also send the URL of your personal Web portfolio. Create a Web portfolio that resembles your paper resume but also has professional looking graphics and additional details about your experience. Remember, this is a reflection of you and it should encourage the visitor to want to hire you.

Project 11.6: Unsolicited Application for Summer Employment

Assume that you are looking for summer employment but few summer jobs have been advertised. Write to H. L. Winman and Associates or Macro Engineering Inc., asking for a summer job. Use your knowledge of the companies, plus your actual background, to write an interesting letter.

Address your letter to the attention of Tanys Young in Cleveland or George Dunn in Phoenix. If there is an H. L. Winman and Associates branch in your area, you may address your letter to branch manager Vern Rogers.

Chapter 12
The Technique of Technical Writing

This chapter concentrates on a few writing techniques that will enable you to convey information both quickly and efficiently. It considers technical writing from five points of view: how to create the whole document, how to structure paragraphs, how to write individual sentences, how to use specific words, and how to create a good technical style. At the end of the chapter you will find several pages of exercises that test your ability to adopt an effective writing style and, in some cases, establish a suitable tone.

We assume you are already proficient in grammar and can recognize and correct basic writing problems. If you need practice in basic writing, we suggest you refer to a textbook such as the *Simon & Schuster Handbook for writers*'[1] You can also refer to the Glossary of Technical Usage (see page 374) for information on how to form abbreviations and compound adjectives, spell problem words, and use numerals or spell out numbers in narrative.

The Whole Document

Three factors affect the whole document: the tone you set, the writing style you adopt, and how you arrange the information on the page. Tone is by far the least tangible: a reader is less likely to be aware of the tone you establish than the writing techniques you use and the arrangement of paragraphs and headings.

Tone

Whether your writing should be formal or informal depends on the situation and your familiarity with the reader. Formal reports should adopt a

1 Lynn Quitman Troyka, *Simon & Schuster Handbook for Writers*, 6th ed. (Upper Saddle River, NJ: Prentice Hall, 2002).

formal tone. (Note, however, that a formal tone is neither stiff nor pompous: there is no room for writing that makes readers feel uncomfortable because they are not as knowledgeable as you are.) Business correspondence is generally less formal, depending on its importance. For example, a management-level letter proposing a joint venture on a major defense project would be formal, whereas letters between engineers discussing mutual technical problems would be informal, and email would be very informal. A memo report also can be informal, since normally it would be an in-plant document written between people who know each other.

You need to know your reader if you are to set the right tone

Varying levels of tone are evident in the following extracts from three separate documents, all written on the same subject.

1. **Extract from a Memo.** John Wood's materials testing laboratory has compression-tested samples of concrete for Karen Woodford of the civil engineering department at H. L. Winman and Associates. In his memo reporting the test results, John writes:

Informal Tone

I have tested the samples of concrete you took from the sixth floor of Tarryton House and none of them meets the 33.25 MPa you specified. The first failed at 28.08 MPa; the second at 26.84 MPa; and the third at 27.95 MPa. Do you want me to send these figures over to the architect, or will you?

2. **Extract from a Letter Report.** Karen Woodford conveys this information to the architect in a brief letter report:

Semiformal Tone

Our tests of three samples taken from the sixth floor of Tarryton House show that the concrete at 52 days still was 5.63 MPa below your specification of 33.25 MPa. We doubt whether further curing will increase the strength of this concrete more than another 1.10 MPa. We suggest, however, that you examine the design specifications before embarking on an expensive and time-consuming remedy.

3. **Extract from a Formal Report.** The architect rechecked the design specifications and decided that 28.50 to 29.00 MPa still would not satisfy the design requirements. He then requested that H. L. Winman and Associates prepare a formal report he could present to the general contractor and the concrete supplier. Karen Woodford's report said, in part:

Formal Tone

At the request of the architect we cut three 0.3 x 0.15 meter diameter cores from the sixth floor of Tarryton House 52 days after the floor had been poured. These cores were subjected to a standard compression test with the following results (detailed calculations are attached at Appendix A):

Imagine you are speaking personally to the reader and adjust your tone accordingly

Core No.	*Location*	*Failed at:*
1	0.46 m W of col 18S	28.08 MPa
2	0.84 m N of col 22E	26.84 MPa
3	1.42 m N of col 46E	27.95 Mpa

The average of 27.62 MPa for the three cores is 5.63 MPa below the design specification of 33.25 MPa. Since further curing will increase the strength of the concrete by no more than 1.10 MPa, we recommend rejecting this concrete pour.

Although the information conveyed by these three examples is similar, the tone the writer adopts varies in response to each situation.

Deal with all the details before you start writing

The sequence in which you write longer reports also affects tone. To set the right tone throughout, write in reverse order, starting with the full development. Writing a report in the order in which it will be read is difficult, if not impossible. An engineering technician who writes the summary before the full development will use too many adjectives and adverbs, big words when shorter words would be more effective, and dull opening statements such as "This report has been written to describe the investigation into defective RL-80 video terminals carried out by H. L. Winman and Associates." He or she will be writing without having established exactly what to say in the full development.

The comments on Karen Woodhouse's formal report in Chapter 6 explain how Karen set about writing her report in reverse order (see pages 148 to 151). In brief, they identify the following writing sequence:

Step 1 Assemble and document all the details and technical data. These will become the appendixes to your report.

Step 2 Write the Discussion, or full development. Direct it to the type of technical reader who will use or analyze your report in depth.

Step 3 Write the Introduction, Conclusions, and Recommendations. Keep them brief and direct them to a semitechnical reader or person in a supervisory or managerial position.

Step 4 Write the Summary. Direct it to a nontechnical reader who has absolutely no knowledge of the project or the contents of your report.

Style

Style is affected by the complexity of the subject you are describing and the technical level of the reader(s) to whom you are writing. Consequently you need to "tailor" your writing style to suit each situation, following these guidelines:

Adjust sentence length to suit the subject and the reader's familiarity with it

1. When presenting low-complexity background information, and descriptions of nontechnical or easy-to-understand processes, write in an easygoing style. That tells readers they are encountering information that does not require total concentration. Use slightly longer

paragraphs and sentences, and insert a few adjectives and adverbs to color the descriptions and make them more interesting.

2. For important or complex data, use short paragraphs and sentences. Present one item of information at a time. Develop it carefully to make sure it will be fully understood before proceeding to the next item. Use simple words. The more punchy style will warn readers that the information demands their full attention.

3. When describing a step-by-step process, start with a narrative-type opening paragraph that introduces the topic and presents any information that the readers should know or would find interesting. Follow it with a series of subparagraphs, each describing a separate step, choosing between two alternative styles:

Style A: Integrated Lead-in Line

In this style the lead-in line becomes part of each subparagraph that follows it, as in this example:

Precede subparagraphs with bullets or sequential numbers...

Let each subparagraph

- develop only one item or aspect of the process,
- be short, and
- be parallel in construction (the importance of parallelism is discussed later in this chapter).

Here, the lead-in words (*Let each subparagraph*) do not form a complete sentence and so do not end with a colon. Consequently the bulleted items each start with a lower-case letter and end with a comma (except the last item), because the lead-in words and the bulleted items really make one long, complete sentence.

Style B: Separate Lead-in Line

In this style the lead-in line stands alone, like this:

If you use subparagraphs to present a series of points, follow these guidelines:

1. Indent each subparagraph as a complete unit of information, to show your readers how you are subordinating your ideas.
2. Precede each subparagraph with either a bullet (as in style A) or a sequential number (as in style B).
3. Number the subparagraphs if you want to identify that the information is presented in a prescribed sequence or in decreasing order of importance, or if you want to refer to the subparagraph later in your letter or report. At all other times use bullets.

Here, the lead-in words (*If you use...follow these guidelines:*) create a complete sentence. Consequently the lead-in words end with a colon, and the subparagraphs each start with a capitalized letter and end with a period. They are complete thoughts in themselves.

Note also the difference in spacing between the subparagraphs in the two examples:

- In style A (and here) there is no spacing between the subparagraphs. This is known as textbook style, in which the publisher wants to economize on space.
- In style B there is additional white space between the subparagraphs. We recommend you insert half a line of white space between the subparagraphs in your technical reports and proposals.

Appearance

If you incorporate Information Design techniques into your writing you can help your readers understand and access information more readily. Information Design can be applied to letters, memos, reports, proposals, instructions, newsletters, and many other documents.

Insert Headings as Signposts

Make your headings *contribute* to the overall appearance

In longer documents, and particularly those discussing several aspects of a situation, you can help your reader by inserting headings. Each heading must be short yet informative, summarizing clearly what is covered in the paragraphs that follow. If, for instance, we had replaced the four-word heading preceding this paragraph with the single word "Headings," we would not have summarized adequately what this paragraph describes. Here are some guidelines:

- Use upper- and lower-case letters rather than all capital letters.
- Use boldface type rather than underlining the headings.
- Keep headings in the same font (typeface) as the main text.
- Use larger point sizes for the principal headings, and progressively smaller point size for each level of subsidiary heading.

Figure 12-1 illustrates how this can be done.

Insert Paragraph Numbers

In some documents—particularly specifications, technical instructions, and military reports, the paragraphs and subparagraphs are numbered. A simple paragraph numbering system starts at 1 and continues consecutively to the end of the document. More complex systems combine numbers, decimals, and letters to allow for subparagraphing, as shown in Figure 12-2. (Note that roman numerals are *not* used.)

Guidelines for Integrating Paragraphs and Headings

The main center heading (above) is set in a larger boldface type than all other headings. Here, it is set in 14 pt Times New Roman, which is the same font we have used for all the text and headings in this example.

Subparagraphing Without Paragraph Numbering

If a subsidiary center heading is used, it is set in 12 pt type (see immediately above), while the side headings and the text in this example are all set in 11 pt type.

Side Heading

A side heading introduces a new section of text and is set flush against the left margin. Paragraphs following the side heading are also typed with all lines flush against the left margin. In technical writing, the first line of each paragraph is seldom indented.

Subparagraph Headings and Subparagraphing

Subparagraph headings are indented about one-third of an inch in from the left margin, as is any text that follows the subparagraph heading.

Each subparagraph is typed as a solid indented block, so that readers can see the subordination of ideas.

Secondary Subparagraphing

If further subparagraphing is necessary, the headings and subparagraphs are indented a further one-third of an inch (i.e. a total of two-thirds of an inch) from the left margin.

Headings Built into the Paragraph. In this lesser-used arrangement, the text continues immediately after the heading. Usually, a paragraph heading applies only to one paragraph of text.

For headings, vary font size and use boldface type rather than underlining the words

Figure 12-1 Guidelines for integrating paragraphs and headings.

Pick Only One Font

The availability of numerous type styles is not an invitation to mix and match

A font such as Century OldStyle or Helvetica is a set of printing type consisting of the same features. Some fonts are called serif (they have a slight finishing stroke - T) and some are called sans serif (they don't have a finishing stroke - **T**). The font you choose will project an image of you, your company and your document. Statistics show that a serif font like Times New Roman is easier to read because the serifs lead the eye from letter to letter, and so are more suitable for longer documents. (This text is printed in a serif type called Sabon.) Sans serif fonts, like Arial, Helvetica or

Subparagraphing Combined with Paragraph Numbering

1. **Side Headings**

1.1 When paragraph numbers are used, side headings normally are assigned simple consecutive paragraph numbers, as has been done here.

1.2 Where only one paragraph follows a side heading, it is not assigned a separate paragraph number and is typed with its left margin level with the side heading, as has been done in the paragraph immediately below heading 1.3.

1.3 **Subparagraph Headings and Subparagraphing**

If more than one paragraph follows a subparagraph heading, each is assigned an identification number or letter:

a) This would be the first subparagraph.

b) This would be the second subparagraph.

c) Each subparagraph can be further subdivided into a series of very short secondary subparagraphs:

(1) Here is a secondary subparagraph.

(2) Ideally, each secondary subparagraph should contain no more than one sentence.

d) We do not recommend using a third level of decimal numbers, such as 1.3.1, 1.3.2, etc.

Ensure your paragraph numbering system is unobtrusive

Figure 12-2 Guidelines for integrating a paragraph-numbering system.

Franklin Gothic are clean and clear, and portray a neat and modern image, yet are not as easy to read and so are suitable for shorter documents and electronic mail.

Daniel Thomashewski chose a sans serif font to type the agenda for the Electronic Facsimile Research Committee meeting in Figure 10-3 on page 273). Morley Wozniak, however, used a serif font for his report evaluating proposed landfill sites in Figure 5-6 on page 112).

Once you decide on a font, stay with it. Don't switch to a different font to show emphasis. Instead, use **bold**, *italic*, or a larger character size to emphasize particular sections of text. Notice, however, that Morley's cover letter and report are printed on H. L. Winman and Associates' company letterhead which uses a sans serif font. Letterhead and logos are excluded from the "maintain one font" guideline.

Make sure that the character size you choose is appropriate for your document and audience. In a one-page letter or memo we sometimes use 10 pt (point size) to help keep the document to only one page. In a longer document we use 12 pt because the type will be slightly larger and the reader's eye won't tire so easily.

Select character size to suit font style and readers' visual acuity

Justify Only on the Left

Word processors make it easy to justify both the left and right margins, which permits you to create lines of exactly the same length. We recommend you justify text only at the left margin and leave the text at the right margin "ragged." Otherwise the computer will generate spaces between words and characters to force the right margin to be straight. Unless you are using very sophisticated word-processing software, such as has been used for this book, the uneven spaces may prove stressful for readers' eyes, since they constantly have to adjust to the unevenness. It may be only a subtle difference, yet it's something you as an author can control and so make the reader's task more pleasant.

Morley Wozniak has used a ragged right margin for his report on landfill sites in Figure 5-6 (see page 112), whereas Karen Woodhouse has used a justified right margin for her formal report on radiant heating in Figure 6-6 (see pages 155 to 168).

Use Wider Margins to Draw Attention

Many technical people hesitate to change the standard settings that come with word-processing packages. Yet once they see the value of being unique, they are easily convinced to try a new way. For example, Anna King, the technical editor at H. L. Winman and Associates, encourages the company's engineers to use a wider left margin in their longer documents and to place the headings all the way to the left. She explains that this helps draw readers' attention to the headings and so helps them retrieve information faster.

Figure 12-3 shows a page from a letter proposal presented with normal paragraphs and headings (a), and a page with a wider left margin and left-justified headings (b). The latter may use more space, but it makes the information much more accessible.

Let careful use of white space focus readers' attention

Adding white space or blank areas in your text is another valuable Information Design technique. You can also use diagrams and figures to break up long passages of text and to complement the message. A simple flow chart or table makes a nice diversion for the reader and makes the information more visual and concrete. For example, Bob Walton's memo-form incident report in Figure 4-3 on page 69 describes an accident in which he was involved. He could have described the positions of the vehicles prior to the accident, but instead he sketched a diagram and attached it to his memo.

Shaping your information can encourage readers to keep reading

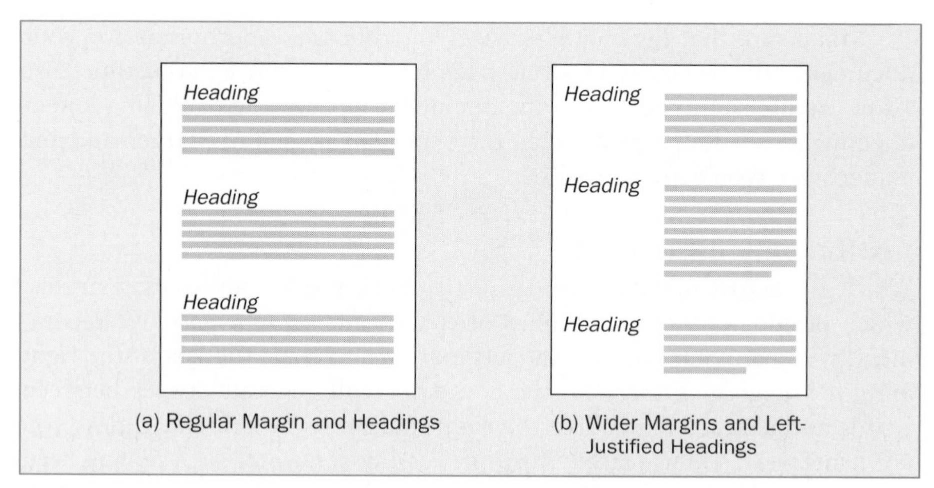

Figure 12-3 Using standard margins and a wide left margin for text.

Use Subparagraphs to Present Ideas

Anna King encourages H. L. Winman engineers to use bulleted lists to break up text and make it visually accessible. Morley Wozniak effectively used white space in his report evaluating proposed landfill sites (see Figure 5-6 on page 112, and particularly the "chunks" of information on its fourth page). Here's an excerpt of a report Morley wrote before he consulted with Anna:

L-o-n-g paragraphs build reader resistance

> I have analyzed our present capabilities and estimate that we can increase our commercial business from $20,000 to $30,000 per month. But to meet this objective we will have to shift the emphasis from purely local customers to clients in major centers. To increase business from local customers alone will require extensive sales effort for only a small increase in revenue, whereas a similar sales effort in a major center will attract a 30% to 40% increase in revenue. We will also have to increase our staff and manufacturing facilities. The cost of additional personnel and new equipment will in turn have to be offset by an even larger increase in business. Properly administered, such a program should result in an ever-increasing workload. And, third, we will have to create a separate department for handling commercial business. If we remove the department from the existing production organization it will carry a lower overhead, which will result in products that are more competitively priced.

Anna made a simple suggestion, "If you break up the second long sentence by inserting *take three steps* and a colon after *we will have to*, and then make a numbered list of the actions you need to take, the information will be much easier to read and understand. Watch what happens visually."

> I have analyzed our present capabilities and estimate that we can increase our commercial business from $20,000 to $30,000 per month. But to meet this objective we will have to take three steps:

Bite-size paragraphs build reader acceptance

> 1. Shift the emphasis from purely local customers to clients in major centers. To increase business from local customers alone will require extensive

sales effort for only a small increase in revenue, whereas a similar sales effort in a major center will attract a 30% to 40% increase in revenue.

2. Increase our staff and manufacturing facilities. The cost of additional personnel and new equipment will in turn have to be offset by an even larger increase in business. Properly administered, such a program should result in an ever-increasing workload.
3. Create a separate department for handling commercial business. If we remove the department from the existing production organization it will carry a lower overhead, which will result in products that are more competitively priced.

Use Tables to Capture Information

Displaying text in a table is an alternative way to design information for maximum impact. Many technicians reserve tables for numerical data, but we suggest you also try using tables for presenting text. A table can compartmentalize information into easy-to-find chunks, as Anna King demonstrates in Table 12-1. She uses the table to describe the writing compartments for a request letter. Another example is in Table 9-3 on page 244.

Use Good Language

It hardly seems necessary to tell you to use good language, but in this case we mean language that you know your readers will understand. Use only technical terms and abbreviations they will recognize immediately. If you are in doubt, define the term or abbreviation, or replace it with a simpler

Table 12-1 Writing plan for a request.

Compartment	What goes in it
Summary	A brief description of your request and a request for approval.
Background or Reason	The circumstances leading up to the request.
Request Details	A detailed explanation of what your request entails, what will be gained if the request is approved, any problems the request may cause, and what the cost will be.
Action	A statement that identifies clearly what you want the reader to do after reading your request.

A table can create distinct bite-size compartments of information

expression. Pages 352 to 353 provide special guidelines for abbreviating technical and nontechnical terms. In comprehensive letters and proposals, you may find it helpful to attach a Glossary of Technical Terms that defines new or unusual terminology.

Paragraphs

The role of the paragraph is complex. It should be able to stand alone but normally is not expected to. It must contribute to the whole document, yet it must not be obtrusive (except when called on to emphasize a specific point). And it should convey only one idea, although made up of several sentences each containing a separate thought.

A topic sentence at the start of the paragraph

Experienced writers construct effective paragraphs almost subconsciously. They adjust length, tone, and emphasis to suit their topic and the atmosphere they want to create, and sometimes even stretch or bend the rules to obtain exactly the right impact. But even they once had to master the techniques of good paragraph writing, although now they let the rhythm of the words guide them far more than the rules.

A topic sentence at the end of the paragraph

We are not so fortunate. We have to learn the rules and apply them consciously. Yet we must not let our approach become too pedantic, or become so bound by the rules that we write in a stilted manner that lacks interest and rhythm. We should consider the rules as building blocks that help create good writing, not as bars that imprison our creativity.

Good paragraph writing depends on three elements:

Unity
Coherence
Adequate Development

These elements cannot stand alone. All three must be present if a paragraph is to be useful to its reader.

Unity

For a paragraph to have unity, it must be built entirely around a central idea. This idea is expressed in a topic sentence—often the first sentence—and developed in supporting sentences. In effect, this permits us to construct paragraphs using the pyramid technique, with the topic sentence taking the place of the summary and the supporting sentences representing the full development (see Figure 12-4).

Although the topic sentence does not always have to be the first sentence in a paragraph, for technical writing we recommend that you consistently place it there.

The following paragraph has the topic sentence right up front. It has strong unity because its topic is clearly expressed and the supporting sentences develop it fully.

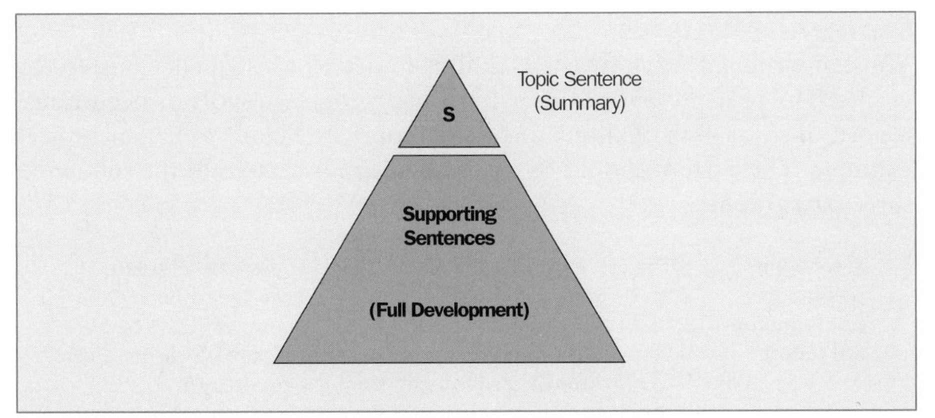

Figure 12-4 Pyramid technique applied to the paragraph.

Every supporting sentence must amplify or evolve from the topic sentence

> Content reuse means writing content once and reusing it many times. Traditional documents are written in files that consist of sections. Reusable content is written as objects or elements, not documents. Documents are therefore made up of content objects that can be mixed and matched to meet specific information needs. For example, a product description (paragraph) could be used in a brochure, on the Web, in a parts catalogue, in product support documentation, or even on a package.[2]

Coherence

Coherence is the ability of a paragraph to hold together as a solid, logical, well-organized block of information. A coherent paragraph is abundantly clear to its readers; they can easily follow the writer's line of reasoning and have no problem progressing from one thought to the next.

Good sequencing and effective transitions help create coherent paragraphs

Most technical people are logical thinkers and should be able to write logical, well-organized paragraphs. But the organization must not be kept a secret; it must be apparent to every reader who encounters their work. Simply summarizing a paragraph in the topic sentence and then following it with a series of supporting sentences does not make a coherent paragraph. The sentences must be arranged in an identifiable order, following a pattern that helps the reader to understand what is being said.

This pattern will depend on the topic and the type of document. Paragraphs describing an event or a process will most likely adopt a sequential pattern; those describing a piece of equipment will probably be patterned on the shape of the equipment or the arrangement of its features.

[2] Ann Rockley, *Managing Enterprise Content: A Unified Strategy* (Indianapolis, Indiana: New Riders, 2003), p. 17.

Sequential Patterns

You can write narrative-type paragraphs to describe a sequence of steps or events. The past-present-future pattern of a progress report or occurrence report, such as Bob Walton's incident report in Figure 4-4, is a typical example. The pattern should be clearly evident, as in two of the following three paragraphs:

A coherent paragraph (in chronological order)

The accident occurred when Dennis Friesen was checking in at the Remick Airlines counter. He placed the company's Nabuchi 300 digital camera on the counter while he completed flight boarding procedure. When the passenger ahead of him lifted a carry-on bag from the counter, its shoulder strap tangled with the carrying strap of the camera and pulled the camera to the floor. Dennis examined the camera and discovered a 40 mm crack across its back. Remick Airlines' representative Kathy Trane took details of the incident and will be calling you to discuss compensation.

Two treatments of the same information

A much less coherent paragraph (containing the same information but not presented in an identifiable pattern)

The accident occurred when Dennis Friesen was checking in at the Remick Airlines counter. Kathy Trane, a Remick Airlines representative, took details of the incident and will be calling you to discuss compensation. The damaged camera received a 40 mm crack across the back. When the passenger ahead of Dennis removed a carry-on bag from the counter, its shoulder strap tangled with the carrying strap of the company's Nabuchi 300 digital camera and pulled it to the floor. Dennis had placed the camera on the counter while he completed flight boarding procedure.

A coherent paragraph (tracing events from evidence to conclusion)

We noticed a mild shimmy at speeds above 65 mph about 10 days after the new tires had been installed. A visual check of all four wheels revealed no obvious defects, so we rotated the four wheels to different positions on the vehicle. This did not eliminate the shimmy but did change its point of origin. To pin down the cause we replaced each wheel in turn with the spare wheel, and found that the shimmy disappeared when the spare was in the left front position. We removed the wheel from that position, tested it, and found that it had been incorrectly balanced.

Descriptive Patterns

You can write descriptive paragraphs to describe scenes, buildings, equipment, and any subject having physical features. This pattern can be defined by the shape of the subject, the order in which parts are operated, the arrangement of parts from smallest to largest, or the importance of the various parts. For example:

Excerpt from paragraph (features in order of importance)

The most important control on the bomb aimer's panel is the firing button, which when not in use is held in the black retaining clip at the bottom left-hand corner. Next in importance is the fusing switch at the top right of the panel; when in the "OFF" position it prevents the bombs from being dropped live. Two safety switches, one immediately above the firing button retaining clip and the other to the right of the bank of selector switches, prevent the firing button from being withdrawn from its clip unless both are in the "LIVE" (up) position.

The topic sentence sets the scene

Continuity

A fully coherent paragraph must also have smooth transitions between its sentences. Smooth transitions give a sense of continuity that makes readers feel comfortable. As they finish one sentence, there is a logical bridge to the next. This can be accomplished by using linking words and by referring back to what has already been said. In the example just quoted, there is a natural flow from "The most important..." in the first sentence to "Next in importance..." in the second. The third sentence then refers back to the firing button and so relates the newly introduced safety switches to the previous information. The transitions are equally good in the first paragraph describing damage to a digital camera, each sentence containing a component that is a development from one of the previous sentences. This is not true of the second camera-damage paragraph, in which each new sentence introduces a new subject with no reference to what has already been said.

Adequate Development

You want to say enough, but not babble on, and on, and on...

Paragraph development demands good judgment. You must identify your readers clearly enough so that you can look at each paragraph from their point of view. Only then can you establish whether your supporting sentences amplify the topic sentence in sufficient detail to satisfy their interest.

Simply inserting additional supporting sentences does not necessarily meet the requirements for adequate development. The supporting sentences must contain just the right amount of pertinent information, all directed to a particular reader. There must never be too little or too much. Too little results in fragmented paragraphs that offer snippets of information that arouse readers' interest but do not satisfy their curiosity. Too much information can lead to long, repetitious paragraphs that annoy readers. Compare the following paragraphs, all describing the result of exploration crews' first venture with machinery across the Peel Plateau east of Alaska, intended for a reader who is interested in the problems of working in the subarctic, but who has never seen what the terrain is like.

Each paragraph has a good topic sentence...

Inadequate development

Trails left by tractors look like long narrow scars cut in the plateau. Many of them have been there for years. All have been caused by permafrost melting. They will stay like this until the vegetation grows in again.

Adequate development

To the visitor viewing this far northern terrain from the air, the trails left by tractors clearing undergrowth for roads across the plateau look like long, narrow scars. Even those that have been there for as long as 28 years are still clearly defined. All have been caused by melting of the permafrost, which started when the surface moss and vegetation were removed and will continue until the vegetation grows in again—perhaps in another 30 years.

...but the development is erratic

Over-development

To conservationists, whose main interest is the protection of the environment, viewing this far northern terrain from the air is a heartrending sight. To them, the trails left by tractors clearing undergrowth for roads across the plateau look like long, narrow scars. The tractors were making way for the first roads to be built by man over an area that until now had been trodden only by Inuit indigenous to the area, and the occasional trapper. Some of these trappers had journeyed from Quebec to seek new sources of revenue for their trade. But now, in the very short time span of 28 years, man has defiled the terrain. With his machines he has cut and gouged his way, thoughtlessly creating havoc that will be visible to those that follow for many decades. Those that preceded him for centuries had trodden carefully on the permafrost, leaving no trace of their presence. The new trails, even those that have been there for 28 years, are visible almost as though they had been cut yesterday. And all were caused by melting of the permafrost...

In these examples, the descriptive pendulum has swung from one extreme to the other. The first paragraph leaves the reader with questions: What were the tractors doing? How many years? How soon will the vegetation grow in again? The second paragraph develops the topic sentence in just enough detail; it explains why the tractors left semipermanent scars and predicts how long they will remain. The third paragraph, although interesting, is filled with irrelevant information (e.g. where the trappers come from) and repetitive statements that detract from the main theme. It might be suitable for a novel but not for a technical report or description.

Correct Length

We recommend that, for a 6 in. typing line, you try to limit paragraph length to no more than 8 printed lines. But also keep these points in mind:

- You can adjust paragraph length to suit the complexity of the topic and the technical level of the reader. Generally, complex topics demand short paragraphs containing small portions of information, while general topics can be covered in longer paragraphs.
- Variety in paragraph length has a lively visual effect. Conversely, if

you write a series of equal-length paragraphs you may create an impression of dullness.

If your work will be printed in narrow, newspaper-like columns, write shorter paragraphs

- Readers attach importance to a paragraph that is clearly longer or shorter than those surrounding it. A very short paragraph among several longer paragraphs particularly attracts attention.
- If you write too many short paragraphs close together, your readers may feel you are providing them with incomplete snippets of information. Conversely, if you write a succession of very long paragraphs, your readers may feel they are facing "heavy going" reading.

Sentences

Although sentences normally form an integral part of a larger unit—the paragraph—they still must be able to stand alone. While helping to develop the whole paragraph, each has to carry a separate thought. In doing so, each plays an important part in placing emphasis—in stressing points that are important and playing down those that are not.

Just as experienced writers first had to learn the elements of good paragraph writing, so they also had to learn the elements of good sentence construction. These are:

Unity
Coherence
Emphasis

Unity and coherence perform a function in the sentence similar to their role in a well-written paragraph.

Unity

The pyramid again!

Although the comparison is not quite as clearly defined, we can still apply the pyramid technique to the sentence in the same way it is applied to the paragraph and the whole document. In this case, the summary is replaced by a primary clause that presents *only one thought*, and the full development by subsidiary clauses and phrases that develop or condition that thought. Thus, a unified sentence, as its name implies, presents and develops only a single thought.

Compare these two sentences:

A unified sentence that expresses one main thought

The Amron Building will make an ideal manufacturing plant because of its convenient location, single-level floor, good access roads, and low rent.

A complicated sentence that tries to express two thoughts	The copier should never have been placed in the general office, where those using it interrupt the work being done by the administrative staff, who have been consistently overworked since the beginning of the year.

Try writing short, simple, uncomplicated sentences. Then later link those that seem to belong together

The first sentence has unity because everything it says relates to only one topic: that the Amron Building will make a good manufacturing plant. The second sentence fails to have unity because readers cannot tell whether they are supposed to be agreeing that the copier should have been placed elsewhere, or sympathizing that the administrative staff have been overworked. No matter how complex a sentence may be, or how many subordinate clauses and phrases it may have, every clause must either be a development of or actively support only one thought, which is expressed in the primary clause.

Coherence

Coherence in the sentence is very similar to coherence in the paragraph. Coherent sentences are continuously clear, even though they may have numerous subordinate clauses, so that the message they convey is apparent throughout. Like the paragraph, they need good continuity, which can be obtained by arranging the clauses in logical sequence, by linking them through direct or indirect reference to the primary clause, and by writing them in the same grammatical form. Parallelism, discussed at the end of this chapter, plays an important role here.

Read examples of good writing in technical journals and magazines (e.g., *Scientific American; IEEE Spectrum*)

The unified sentence discussing the Amron Building has good coherence because its purpose is continuously clear and each of its subordinate clauses links comfortably back to the lead-in statement (through the phrase "because of its…"). The clauses are written in parallel form (i.e. they have the same grammatical pattern), which is the best way to carry the reader smoothly from point to point.

The first sentence below lacks coherence because there is no logic to the arrangement or form of the subordinate clauses. Compare it with the second sentence, which despite its greater length is still coherent because it continuously develops the thought of "late" and "damaged" expressed in the primary clause.

An incoherent sentence	The Amron Building will make an ideal manufacturing plant because of its convenient location, which also should have good access roads, the advantage of its one-level floor, and it commands a low rent. **(34 incoherent words)**
A coherent sentence	Many of the 61 samples shipped in December either arrived late or were damaged in transit, even though they were shipped one week earlier than usual to avoid the Christmas mail tie-up, and were packed in polyurethane as an extra precaution against rough handling. **(45 coherent words)**

Emphasis

Whereas unity and coherence ensure that a sentence is clear and uncluttered, properly placed emphasis helps readers identify the sentence's important parts. By arranging a sentence effectively you can attach importance to the whole sentence, to a clause or phrase, or even to a single word.

Emphasis on the Whole Sentence

You can give a sentence more emphasis than sentences that precede and follow it by manipulating its length or by stressing or repeating certain words. You can also imply to the reader that all the clauses within a sentence are equally important (without affecting its relationship with surrounding sentences) by balancing its parts.

Too many sentences all the same length imply the information is dull!

Although you should aim for variety in sentence length, there may be occasions when you want to adjust the length of a particular sentence to give it greater emphasis. Readers will attach importance to a short sentence placed among several longer sentences, or to a long sentence among predominantly short ones. They will also detect a sense of urgency in a series of short sentences that carry them quickly from point to point. This technique is used effectively by story tellers:

> The prisoner huddled against the wall, alone in the dark. He listed intently. He could hear the guards, stomping and muttering. Cursing the cold, probably.

In technical writing we have little occasion to write very short sentences, except perhaps to impart urgency to a warning of a potentially dangerous situation:

> Dangerously high voltages are present on exposed terminals. Before opening the doors,
>
> 1. set the master control switch to "OFF," and
> 2. hang the red "NO" flag on the operator's panel.
>
> Never cheat the interlocks.

Neither must we err in the opposite direction and write overly long sentences that are confusing. The rule that applies to paragraph writing—adjust the length to suit the complexity of topic and the technical level of the reader—applies equally to sentence writing. If, on average, your sentences exceed 22 words, they are probably too long.

Similarity of shape can signify that all parts of a sentence are of equal importance. Clauses separated by a coordinate conjunction (mainly *and, or, but,* and sometimes a comma) tell a reader that they have equal emphasis. The following sentences are "balanced" in this way:

> Eight test instruments were used for the rehabilitation project, *and* were supplied free of charge by the Dere Instrument Company.

Keep sentence parts parallel

The gas pipeline will be 300 miles shorter than the oil pipeline, *but* will have to cross much more difficult terrain.

The upper knob adjusts the instrument in the vertical plane, *and* the lower knob adjusts it in the horizontal plane.

Emphasis on Part of a Sentence

Coordination means giving equal weight to all parts of a sentence; subordination means emphasizing a specific part and de-emphasizing all other parts. It is effected by placing the most important information in the primary clause and placing less important information in subordinate clauses. In each of the following sentences, the main thought is italicized to identify the primary clause:

When the technician momentarily released his grip, *the control slipped out of reach.*

The bridge over the underpass was built on a compacted gravel base, partly to save time, partly to save money, and partly because materials were available on site.

He lost control of the vehicle when the wasp stung him.

When the wasp stung him, *he lost control of the vehicle*.

Emphasis on Specific Words

Where we place individual words in a sentence has a direct bearing on their emphasis. Readers automatically tend to place emphasis on the first and last words in a sentence. If we place unimportant words in either of these impact-bearing positions, they can rob a sentence of its emphasis:

Let a noun have the last word!

Emphasis misplaced	Such matters as equipment calibration will be handled by the standards laboratory however.
Emphasis restored	Equipment calibration, however, will be handled by the standards laboratory.
Emphasis misplaced	Without exception change the oil every six days at least.
Emphasis restored	Change the oil at least every six days.

The verbs we use have a powerful influence on emphasis. Strong verbs attract the reader's attention, whereas weak verbs tend to divert it. Verbs in the active voice are strong because they tell *who did what*. Verbs in the passive voice are weak because they merely pass along information; they describe *what was done by whom*. The sentences below are written using both active and passive voice. Note that the versions written in the active voice are consistently shorter and more direct:

Passive Voice	Active Voice
Elapsed time is indicated by a pointer.	A pointer indicates elapsed time.
The project was completed by the installation crew on May 2.	The installation crew completed the project on May 2.
It is suggested that meter readings be recorded hourly.	I suggest you record meter readings hourly.
The samples were passed first to quality control for inspection, and then to the shipping department where they were packed in polyurethane.	Quality control inspected the samples and then the shipping department packed them in polyurethane.

The passive voice lacks emphasis

The active voice implies action

There are occasions when you will have to use the passive voice because you are reporting an event without knowing who took the action, or prefer not to name a person. For instance, you may prefer to write:

The strain gauge was read at 10-minute intervals.

rather than:

Kevin McCaughan read the strain gauge at 10-minute intervals.

Completeness

Every day we see examples of incomplete sentences—on television, in magazines, and especially in advertising. Media writers use them to create a crisp, intentionally choppy effect:

SHEER COMFORT!

33,000 feet high. Wide seats, just like your living room.
Tempting meals. Complimentary refreshments.
Only on Remick Airlines. Our Business Class. Try us!

But if we do the same in our business letters and reports, our sentences are likely to be read with raised eyebrows.

Avoid Writing Sentence Fragments

It's easy to form a sentence fragment. Normally you correct it later, when you are checking what you have written, but sometimes your familiarity with the information may cause you not to notice it. Consider these two pieces of information:

- The meeting achieved its objective. Even though three members were absent.
- The staff were allowed to leave at 3 p.m. Seeing the air-conditioning had failed.

All right for a first rough draft...

The first sentence in each of these examples is complete (it has a subject-verb-object construction), but the two second sentences are incomplete because, to be understood, each depends on information in the first sentence.

A useful way to check whether or not a sentence is complete is to read it aloud entirely on its own. If it contains a complete thought it will be understood just as it stands. For example, *The meeting achieved its objective* is a complete thought because you do not need additional information to understand it. *The staff were allowed to leave at 3 p.m.* is also complete. But you cannot say the same when you read these aloud:

- Even though three members were absent.
- Seeing the air-conditioning had failed.

In most cases a sentence fragment can be corrected by removing the period that separates it from the sentence it depends on, inserting a comma in place of the period, and adding a conjunction or connecting word such as *and, but, which, who,* or *because*:

...but it should be corrected by the second draft

- The meeting achieved its objective, even though three members were absent.
- The staff were allowed to leave at 3 p.m. *because* the air-conditioning had failed. ("*Seeing*" has been changed to "*because*.")

Here are two others:

- Staff will have to bring bag lunches or go out for lunch from October 6 to 10. While the lunchroom is being renovated. *(Change the period to a comma.)*
- All 20-year employees are to be presented with long-service awards. Including three who retired earlier in the year. At the company's annual banquet. *(Change both periods to commas.)*

In particular, check sentences that start with a word that ends in "-ing" (e.g. referr*ing*, answer*ing*, be*ing*) or an expression that ends in "to" (e.g. with reference *to*):

- With reference to your letter of June 6. We have considered your request and will be sending you a check. *(Change the period to a comma.)*
- Referring to the problem of vandalism to employees' automobiles in the parking lot. We will be hiring a security guard to patrol the area from 8 a.m. to 6 p.m., Monday through Friday. *(Although the period could be changed to a comma, a better sentence could be formed by reconstructing the fragment:)*
- To resolve the problem of vandalism to employees' automobiles in the parking lot, we will be hiring...

Avoid Forming Run-on Sentences

Your reader will know what you are saying, but will feel uncomfortable reading your words

A similar sentence error occurs if you link two separate thoughts in a single sentence, joining them with only a comma or even no punctuation. The effect can jar a reader uncomfortably. For example:

- Ms. Solvason has been selected for the word-processing seminar on March 11, she is not eager to attend.

This awkward construction is known as a run-on sentence. It can be corrected by

- replacing the comma with a period, to form two complete sentences:

  ...on March 11. She is not eager to attend.

- or retaining the comma and following it with *which* or *but:*

  ...on March 11, which she is...
  ...on March 11, but she is...

A run-on sentence with no punctuation is even more noticeable:

The customer said he never received an invoice I made up a new one.

Either a period, or a comma and a linking word, must be inserted between the words *invoice* and *I*:

...an invoice. I made up...
...an invoice, so I made up...

Positioning End Punctuation Correctly

There are particular rules for inserting punctuation after quotation marks and closing parentheses:

- If a sentence ends with a quotation mark, place the period *inside* the quotation mark:

  "That's the information we need," the chief engineer remarked.
  ↑
  "Now we can start the project."
  ↑

(The above example shows that when a comma ends an introductory statement, it is also placed *inside* the closing quotation mark.)

These are North American rules; they differ in the UK

- If a sentence is enclosed within parentheses, place the period *outside* the closing parenthesis if the words within the parentheses do not form a complete sentence:

  Sound levels measured in the laboratory exceeded the tolerance specification (as shown in Table 2).
  ↑

- However, if the parentheses are preceded by a period, and the words within the parentheses create a complete sentence, then place the period *inside* the closing parenthesis:

  Sound levels measured in the laboratory exceeded the tolerance specification. (They were above 82 dBA for more than two hours per day.)
  ↑

Words

The right words in the right place at the right moment can greatly influence your readers. A heavy, ponderous word will slow them down; an

The keyword here is "specific"...

overused expression will make them doubt your sincerity; a complex word they do not recognize will annoy them; and a weak or vague word will make them think of you as indefinite. But the right word—short, clear, specific, and necessary—will help them understand your message quickly and easily.

Words That Tell a Story

Words should convey images. We have many strong, descriptive words in our individual vocabularies, but most of the time we are too lazy to use them because the same old routine words spring easily to mind. We write "put" when we would do better to write "position," "insert," "drop," "slide," or any one of the numerous descriptive verbs that better describe the action. Compare these examples of vague and descriptive words:

Whenever possible, insert specific words rather than generalizations

Vague Words	Descriptive Words
While the crew was in the town they *got* some spare parts.	they *bought* they *purchased* they *borrowed* they *requisitioned*
We have *contacted* the site.	We have *telephoned...* We have *visited...* We have *written to...* We have *spoken to...* We have *faxed...* We have *emailed...*
The project will *take a long time.*	will *last four months* will *require 300 work hours* will *employ two installers for three weeks*

Story writers use descriptive words to convey active images to their readers. Because we are concerned with *technical* writing does not mean we should avoid seeking colorful words. One descriptive word that defines size, shape, color, smell, texture, or taste is much more valuable than a dozen words that only generalize. (But a word of warning: reserve most of these colorful, descriptive words for your verbs and nouns rather than for adverbs and adjectives.)

Analogies can offer a useful means for describing an unfamiliar item in terms a nontechnical reader will recognize:

Relate a complex concept to a well-known idea or fact

> A resistor is a piece of ceramic-covered carbon about the size of a cribbage peg, with a 2 inch length of wire protruding from each end.

Specific words tell the reader that you are a definite, purposeful individual. Vague generalities imply that you are unsure of yourself. If you write

> It is considered that a fair percentage of the samples received from one of our suppliers during the preceding months contained a contaminant.

you give your reader four opportunities to wonder whether you really know much about the topic:

1. "It is considered"	Who has voiced this opinion?
2. "a fair percentage"	How many?
3. "one of our suppliers"	Who? One in how many?
4. "contained a contaminant"	What contaminant? In how strong a concentration?

All these generalizations can be avoided in a shorter, more specific sentence:

> We estimate that 60% of the samples received from RamSort Chemicals last June were contaminated with 0.5% to 0.8% mercuric chloride.

This statement tells the reader that you know exactly what you are talking about. As a technical person, you should never create any other impression.

Combining Words into Compound Terms

The trend is to compound multiword expressions into a single word

One of the biggest problems for technical writers is knowing whether multiword expressions should be compounded fully, joined by hyphens, or allowed to stand as two or more separate words. For example, should you write

cross check, cross-check, or crosscheck?
counter clockwise, counter-clockwise, or counterclockwise?
change over, change-over, or changeover?

The tendency today is to compound a multiword expression into a single term. But this bare statement cannot be applied as a general rule because there are too many variations, some of which appear in the glossary.

Most multiword expressions are compound adjectives. When two words combine to form an adjective they are either joined by a hyphen or compounded to form one word. They are usually joined by a hyphen if they are formed from an adjective-noun expression:

Adjective + Noun	*As a Compound Adjective*
heavy water	heavy-water production
four channels	four-channel receiver
high frequency	high-frequency oscillator

But when one of the combining words is a verb, they often combine into a one-word adjective. Under these conditions they will normally also compound into a single-word noun:

Two Words	*As a Noun*	*As an Adjective*
lock out	lockout	lockout voltage
shake down	shakedown	shakedown test
cross over	crossover	crossover network

Refer to the Glossary for the more common multi-word expressions

Three or more words that combine to form an adjective in most cases are joined by hyphens. For example, *lock test pulse* becomes *lock-test-pulse generator*. Occasionally, however, they are compounded into a single term, as in *counterelectromotive force*. Specific examples are listed in the glossary.

Obviously, these "rules" cannot be taken at full face value because there are occasions when they do not apply. Useful guides for doubtful combinations are contained in many dictionaries, and in the *Simon & Schuster Handbook for Writers* referenced earlier.

Long Versus Short Words

Avoid using an 89 cent word when an equally suitable 25 cent word is available

Big words create a barrier between writer and reader. There are many long scientific words that we have to use in technical writing; we should surround them with short words whenever possible so our writing will not become ponderous and overly complex.

Low-Information-Content Expressions

Words and expressions of low information content (LIC) contribute little or nothing to the facts conveyed by a sentence. Remove them and the sentence appears neater and says just as much. The problem is that practically everyone inserts LIC words into sentences, and we become so accustomed to them that we do not notice how they fill up space without adding any information. For example:

Vague and Wordy Pursuant to the client's original suggestion, Mr. Richards is of the opinion that the structure planned for the client would be most suitable for erection on the site until recently occupied by the old established costume manufacturer known as Garrick Garments. In accordance with the client's anticipated approval of this site, Mr. Richards has taken great pains to design a multi-level building that can be considered to use the property to an optimum extent.

"Waffle" is an ideal word to describe such cumbersome writing. By eliminating unnecessary expressions (such as *pursuant to; of the opinion that; for immediate erection on; in accordance with; can be considered to use; an optimum extent*), we can cut the original 74 words to a much more effective 40 words:

Clear and Direct	Mr. Richards believes the building planned for the client should be erected on the site previously occupied by Garrick Garments. He has assumed the client will approve this site, and has designed a multi-level building that fully develops the property.

Low-information-content words are untidy lodgers...

Table 12-2 contains some of the words and phrases you should delete from your writing. They are difficult to identify because they often sound like good prose. In the following sentences, the LIC words have been ital-

Table 12-2 Examples of low-information-content (LIC) words and phrases.

This is only a partial list: there are many more LIC expressions

The LIC words and phrases in this partial list are followed by an expression in parentheses (to illustrate a better way to write the phrase) or by an (X), which means that it should be dropped entirely.

actually (X)
a majority of (most)
a number of (many; several)
as a means of (for; to)
as a result (so)
as necessary (X)
at present (X)
at the rate of (at)
at the same time as (while)
at this time (X)
bring to a conclusion (conclude)
by means of (by)
by use of (by)
communicate with (talk to; telephone; write to)
connected together (connected)
contact (talk to; telephone; write to)
due to the fact that (because)
during the course of, during the time that (while)
end result (result)
exhibit a tendency (tend)
for a period of (for)
for the purpose of (for; to)
for this reason (because)
in all probability (probably)
in an area where (where)
in an effort to (to)
in close proximity to (close to; near)
in color, in length, in number, in size (X)
in connection with (about)
in fact, in point of fact (X)
in order to (to)
in such a manner as to (to)
in terms of (in; for)
in the course of (during)
in the direction of (toward)
in the event that (if)
in the form of (as)
in the light of (X)
in the neighborhood of; in the vicinity of (about; approximately; near)
involves the use of (employs; uses)
involves the necessity of (demands; requires)
is a person who (X)
is designed to be (is)
it can be seen that (thus; so)
it is considered desirable (I or we want to)
it will be necessary to (I, you, or we must)
of considerable magnitude (large)
on account of (because)
previous to, prior to (before)
subsequent to (after)
with the aid of (with; assisted by)
with the result that (so, therefore)

Note: Many of these phrases start and end with words such as *as, at, by, for, in, is, it, of, to,* and *with*. This knowledge can help you identify LIC words and phrases in your writing.

icized; they should be either deleted or replaced, as indicated by the notes in brackets.

- Flow is controlled by *means of* No. 3 valve. (delete) Or: No. 3 valve controls the flow. (active voice)
- Adjust the control *as necessary* to obtain maximum deflection. (delete)
- Tests were run for *a period of* three weeks. (delete)
- If the project drops behind schedule *it will be necessary to* bring in extra help. (*I, you, he, we,* or *they will* bring in extra help)
- By Wednesday we had a backlog of 632 units, *and for this reason* we adopted a two-shift operation. (replace with *so we adopted...*)
- A new store will be opened *in an area* where market research has *given an indication that there actually is a* need for more retail outlets. (delete both expressions; replace the second one with *research has indicated that...*)

Clichés and hackneyed expressions are similar to LIC words and phrases, except that their presence is more obvious and their effect can be more damaging. If you refer to yourself as "the writer," start and end letters with overworked phrases such as "We are in acknowledgment of..." and "...please feel free to call me," or use semilegal jargon such as "the aforementioned discussion," you will be considered garrulous and insincere. Further examples are listed in Table 12-3.

Some Fine Points

Using Parallelism to Good Effect

Good parallelism has a subtle, mostly hidden effect on readers

Parallelism in writing means "similarity of shape." It is applied loosely to whole documents, more firmly to paragraphs, sentences, and lists, and tightly to grammatical form. Readers generally do not notice that parallelism is present in a good piece of writing—they only know that the sentences read smoothly. But they are certainly aware that something is wrong when parallelism is lacking. Good parallelism makes readers feel comfortable, so that even in long or complex sentences they never lose their way. Its ability to help readers through difficult passages makes parallelism particularly applicable to technical writing.

The Grammatical Aspects

If you keep your verb forms similar throughout a sentence in which all parts are of equal importance (i.e. in which the sentence has coordination, or is balanced), you will have taken a major step toward preserving parallelism. For example, if you write "unable to predict" in the early part of a sentence, you should write "able to convince" rather than "successful in convincing" later in the sentence:

Table 12-3 Typical clichés and hackneyed expressions.

all things being equal	in the foreseeable future
a matter of concern	in the long run
and/or	in the matter of
as a matter of fact	last but not least
as per	many and diverse
attached hereto	needless to say
at this point in time	please feel free to
enclosed herewith	pursuant to your request
for your information (as an introductory phrase)	regarding the matter of
if and when	this will acknowledge
in our opinion	we are pleased to advise
in reference to	we wish to state
in short supply	with reference to
	you are hereby advised

Readers feel uncomfortable when parallelism is violated

Parallelism violated Mr. Johnson was *unable to predict* the job completion date, but was *successful in convincing* management that the job was under control.

Parallelism restored (A) Mr. Johnson was *unable to predict* the job completion date, but was *able to convince* management that the job was under control.

Parallelism restored (B) (more direct alternative) Mr. Johnson *predicted* no completion date, but *convinced* management that the job was under control.

The rhythm of the words is much more evident in the latter two sentences. In sentence (B) particularly, the parallelism has been stressed by converting the still awkward "...unable...able..." of sentence (A) to the positively parallel "predicted...convinced...."

Two other examples follow:

Parallelism violated Pete Hansk likes surveying airports and to study new construction techniques.

Parallelism restored Pete Hansk likes to survey airports and to study new construction techniques.

or

Pete Hansk likes surveying airports and studying new construction techniques.

Parallelism violated His hobbies are developing new software programs and stereo component construction.

Parallelism restored His hobbies are developing new software programs and constructing stereo components.

or

His hobbies are new software development and stereo component construction.

Parallelism is particularly important when you are joining sentence parts with the coordinating conjunctions *and, or,* and *but* (as in the examples above), with a comma, or with correlatives such as

either...or
neither...nor

Be particularly careful when using the expression *Not only...but also*

Each part of a correlative must be followed by an expression in the same grammatical form. That is, if *either* is followed by a verb, then *or* must also be followed by the same form of verb:

Parallelism violated	You may either repair the test set or it may be replaced under the warranty agreement.
Parallelism restored	You may either *repair* the test set or *replace* it under the warranty agreement.

Application to Technical Writing

Although parallelism is a useful means for maintaining continuity in general writing, it has special application in technical writing. Parallelism can clarify difficult passages and give rhythm to what otherwise might be dull material. When building sentences that have a series of clauses, you can help the reader see the connection between elements by molding them in the same shape throughout the sentence. There is complete loss of continuity in the following sentence because it lacks parallelism:

Parallelism violated	In our first list we inadvertently omitted the 7 lathes in room B101, 5 milling machines in room B117, and from the next room, B118, we also forgot to include 16 shapers.

When the sentence is written so that all the item descriptions have a similar shape, the clarity is restored:

Maintain similarity of shape within each sentence...

Parallelism restored	In our first list we inadvertently omitted 7 lathes in room B101, 5 milling machines in room B117, and 16 shapers in room B118.

Within the paragraph, parallelism has to be applied more subtly. If it is too obvious, the similarity in construction can be dull and repetitive. The verbs are often the key: keep them generally in the same mood and they will help bind the paragraph into one cohesive unit (this is closely tied in with coherence, discussed under "Paragraphs"). This paragraph has good parallelism:

...and within each paragraph

> The effects of sound are difficult to measure. What to some people is simply background noise, to others may be ear-shattering, peace-destroying drumming. The roar of a jet engine, the squeal of tires, the clatter of machinery, the hiss of air-conditioning, the chatter of children, and even the repetitive squeak of an unoiled door hinge, can seriously affect them and create a distinct feeling of uneasiness.

Similarity of shape is most obvious in the third sentence, with its rhythmic use of "sound" words:

roar of a jet engine	*hiss* of air-conditioning
squeal of tires	*chatter* of children
clatter of machinery	*squeak* of an unoiled door hinge

Here we have words that paint strong images. They not only have the same grammatical form but are also bound together because they relate to the same sense: hearing.

Even if the subject has less noticeable impact, you should still try to use parallelism to carry your reader smoothly through your description, as in this description of a surveyor's transit:

> Two sets of clamps and tangent screws are used to adjust the leveling head. The upper clamp fastens the upper and lower plates together, while the upper tangent screw permits a small differential movement between them. The lower clamp fastens the lower plate to the socket, while the lower tangent screw turns the plate through a small angle. When the upper and lower plates are clamped together they can be moved freely as a unit; but when both the upper and lower clamps are tightened the plates cannot be moved in any plane.

Very technical, yet the rhythm binds the parts together

Application to Subparagraphing

Subparagraphing seldom occurs in literature, but is used frequently in technical writing to separate events or steps, describe an operation or procedure, or list parts or components. Subparagraphing always demands good parallelism.

In the following example, a description of three tests has been divided into subparagraphs (for clarity, only the initial words of each test are shown here):

> Three tests were conducted to isolate the fault:
> - In the first test a matrix was imposed upon the video screen and...
> - For the second test, voltage measurements were taken at...
> - A continuity tester was connected to the unit for test 3 and...

The last subparagraph is not parallel with the first two. To be parallel it must adopt the same approach as the others (i.e. it should first mention the test number and then say what was done):

It's when writing point form that parallelism becomes particularly important

> - For the third test, a continuity tester was connected to the unit and...

To be truly parallel the subparagraphs should start *exactly* the same way:

> - In the first test a matrix was...
> - In the second test voltage measurements were...
> - In the third test a continuity tester was...

But this would be too repetitive. The slight variations in the original version make the parallelism more palatable.

If more than three tests have to be described, a different approach is necessary. To continue with "A fourth test showed...," "For the fifth test...," and so on, would be dull and unimaginative. A better method is to insert a number in front of each paragraph:

Seven tests were conducted to isolate the fault:

1. A matrix was imposed upon the video screen and...
2. Voltage measurements were taken at...
3. A continuity tester was connected to the unit and...

Parallelism has been retained and we now have a more emphatic tone that lends itself to technical reporting.

A third version employs active verbs together with parallelism to build a strong, emphatic description that can be written in either the first or third person:

A running lead-in line and imperative-mood verbs create a convincing effect

In laboratory tests conducted to isolate the fault we

1. *imposed* a matrix upon the video screen and...
2. *measured* voltage at...
3. *connected* a continuity tester to the unit and...

To change from the first person to the third person, only the lead-in sentence has to be rewritten:

In tests conducted to isolate the fault, the laboratory

1. *imposed...*
2. *measured...*
3. *connected...*

See page 325 for more information on inserting bullets and paragraph numbers.

Abbreviating Technical and Nontechnical Terms

You may abbreviate any term you like, and in any form you like, providing you indicate clearly to the reader how you intend to abbreviate it. This can be done by stating the term in full, then showing the abbreviation in parentheses to indicate that from now on you intend to use the abbreviation. Here is an example:

In technical narrative, spell out single-digit numbers (sdn). However, when sdn are being inserted into a series of numbers, write them as numerals.

When forming abbreviations, observe these three basic rules:

1. **Use lower case letters,** unless the abbreviation is formed from a person's name:

centimeter	cm
kilogram	kg
approximately	approx
decibel	dB (the *B* represents *Bell* [Alexander Graham Bell])

2. **Omit all periods,** unless the abbreviation forms another word:

These guidelines are recognized worldwide

horsepower	hp
cubic centimeter	cm^3
cathode-ray tube	crt

meter	m
pascal	Pa
singular	sing.

3. **Write plural abbreviations in the same form as the singular abbreviation:**

meters	m
pascals	Pa
kilograms	kg
hours	h *or* hr

There are, however, exceptions, which have grown as part of our language. Through continued use, these unnatural abbreviations have been generally accepted as the correct form. A few examples follow:

The Glossary is there to help you

for example *(exempli gratia)*	e.g.	*(There is a slowly growing trend to write these as* eg *and* ie*)*
that is *(id est)*	i.e.	
morning *(ante meridiem)*	a.m.	
afternoon *(post meridiem)*	p.m.	
inside diameter	ID	
number(s)	No.	

The Glossary of Technical Usage offers a reasonably comprehensive list of standard abbreviations and some technical abbreviations. For specific technical terms, you may need to refer to a list of abbreviations compiled by one of the technical societies in your discipline.

Writing Numbers in Narrative

The conventions that dictate whether a number should be written out or expressed in figures differ between ordinary writing and technical writing. In technical writing you are much more likely to express numbers in figures.

The rules listed below are intended mainly as a guide. They will apply most of the time, but there will be occasions when you will have to make a decision between two rules that conflict. Your decision should then be based on three criteria:

- Which method will be most readable.
- Which method will be simplest to type.
- Which method you used previously, under similar circumstances.

Good judgment and a desire to be consistent will help you select the best method each time.

The basic rule for writing numbers in technical narrative is

- spell out single-digit numbers (one to nine inclusive), and

- use figures for multiple-digit numbers (10 and above).

However, there are exceptions to this rule:

Always use figures

Numerals are more common than spelled-out numbers in most technical narrative

- when writing specific technical information, such as test results, dimensions, tolerances, temperatures, statistics, and quotations from tabular data,
- when writing any number that precedes a unit of measurement: 3 mm; 7 kg; 121.5 MHz,
- when writing a series of both large and small numbers in one passage: During the week ending May 27 we tested 7 transmitters, 49 receivers, and 38 power supplies,
- when referring to section, chapter, page, figure (illustration), and table numbers: Chapter 7; Figure 4,
- for numbers that contain fractions or decimals: 7 1/4, 7.25,
- for percentages: 3% gain; 11% sales tax,
- for years, dates, and times: At 3 p.m. on January 9, 2004; 08:17, 20 Feb 04,
- for sums of money: $2000; $28.50; $20; 27 cents or $0.27 (preferred), and
- for ages of persons.

Always spell out

- round numbers that are generalizations: about five hundred; approximately forty thousand,
- fractions that stand alone: repairs were made in less than three-quarters of an hour, and
- numbers that start a sentence (better still, rewrite the sentence so that the number is not at the beginning).

There are five additional rules:

- Spell out one of the numbers when two numbers are written consecutively and are not separated by punctuation: 36 fifty-watt amplifiers or thirty-six 50-watt amplifiers. (Generally, spell out whichever number will result in the simplest or shortest expression.)

Always place an "0" in front of an open decimal

- Insert a zero before the decimal point of numbers less than one: 0.75; 0.0037.
- Use decimals rather than fractions (they are easier to type), except when writing numbers that are customarily written as fractions.
- Insert commas in large numbers containing five or more digits: 1,275,000; 27,291; 4056. (Insert a comma in four-digit numbers only when they appear as part of a column of numbers.)
- Write numbers that denote position in a sequence as 1st, 2nd, 3rd, 4th...31st...42nd...103rd...124th...

When writing and abbreviating numerical prefixes such as "giga" and "kilo," follow the guidelines in Table 12-4.

Writing Metric Units and Symbols (SI)

The Glossary of Technical Usage includes terms and symbols prescribed by the International System of Units (SI). The trend toward worldwide adoption of metric units of measurement means that for some time both the imperial inch/pound system and the metric (SI) system will be in use concurrently. The terms and symbols introduced here are those you are most likely to encounter in your technical reading or may use in your technical writing.

The acronym "SI" represents the name "Système International d'Unités." Both the acronym and the name were adopted for universal usage in 1960 by the eleventh Conférence Générale des Poids et Mesures (CGPM), which is the international authority on metrication. Since then, many of the metric terms the conference established have crept into our language. For example, *Hertz*, the unit of frequency measurement, was first introduced as a replacement for *cycles per second* in the early 1960s; now it is used universally, both in the technical world and by the general public. Other terms already in use are:

	non-SI	*SI*
Temperature:	degrees Fahrenheit	degrees Celsius
Length:	miles, yards, feet, inches	kilometers, meters, millimeters

In SI, the "-re" spelling is correct for *litre* and *metre*, although only the "-er" spelling is recognized in the US

Table 12-4 Numerical prefixes and abbreviations.

Multiple/ Submultiple	Prefix	Symbol	Multiple/ Submultiple	Prefix	Symbol
10^{24}	yotta	Y	10^{-1}	deci	d
10^{21}	zetta	Z	10^{-2}	centi	c
10^{18}	exa	E	10^{-3}	milli	m
10^{15}	peta	P	10^{-6}	micro	µ
10^{12}	tera	T	10^{-9}	nano	n
10^{9}	giga	G	10^{-12}	pico	p
10^{6}	mega	M	10^{-15}	femto	f
10^{3}	kilo	k	10^{-18}	atto	a
10^{2}	hecto	h	10^{-21}	zepto	z
10	deca	da	10^{-24}	yocto	y

This guideline is also recognized worldwide

Weight:	tons, pounds, ounces	tonnes, kilograms, grams, milligrams
Liquid volume:	gallons, quarts	kiloliters, liters

There are nine general guidelines for writing SI symbols, and these are applied consistently to the entries in the Glossary of Technical Usage. SI symbols must be written, typed, or printed

The abbreviation for *liter* is L, because a lower case letter l looks like the numeral 1

1. in upright type, even if the surrounding type slopes or is in italic letters,
2. in lower case letters, except when the name of the unit is derived from a person's name (e.g. the symbol F for *farad* is derived from *Faraday*), in which case the first letter of the symbol is capitalized (e.g. Wb for *weber*),
3. with a space between the last numeral and the first letter of the symbol: **355 V, 27 km** (not 355V, 27km),
4. with no "s" added to a plural: **1 g, 236 kg,**
5. with no period after the symbol, unless it forms the last word in a sentence,
6. with no space between the multiple or submultiple symbol and the SI symbol: **3.6 kg, 150 mm, 960 kHz,**
7. with a solidus (oblique stroke: /) to represent the word *per*: **m/s** (meters per second), and with only one solidus used in each expression,
8. with a dot at midletter height (·) to represent that symbols are multiplied: **lm·s** (lumen second), and
9. always as a symbol, when a number is used with the SI unit (e.g."the tank holds 400 L"), but spelled out when no number is used with the unit (e.g "capacity is measured in liters" [*not* "capacity is measured in L"]).

Writing Non-Gender-Specific Language

History has provided us with a scenario in which men were the warriors and hunters, and subsequently the breadwinners, and women were the home bodies who cooked and reared children and catered to their men's needs. Today, all that has changed and it is universally recognized in developed countries that women and men are equal and can for the most part have equal occupations and equal roles. Consequently, we now see men as administrative assistants, nurses, and child care workers, and women as airline captains, engineers, truck drivers, and backhoe operators.

Unfortunately, our language has not kept pace and we still see some people who write like this:

Awareness is the key to writing non-gender-specific language

> An administrative assistant will be brought in to record the minutes of the client/contractor project meeting. *She* will be responsible for making travel arrangements for all meeting participants.

After much deliberation, the committee decided to hire an engineer to look into the problem. *He* will evaluate the extent of erosion that occurred when the river overflowed its banks.

Neither of these writers knew whether the administrative assistant and the engineer were going to be male or female, yet *they automatically assumed* that the administrative assistant would be a woman and the engineer would be a man. Traditionally, that was what they and their parents, and their grandparents, and their parents before them, were accustomed to.

It's our job to eradicate gender-specific references like these from our writing, until it becomes *automatic* always to write non-gender-specific references. For example:

An administrative assistant will be brought in to record the minutes of the client/contractor project meeting, and to make travel arrangements for all meeting participants.

After much deliberation, the committee decided to hire an engineer to evaluate the extent of erosion that occurred when the river overflowed its banks.

Interestingly, these examples are shorter than the original sentences

Eliminate Masculine Pronouns

When describing engineers, scientists, architects, managers, supervisors, technical people, and even accountants and lawyers, historically our language has abounded with masculine pronouns. The engineer described earlier is a typical example. Here is another, this time an excerpt from a company's operating procedures:

4.3 **Senior Systems Engineer.** *His* primary role is to plan, schedule, manage, and coordinate the activities of the engineers within the Systems Engineering Department. *He* also is responsible for preparing budgets and maintaining fiscal control of operations performed by the department, and for maintaining liaison with and reporting progress to clients.

There are several ways you can remove the male pronouns:

1. Repeat the job title, and abbreviate it:

4.3 **Senior Systems Engineer (SSE).** The SSE's primary role is to plan, schedule, manage, and coordinate the activities of the engineers within the Systems Engineering Department. The SSE also is responsible for preparing budgets and maintaining fiscal control of operations performed by the department, and for maintaining liaison with and reporting progress to clients.

2. Use a bulleted list:

4.3 **Senior Systems Engineer.** The Senior Systems Engineer is responsible for

- planning, scheduling, managing, and coordinating the activities of the engineers within the Systems Engineering Department,

This is probably the clearest and most comfortable revision

- preparing budgets and maintaining fiscal control of operations performed by the department, and
- maintaining liaison with and reporting progress to clients.

3. Create a table:

4.3 Senior Systems Engineer

This revision employs good information design principles

Primary Responsibility	Secondary Responsibilities
To plan, schedule, manage, and coordinate the activities of the engineers within the Systems Engineering Department.	To prepare budgets and maintain fiscal control of operations performed by the department. To maintain liaison with and report progress to clients.

4. Replace the male pronoun with "you" and "your":

4.3 Senior Systems Engineer. *Your* primary role is to plan, schedule, manage, and coordinate the activities of the engineers within the Systems Engineering Department. *You* also are responsible for preparing budgets and maintaining fiscal control of operations performed by the department, and for maintaining liaison with and reporting progress to clients.

(Note: If you use "you" in one part of a document, be consistent and use it throughout the document. Avoid bouncing back and forth between "you" and "he" or "she".)

5. Replace the male pronoun with "he or she":

Suggestions 4, 5, and 6 are less comfortable revisions

4.3 Senior Systems Engineer. *His or her* primary role is to plan, schedule, manage, and coordinate the activities of the engineers within the Systems Engineering Department. *He or she* also is responsible for preparing budgets and maintaining fiscal control of operations performed by the department, and for maintaining liaison with and reporting progress to clients.

(Note: This is the least recommended method.)

6. Change singular pronouns to plural pronouns:

4.3 Senior Systems Engineers. *Their* primary role is to plan, schedule, manage, and coordinate the activities of the engineers within the Systems Engineering Department. *They* also are responsible for preparing budgets and maintaining fiscal control of operations performed by the department, and for maintaining liaison with and reporting progress to clients.

(Note: This method can be used only when the description lends itself to using plural nouns and pronouns; i.e. there must be more than one Senior Systems Engineer.)

Replace Gender-Specific Nouns

Each state has its Workers Compensation Board, an organization that provides financial help to employees who are injured at work. Yet, not many years ago, all Workers Compensation Boards were known as *Workman's* Compensation Boards. The previous title seemed to imply that the board provided help *only* to male workers, which was not true. Similarly, until about 15 years ago, flight attendants on airlines were known as stewardesses, implying that the job was held only by females. Again, particularly today, this is plainly inaccurate.

There are many other job titles that are equally gender-specific and predominantly male-oriented. These have been changed in recent years to reflect that the title refers to both male and female employees. Table 12-5 lists gender-specific titles and suggests better alternatives.

The term "man-hours" previously was used to define the time that would be expended on a particular job. Today, we write *work-hours* or *staff hours*.

Be Consistent When Referring to Men and Women

Men throughout recent history have been given the courtesy title *Mr.* before their names. Until 20 years ago, women had two courtesy titles, to denote whether they were married or single: *Mrs.* and *Miss*. Today, a woman's marital status is *never* implied in the title: all women should be referred to as *Ms.* (English is not the only language to have created this anomaly. For example, in France men are referred to as *Monsieur* and women as *Madame* or *Mademoiselle*. In Russia, a sex-identifying title is not placed before a per-

Titles such as *fireman*, *foreman*, and *salesman* are deeply entrenched, and so are difficult to eradicate

Table 12-5 Preferred names for gender-specific titles.

If you are tempted to write:	Consider replacing it with:
actor; actress	actor (for both sexes)
chairman	chairperson (*or* chair)
cowboy	cattle rancher
fireman	firefighter
foreman	supervisor
newsman	reporter
policeman; policewoman	police officer
postman	letter (*or* mail) carrier
repairman	service technician
salesman	sales representative
spokesman	spokesperson
workman	worker; employee
waiter; waitress	server (or "waiter" for both sexes)

son's name, but a woman's family name has an "a" added to the end to denote the person is female: for example, Boris Serov; Svetlana Serova.)

Be particularly careful when addressing letters and writing the salutation

Never address a letter to "Dear Sir or Madam." If you are replying to a letter signed by A. J. Winters, and you don't know if A. J. Winters is male or female, then in the salutation write "Dear A. J. Winters:". If you don't know the person's name, then address the letter to *the position the person holds*: "Dear Customer Services Manager:" or "Dear Manager, Human Resources:".

To sum up: we need to be continually vigilant and check that we never use gender-specific expressions in our writing.

Writing for an International Audience

Three years ago, when Macro Engineering Inc. started doing business with companies in Eastern Europe, Senior Engineer Sharleen Burton had to learn a different set of rules for communicating with the company's new customers and business associates. She found it strange that she had to change the way she wrote her letters and memos.

In a growing global economy, we have to be aware of and adapt to cultural differences

"I took it for granted that everyone used the pyramid method to construct their letters," she explained. "But then I found that in many Eastern European and Asian countries, starting immediately with the main message—getting right down to business—is considered downright rude."

Sharleen also discovered that people in many European and Asian countries are much more formal in their greetings than we tend to be in North America. Where an American writing to Paul Villeneuve would write *Dear Paul* quite early in an exchange of letters, an Italian or Norwegian (for example) would continue to write *Dear Mr. Villeneuve* for much longer.

Writing Business Correspondence

At first, Sharleen thought it would be just a matter of explaining to people she corresponded with in Russia, Lithuania, and the Ukraine, that she wrote this way because in the West we had discovered it was much more efficient. She thought it would help their economy if they adopted our methods. But she had forgotten the hundreds of years of history that has fashioned the Eastern and Asian cultures, and found that she had to adapt her methods to suit them. "You have to understand the culture prevalent in each society and adjust to it," Sharleen says now.

She did not entirely discard the pyramid method. She decided she could still use the pyramid for the central part of her letters, but that she would have to precede it with a personal greeting and polite remarks concerning the health and happiness of her reader (and, often, also of her reader's family). And she would have to follow the pyramid with a polite closing remark, such as wishing the reader continuing good health and prosperity in the months and years ahead.

So at first she constructed her letters to Eastern Europe as shown in Figure 12-5. "Yet even then I had to be cautious," Sharleen continued. "I learned fairly quickly that this arrangement was fine for readers I had corresponded with before, but to new readers—particularly older, more traditional readers—it still seemed too abrupt. For them, I had to move the Summary Statement further down in the letter."

Adapt the pyramid to suit the reader's culture

Revising the Writing Plan

Sharleen's revised writing plan looked like this:

1. Greeting
2. Background
3. Details
4. Outcome and Summary Statement (combined)
5. Complimentary Close

This, of course, is the inverted pyramid, which is the reverse of the sequence promoted in previous chapters. And it applies not only to formal business letters but also to memos and electronic means of communication: faxes and particularly email.

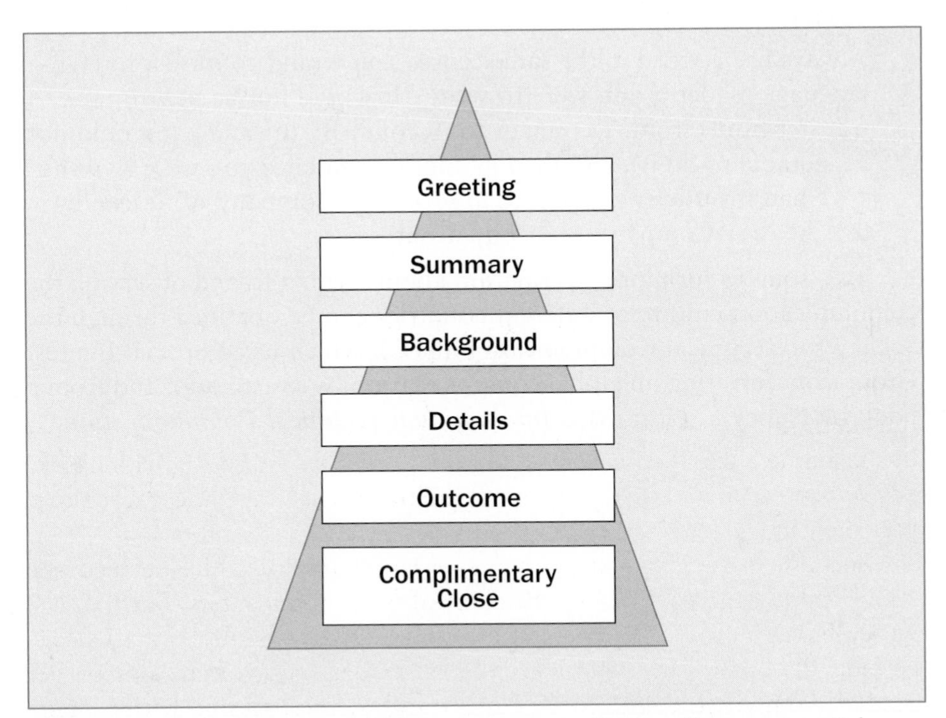

Figure 12-5 "Adapted" pyramid for a letter to an Eastern European or Asian reader.

Avoid jargon: choose words that will be understood by both writer and reader

However, changing the focus is not the only part of a message to an Eastern or Asian country that requires attention. When you write in English to readers who normally speak another language—German, French, Italian, Spanish, Malay, or Chinese, for example—you have to choose words that will be clearly understood. (This also holds true for different cultures who speak the same language. In Great Britain, for instance, the word "fortnight" is commonly used to mean "two weeks," yet in the United States it would not be understood. Similarly, the word "presently" means "shortly" in Great Britain but "right away" in the US.

Writing Guidelines

Here are some guidelines to follow when writing to an international audience:

- Avoid long, complex sentences.
- Avoid long, complex words. If you have a choice between two or more words or expressions that have roughly the same meaning, choose the simpler one. For example, write "pay" rather than "salary" or "remuneration."
- Use the same word to describe the same action or product consistently throughout your letter. Decide, for example, whether you will refer to money in the bank as *funds, currency, deposits, capital,* or *money*.
- Always use a word in the same sense. You would confuse a foreign-language reader if you were to write "It would not be *appropriate* to transfer funds from Account A to Account B" (meaning it would not be suitable to do it), and then in another sentence you were to write "We had insufficient capital to *appropriate* Company A" (meaning to take over Company A, or buy it out).

Two sources for more information about writing to and observing the communication culture of different countries can be obtained through the Society for Technical Communication (STC), which has a Special Interest Group concentrating on global communication (www.stc.org), and from a book by Nancy L. Hoft titled *International Technical Communication.*[3]

---

[3] Nancy L. Hoft, *International Technical Communication: How to Export Information About High Technology* (New York: John Wiley & Sons, Inc., 1995).

ASSIGNMENTS

Exercise 12.1

The following sentences and short passages lack compactness, simplicity, or clarity, and particularly contain low-information-content (LIC) expressions. Improve them by deleting unnecessary words, or by partial or complete rewriting.

Simple ideas confused by wordy, rambling sentences

1. A gas leak was the cause of a one-day delay in production.
2. The company has installed a microprocessor-controlled scanner in an effort to increase quality control of assembled components.
3. It's our considered opinion that you will be able to start beta-testing the online preventative maintenance training program in the neighborhood of eight days before the scheduled completion date.
4. If you have experience with any further problems with the Model 17 LCD panel, please do not hesitate to contact me at any time.
5. The *4Tell* software will be initiated as a means of preventing slippage of the schedule.
6. The Freeling Lake Mine survey project has, at this point in time, reached the point of being 3.5 days behind schedule.
7. It is a possibility that the technician may or may not have read the addendum correctly.
8. Company tools may be borrowed for home use for a time that is not in excess of 48 hours.
9. You can make arrangements for the borrowing of company equipment by the completion of Form 210A.
10. For your information, it was decided by the executive committee that a sum in the amount of $240,000 be set aside in next year's budget for the purpose of renovating the metrology lab.
11. The end result was a 22% decrease in acidity following the introduction of Limasol Plus into the solution.
12. Any attempt to operate the engine in excess of 1500 rpm is likely to result in and be the cause of accelerated bearing failure.
13. If it is your intention to effect repairs to your model 1800 scanner, it would be wise first to assure the availability of parts before starting work.
14. Check for grain temperature with the use of a probe 5 feet in length inserted vertically downward into the depths of the grain.
15. Registration for next year's Institute in Technical Management should be made before or no later than December 15.
16. Between a speed of 65 and 70 mph the wheel at the left front exhibits a tendency to shimmy.
17. It is with considerable concern that we have noticed a decrease in production in the region of 3% to 5% for last year.
18. Overtime will be worked for a period of two weeks in order to bring the project back on schedule by no later than the end of next week.

19. Installation and testing of the remote sensing unit will bring the Fairview research program to a conclusion.
20. The purchase of a replacement vehicle for the survey crew will be completed in short time so the vehicle will be ready in good time for the start of the survey season.

Exercise 12.2

The following sentences offer choices between words that sound similar or are frequently misused. Select the correct word in each case.

These word choices offer unexpected surprises!

1. Signals from space probe 811 traveled 86 billion miles (farther/further) than signals from space probe 260.
2. Three attempts were made to (elicit/illicit) a response from station K-2.
3. Although we have developed a unique software product, our clients apparently are (disinterested/uninterested) in purchasing it.
4. Before clamping the scale onto its base, (orient/orientate) the 0° point on the scale so that it is opposite the "N" mark.
5. Our vehicle was (stationary/stationery) when truck T48106 skidded into it.
6. The inspection team calibrated 38 of the 47 test instruments between April 28 (and/to) May 6. The (balance/remainder) will be calibrated on May 13 and 14.
7. Fuel consumption is (affected/effected) by driving speed.
8. The quality control inspection team recommended instituting a (preventive/preventative) maintenance program to reduce system outages.
9. We could not complete the modifications on schedule (as/for/because) the remote control unit had not yet been shipped to the site.
10. Although the report (seemed/appeared) to document the investigation results accurately, its executive summary (implied/inferred) there were hidden costs.
11. When compared (to/with) the previous period, the (amount/number) of products rejected by the quality control inspectors decreased from 237 to 189.
12. Under current human rights legislation, employers have to be particularly (discreet/discrete) when making personal enquiries about potential employees.

Some words are so similar it can be difficult to choose between them

13. Analyses of the data (is/are) being delayed until all the data (has/have) been received.
14. The inspection team will submit (its/it's/their) final report on October 26.
15. Improved maintenance in 2003 resulted in 27 (fewer/less) service interruptions than in 2002.
16. When the temperature within the cabinet rises above 28.5°C, the blower motor cuts in automatically and operates (continually/continuously) until the cabinet temperature drops to 25°C.

17. Before accepting each job lot of steel from the foundry, samples are tested mechanically and electronically to ensure the steel is (free from/free of) flaws.
18. The chief engineer agreed that the agenda (was/were) too comprehensive for a one-hour meeting.
19. When US Air had to cancel my flight from New York to Milan because the aircraft had technical problems, the gate agent had to find an (alternate/alternative) route for me to fly on another airline.
20. The (principal/principle) reason for including this exercise in *Technically-Write!* is to draw your attention to the glossary of technical usage.

Exercise 12.3

Rewrite the following sentences to make them more emphatic (in many cases, change them from the passive to active voice). Create a "doer" if one is not identified.

1. The minutes of the November 17 project meeting were recorded by technician Jean Melnyk.
2. It was recommended by the quality control inspector that production lots 185 and 224 be reworked.
3. Yesterday's power failure was caused by a lightning strike at Westbourne power station.
4. The fractured rotor arm was repaired by Roger Cormier.
5. The *Amaze* software was shipped by express post on Thursday, July 12, but was not received by the customer until Tuesday, July 17.
6. Although Whistler Mountain's Olympic Station was obscured by fog for 24 hours, work by the crew repairing the chairlift continued throughout the night.
7. Safety boots and hardhats are to be worn by all technicians working within 100 yards of the construction site.
8. The permafrost north of Lac la Biche has been damaged irreparably by SUVs driven by oil prospectors.
9. Due to poor visibility and drifting snow caused by a blizzard on January 15, the highway to Montrose International Airport was closed by the State Police from 4 p.m. to 9 a.m. the following day.
10. As a result of the high sound levels recorded by Murray Walsh, it was decided by management that the machine shop should be operated on a reduced scale until sound-reduction measures could be implemented.
11. When toxicity levels in No. 3 tank were measured by Fran Wheeler, it was evident from her report that the requirements of specification DZ0286 had been exceeded by 18%.
12. It was stipulated by the accounting department that all requests for international travel must first be approved by the manager of admin-

Changing from passive to active voice will shorten these sentences

A passive-voice sentence is longer than an active-voice sentence

istration, and then all arrangements for international travel are to be made by Columbia Travel at 3030 Windmark Drive.

13. The opportunity to prepare a proposal for installation and maintenance of the ISDN network for Regent Courier Systems was missed because their Request for Proposal was sent to our old address. It was forwarded by the post office too late for us to meet the submission deadline.
14. When the *4Tell* software was evaluated, it was discovered by Sheila Fieldstone that the program was only 80% compatible with the operating system used by the company for its mainframe computer.
15. In a memo dated February 10 from Luis Cruz, Fern Wilshareen was asked to recommend an engineering technologist from her department to attend the Engineering Technology Symposium in Chicago on April 3 and 4.

Exercise 12.4

Improve the parallelism in the following sentences.

1. A survey was carried out to define property boundaries between 276 and 278 Lawson Avenue and as a means for settling disputes between the property owners.
2. Write your name, address, postal code, and telephone number into spaces 1, 2, 4, and 6 of the application form, and in space 8 list your email address.
3. Multiple Industries' metal shears failed on January 12, were repaired on January 15, and then on January 22 they failed again.
4. Getting approval to purchase a replacement control unit will not be a problem, but there will be difficulty in finding a local supplier who can deliver it by June 16.
5. If you decide to order three Nabuchi 310 portable computers, the price will be $1995 each, but the price will be $2250 each if you order only one or two.

Keep the ideas—like trains—on parallel tracks

6. When the Montrose branch office closed, two staff members were transferred to the Syracuse office and a generous severance package was arranged for five who were laid off.
7. We selected the Nabuchi 700 inkjet printer because it is fast, moderately priced, readily available, and its scalable fonts can be used directly with our word-processing software.
8. The accident delayed installation work for three days and was the cause of increased operating costs.
9. Inspection of the air extraction unit revealed
   - rust on the elements,
   - a crack in the inlet bellows,
   - someone had deposited chewing gum on the fan blades, and
   - loose clamps on the fresh air inlet and the outlet pipes.

10. Management hired a consultant to conduct an independent study for three reasons: (1) to determine the cause of the accident; (2) as a means for demonstrating to staff that the company has taken the problem very seriously; and (3) to find a means for preventing a recurrence.
11. The supplier explained that version 8.1 of the *4Tell* 2000 software not only has made the program more user-friendly, but also the severe tendency for the version 7.4 software to cause "mouse freeze" has been corrected.
12. A revised procedure for submitting expense accounts is to be followed by all field staff:
    1. They are to be submitted weekly.
    2. Complete them on Monday for the preceding week.
    3. They should be completed in quadruplicate.
    4. Mail copies 1, 2, and 3 to head office, retaining copy 4 for your files.
    5. Be sure to mail them no later than noon on Tuesday.

Exercise 12.5

Abbreviate the terms shown in italics in the following sentences. In some cases you will also have to express numerals in the proper form. (Guidelines for forming abbreviations and writing numbers in narrative are on pages 352 to 355. The Glossary of Technical Usage also contains many technical abbreviations.)

1. A 24-*inch diameter* culvert, *approximately* 330 *yards* long, will connect the proposed development to the main drainage channel.
2. The accident damaged three Nabuchi model 6200 portable computers, *serial numbers* 0087, 0192, and 0698, that were packed in carton *numbers* three, seven, and 14.
3. The sound level recorded six *feet* from the south wall of the office was 58.6 *decibels*.
4. 8,000 of the 68 656 cylinders for the Norland contract were manufactured with a *22.5 millimeter inside diameter.*
5. Rotate the antenna in a *counterclockwise* direction until the needle on the dial dips to between the .6 and 1.4 markings.
6. Vibration became severe at speeds above 54 *miles per hour*, which forced us to drive more slowly than anticipated. However, this enabled us to achieve a fuel consumption of 21.7 *miles per gallon.*
7. Radio station DMON's two operating frequencies are 88.1 *kilohertz* and 103.6 *megahertz.*
8. Paragraph three on page seven of the proposal quoted an additional six hundred and fifty-seven dollars ($657.00), for repairs to the damaged roof.
9. One degree of latitude is the equivalent of sixty *nautical miles*, so each minute of latitude equals one *nautical mile* or 6080 feet (or 1.609 *kilometers*).

Refer to the guidelines earlier in this chapter, or turn directly to the Glossary

10. The new furnace produces 56,000 *British thermal units*, which is seven *per cent* more than the unit it replaces.
11. Comparing temperatures is not difficult if you remember that zero *degrees Celsius* is the same as thirty-two *degrees Fahrenheit*, twenty-two *degrees Celsius* is equal to seventy-two *degrees Fahrenheit*, and that both *temperatures* are the same at minus forty *degrees*.
12. Tank *serial number* 2821 holds one thousand *liters* of oil, which is equivalent to 264 *United States gallons* or 220 *imperial gallons*.

Exercise 12.6

Correct any of the following sentences that are not complete or have not been punctuated properly.

An exercise in using appropriate punctuation

1. We have examined your bubble-jet printer, repairs will cost $148.00 plus tax, shall we go ahead?
2. The hard disk crashed at 4:15 p.m. Before I had time to copy today's work onto a safety disk.
3. Do you have a Hama A16 Electronic Camera Flash in your department, you can recognize it by the words *Sorte mit Automatik* on the base.
4. With effect from October 31 we will assign Myra Weiss and Dan Helwig to your Division for six months, this confirms our telephone conversation of October 17.
5. The safety label warned: "Toxic solution, handle with great care".
6. Progressive corrosion inside the pipes has reduced liquid flow by 21% since 1996. A condition which, if not corrected, could cause system shutdown in less than 12 months.
7. Work on the Feldstet contract was completed on February 16 three days ahead of the February 19 scheduled completion date, a cause for celebration.
8. When the digital exchange was installed at Multiple Industries, eight lines were left unused for anticipated staff expansion. Also for providing dedicated lines for a planned facsimile transmission network between branches.
9. The overhead steam pipe ruptured at 10:10 a.m., fortunately the office was empty as everyone had gone down to the cafeteria for coffee break. Although damage was effected to the computer equipment and oak furniture.
10. May I have company approval to attend the course "Preparing for ISO 2001 Approval," to be held at Oklahoma State University May 16–20. And as attendance is limited may I have a reply by April 28.
11. Your request to attend the May course on "Preparing for ISO 2001 Approval" as requested in your memo dated April 10 was approved at the April 15 meeting of the Executive Committee, Francine Williams who attended the same course in January was particularly helpful in recommending that you attend.

12. Before submitting purchase requests for new or replacement equipment, ensure that
    - manufacturer details are complete,
    - price quotations are attached,
    - The appropriate specifications are quoted,
    - An alternative supplier is listed, and
    - The divisional manager approves the request.

Exercise 12.7

Select the correctly spelled words among the choices offered below.

Well? How good are you at spelling?

1. The (computer/computor) is supplied with a built-in 99-year (calendar/calender).
2. A (coarse/course)-grained (aggregate/agreggate) is used as a base before pouring the concrete.
3. Profits in the (forth/fourth) quarter increased by (forty/fourty) percent.
4. The (affluent/effluent) produced by the paper mill is (enviromentally/environmentaly/environmentally) sound.
5. The preface to a book or report is sometimes called a (forward/foreward/foreword).
6. The (cite/site) is (inaccessable/inaccessible) except by helicopter.
7. Version 6.0 of *4Tell for Windows 05* (supercedes, supersedes) version 5.5.
8. The tests show that the materials have (similar/similiar) properties.
9. To a young business owner seeking a cash flow loan, the (colatteral/collatteral/collateral) demanded by the bank may seem (exhorbitant/exorbitant).
10. When Multiple Industries bought all the outstanding shares of Torrance Electronics, the latter company became a (wholely/wholly)-owned subsidiary.
11. Its better to edit your own writing on (hard copy/hardcopy) rather than (online/on line).
12. Silica gel is a drying agent, or (desiccant/dessiccant/dessicant), that is packed with electronic equipment before shipment.
13. Software designers who have (entepreneurial/entrepeneurial/entrepreneurial) drive do not (necessarily/neccessarily) have good management expertise.
14. After (lengthy/lengthly) deliberation, the executive committee admitted that Ken Wynne's innovative design was indeed (ingenious/ingenuous).
15. After we have (accumulated/accummulated) all the results from product tests, we will (prescribe/proscribe) definitive purchase specifications.
16. An (auxiliary/auxilliary) heater cuts in when temperature drops below 3°C.

17. The (eigth/eighth) test demonstrated that the process is (feasable/feasible).
18. The incandescent lamps have been replaced with (flourescent/fluorescent/fluourescent) lamps.
19. The sales manager was (embarassed/embarrassed/embarrased) that customers were being (harassed/harrassed/harrased) by overly zealous sales staff.
20. Well? How many words did you (mispell/misspell)?

All these words are in the Glossary

Exercise 12.8

Describe why gender-specific terms are inappropriate in the letters and reports you write. Identify five additional gender-specific terms not listed in this chapter and, for each, provide a non-gender-specific alternative.

Exercise 12.9

Improve the following passages so they contain no gender-specific language and (where appropriate) are better conveyors of information:

1. Memo to all lab technicians:

   The Department of Defense has informed us that a D.O.D. inspector will visit our calibration lab on May 17. He will be evaluating our equipment and calibration hierarchy to determine whether our lab meets MIL-STD-202 specifications. Please extend him every courtesy and your cooperation.

2. To: Andy Rittman:

   Please inform each field technician that from April 1 he will be covered by company-sponsored travel insurance arranged through Tri-State Assurance Corporation. They won't have to pay for it, but belonging to the scheme won't be automatic, they have to apply for it. I suggest you write a personal memo to each technician and enclose a copy of the enclosed application form and explanatory brochure. In each case be sure to remind him to apply by March 25, otherwise he'll have to wait until May 1 for his coverage to start.

Correcting gender-specific terms can be more subtle than is immediately evident

3. From a company notice board:

Christmas Cheer is About to Start!

Are you planning to attend the Christmas Party? Tell your most significant other it will be at O'Halloran's on December 15 and to expect a royal feast!

Line up a babysitter today and tell her to expect to stay late...!

Call Dave Michaelson at extension 207 if you want to announce your intentions or to reserve a table, or if you want more information.

Exercise 12.10

Some research is needed here

Assume that you are employed by H. L. Winman and Associates and that you have been selected to be sent overseas in six weeks time, as part of a team of five engineers and technicians who will be working on a project in ________________ *(your instructor will tell you which country)*. From your local library, research information on customs that will influence how you write, speak, and handle yourself in the host country. Write three short paragraphs describing the factors you will have to consider when you

1. write letters to people in the host country,
2. speak to your hosts at a conference, and
3. attend meetings or business lunches.

Exercise 12.11

Rewrite the following letters and memos to improve their effectiveness.

An overwritten letter

1. Customer Service Manager

   Emerald Air Express

   To whomever it may concern:

   I am writing with reference to a shipment one of your drivers delivered to me on August 17. This particular shipment was an envelope containing 8 papers that I needed urgently. For your records, your waybill number 7284 06292 36 is in reference to this shipment. The documents originated in Toronto, Canada, and the package was picked up by one of your drivers and delivered to our Poughkeepsie, N.Y. office on August 13. The shipping envelope was marked "Next Morning Delivery."

   Your driver—here, the one who delivered the package to me—insisted that the shipment had been sent C.O.D., and I had to pay $48.76 for it before he would hand over the envelope. Only after the driver had gone did I discover that the shipment had been traveling for 4 days!!!

   What I want to know is this: Why was I charged such an outrageous sum for documents that were time-sensitive and that, by the time it reached me, was of no value? I look forward to receiving your check at your earliest convenience.

   I remain (annoyed),

   Peter LeMay

A fragmented, underwritten letter

2. Dear Mr. Shasta:

   I am in receipt of your letter of Nov 17, you'll be glad to hear the problem you outlined is under consideration. A defective chip has been discovered in the output stage. Replacements are hard to get, I phoned around but no one has one locally. So I've ordered one directly from the manufacturer—Mansell Microprocessors—and asked them to ship it air express. I'll telephone you immediately it comes in. But it won't be for three weeks, too many are back-ordered, they can't ship before Dec 12.

   Sincerely,

   T. L. Pedersen

Another wordy, over-written letter

3. Dear Mr. Reimer:

   It is with the sincerest regret that Vancourt Computers Inc. has to inform you that there will be an unfortunate delay of about three weeks in filling your order. (Ref. your P.O. 2863 dated April 29.) Due to measures entirely beyond our control the ship carrying a shipment of 1100 portable model 7000 computers from Nabuchi Electronics in Taiwan developed engine trouble in mid-Pacific and had to limp back to its home port. Your 3 computers are among the 1100. The shipping company has recently informed us that the ship will sail on June 6 and arrive in Vancouver on June 19. We will air express your 3 units as soon as they have cleared customs and absorb the additional expense ourselves. We hope this will help you understand our position.

   Yours very truly,

   Vern Kerpov

A nearly-right memo

4. To: All H. L. Winman staff, Calgary

   From: Tanys Young

   Date: August 10

   Ref: New facility

   H. L. Winman and Associates is pleased to announce that construction will start shortly in room B101, which is to become a Child Care Center that will accommodate 20 children between 1 and 5. It will open on November 1 and will be open only to children of H. L. Winman employees.

   If you have preschoolers and would like to take advantage of this unusual new in-house service, you are invited to procure form CCC01 from Rick Davis in Personnel. It is our expectation that the CC Center will be subject to over-subscription, hence we suggest your application be placed early because we will evaluate them in the order in which they are received.

   Additionally, we will be looking for three experienced child care workers to man the facility. If you or your spouse knows of someone who might be suitable, ask her to telephone Rick for information and an application form.

A discontinuous, under-written letter

5. Dear Ms. Sorchan:

   Re: Inspection of your residential lot at 2127 Victoria Street. My survey shows your neighbor's fence is 3.7 inches inside your property (your neighbor to the south, that is, at #2123). Fence to the north is okay: its yours and its 1.2 inches inside your property. Does your neighbor at #2123 know about this? You have basis for watertight legal action if that's the rout you want to take. Our survey report and invoice #236 is enclosed.

   Regards,

   Wilton Candrow

6. To: Annette Lesk

   From: Mark Hoylan

   Date: 09/07/03

   Subject: Progress at MMW

   As you are well aware, Ken Poitras and I have been at Morriss Machine Works for two and a half weeks now, where we are shoring up the flooring for the NCR machine to be installed this week, and have had to build a 10 ft extention along the width of the north wall.

   All this work is now complete and we should've been heading back to the office by now except theres a problem: the NCR machine arrived today and instead of being installed its sitting outside under a tarp. Why? Because no one seems to of calculated that a 42 in. wide machine (which is it's narrowest dimention) can't be greased through a 36 in. wide door!

   So...Mr. Grindelbauer who is the Machine Works manager (actually, he's the owner) has asked Ken and yours truly to stay behind and tear out part of the wall to make the door wider (which we're doing now) so the machine can be put in tomorrow, and then for us to rebuild the wall and reinstal the door which I reckon will take two extra days. I tried to explain to him there would be an extra charge for doing all this as its not in the contract, and he said for you to call him. Will you do that? Thanks.

   Mark

   P.S. I reckon the additional time will be 32 hours and there will be two additional nights accomodation and per diem, plus some extra lumber and wallboard at a cost that will probably come in at about $170.

   PPS We'll be back in the office on the tenth. OK?

Chatty and friendly; but are there too many words?

Exercise 12.12

Rewrite this one-paragraph notice and use information design to make it clearer, more personal, and more likely to encourage readers to do as it requests.

PROCEDURE RE EXPENSE CLAIMS

Expense claims must be handed to the Accounts Section before 10:30 a.m. on Wednesday for payment on Friday. Personnel failing to hand in their forms at the proper time, may do so at any time until 4:30 p.m. on Thursday but must wait until Monday for payment. Under no account will a late claim be paid in the same week that it was filed. Claims handed in after Thursday will be processed with the following week's claims and will be paid on the next Friday.

You can use your imagination when revising this one!

Glossary of Technical Usage

A standard glossary of usage contains rules for combining words into compound terms, for forming abbreviations, for capitalizing, and for spelling unusual or difficult words. The glossary in *Technically-Write!* also offers suggestions for handling many of the technical terms peculiar to industry. Hence, it is oriented toward the technical rather than the literary writer.

The entries in the glossary are arranged alphabetically. Among them are words that are likely to be misused or misspelled, such as

- words that are similar and frequently confused with one another; e.g. *imply* and *infer*; *diplex* and *duplex*; *principal* and *principle*,
- common minor errors of grammar, such as *comprised of* (should be *comprises*), *most unique* (*unique* should not be compared), *liaise* (an unnatural verb formed from *liaison*), and
- words that are particularly prone to misspelling, e.g. *desiccant, oriented, immitance*, and words for which there may be more than one "correct" spelling; e.g. *sulfur* or *sulphur*.

Where two spellings of a word are in general use (e.g. *symposiums* and *symposia*), the glossary lists both and states which is preferred.

Definitions have been included when they will help you select the correct word for a given purpose, or to differentiate between similar words having different meanings. These definitions are intentionally brief and are intended only as a guide; for more comprehensive definitions, consult an authoritative dictionary.

All entries in the glossary are in lower case letters. Capital letters are used where capitals are recommended for a specific word, phrase, or abbreviation. Similarly, periods have been eliminated except where they form part of a specific entry. For example, the abbreviation for "inch" is *in.*, and the period that follows it is inserted intentionally to distinguish it from the word "in".

Finally, think of the glossary as a guide rather than a collection of hard and fast rules. Our language is continually changing, so that what was fashionable yesterday may seem pedantic today and a cliché tomorrow. We expect that in some cases your views will differ from ours. Where they do, we hope that the comments and suggestions we offer will help you to choose the right expression, word, abbreviation, or symbol, and that you will be able to do so both consistently and logically.

The Glossary

General abbreviations used throughout the glossary:

abbr	abbreviate(d); abbreviation
adj	adjective
Br	Britain; British
def	definition
lc	lower case
n	noun
pl	plural
pref	prefer; preferred; preference
rec	recommend(ed)
SI	International System of Units
US	United States
v	verb

A

a; an use *an* before words that begin with a silent *h* or a vowel; use *a* when the *h* is sounded or if the vowel is sounded as *w* or *y*; *an hour* but *a hotel*, *an onion* but *a European*

aberration

above - as a prefix, *above-* combines erratically: *aboveboard, above-cited, aboveground, above-mentioned*

abrasion

abscess

abscissa

absence

absolute abbr: **abs**

absorb(ent); adsorb(ent) *absorb* means to swallow up completely (as a sponge absorbs moisture); *adsorb* means to hold on the surface, as if by adhesion

abut; abutted; abutting; abutment

ac abbr for alternating current

accelerate; accelerator; accelerometer

accept; except *accept* means to receive (normally willingly), as in *he accepted the company's offer of employment*; *except* generally means exclude: *the night crew completed all the repairs except rewiring of the control panel*

access; accessed; accessible

accessory; accessories abbr: **accy**

accidental(ly)

accommodate; accommodation

account abbr: **acct**

accumulate; accumulator

acetaminophen

achieve means to conclude successfully, usually after considerable effort; avoid using *achieve* when the intended meaning is simply to reach or to get

acknowledg(e)ment *acknowledgment* pref

acquiesce def: agree to

acquire; acquisition

across not *accross*

actually omit this word: it is seldom necessary in technical writing

actuator

adapt; adept; adopt *adapt* means to adjust to; *adept* means clever, proficient; *adopt* means to acquire and use

adapter; adaptor *adapter* pref

addendum pl: *addenda*

adhere to never use *adhere by*

ad hoc def: set up for one occasion

adjective (compound) two or more words that combine to form an adjective are either joined by a hyphen or compounded into a single word; see page 345; abbr: **adj**

adsorb(ent) see **absorb**

advanced power manager abbr: **apm** (pref) or **APM**

advantageous

adverse; averse *adverse* means unsatisfactory or unsuitable, as in *driving speed was adversely affected by the 4 inches of snow on the highway*; *averse* means in opposition to, as in *the staff were averse to working overtime* (they didn't want to)

advice; advise use *advice* as a noun and *advise* as a verb: *the engineer's advice was sound; the technician advised the driver to take an alternative route*; spell: **adviser, advisable**

ae; e *ae* is pref in Br as in *aesthetic* and *anaemic*; *e* is pref in US as in esthetic and anemic

aerate

aerial see **antenna**

aero- a prefix meaning of the air; it combines to form one word; *aerodynamics, aeronautical*; in some instances it has been replaced by *air: airplane, aircraft*

aesthetic; esthetic *esthetic* pref; see **ae**

affect; effect *affect* is used only as a verb, never as a noun; it means to produce an effect upon or to influence (*the potential difference affects the transit time*); *effect* can be used either as a verb or as a noun; as a verb it means to cause or to accomplish (*to effect a change*); as a noun it means the consequences or result of an occurrence (as in *the detrimental effect upon the environment*), or it refers to property, such as *personal effects*

aforementioned; aforesaid avoid using these ambiguous expressions

after- as a prefix, usually combines to form one word: *afteracceleration,* *afterburner, afterglow, afterheat, afterimage*; but *after-hours*

agenda although plural, *agenda* is generally treated as singular; *the agenda is complete*

aggravate the correct definition of *aggravate* is to increase or intensify (worsen) a situation; try not to use it when the meaning is *annoy* or *irritate*

aggregate

aging; ageing *aging* pref

agree to; agree with to be correct, you should *agree to* a suggestion or proposal, but *agree with* another person

air- as a prefix, normally combines to form one word: *airborne, airfield, airflow, airlift, airmail*; exceptions: *air-condition(ed) (er) (ing), air-cool(ed) (ing), air strike*

air horsepower abbr: **ahp**

airline; air line an *airline* provides aviation services; an *air line* is a line or pipe that carries air

algae

algorithm

alkali; alkaline pl: *alkalis* (pref) or *alkalies*

allot(ted)

all ready; already *all ready* means that all (everyone or everything) is ready: *already* means by this time: *the samples are all ready to be tested; the samples have already been tested*

all right def: everything is satisfactory; never use *alright*

all together; altogether *all together* means all collectively, as a group; *altogether* means completely, entirely: *the samples have been gathered all together, ready for testing; the samples are altogether useless*

allude; elude *allude* means refer to; *elude* means avoid

almost never contract *almost* to *most*; it is correct to write *most of the software has been tested*, but wrong to write *the software is most ready*

alphanumeric def: in alphabetical, then numerical, sequence

alternate; alternative *alternate(ly)* means by turn and turn about: *the inspector alternated between the two construction sites*; *alternative(ly)* offers a choice between two or more things: *the alternative is to find a replacement speaker, change the meeting into a workshop, or cancel the event.*

alternating current abbr: **ac**

alternator

altitude abbr: **alt**

AM abbr for audio modulation

a.m. def: before noon (*ante meridiem*)

amateur

ambience; ambient abbr: **amb**

ambiguous; ambiguity

American standard code for information exchange; abbr: **ASCII**

American Wire Gauge abbr: **AWG**

among; between use *among* when referring to three or more items; use *between* when referring to only two; avoid using *amongst*

amount; number use *amount* to refer to a general quantity: *the amount of time taken as sick leave has decreased*; use *number* to refer to items that can be counted: *the number of applicants to be interviewed was reduced to six*

ampere(s) abbr: **A** (pref) or **amp;** other abbr: **kA, mA, μA, nA, pA, A/m** (amperes per minute)

ampere-hour(s) abbr: **Ah** (pref) or **amp-hr** (more common)

amplitude modulation abbr: **AM**

an see **a**

anaemic; anemic *anemic* pref; see **ae**

anaesthetic; anesthetic *anesthetic* pref; see **ae**

analog

analogous

analogy

anesthetic; anesthesia

AND-gate

and/or avoid using this term; in most cases it can be replaced by either *and* or *or*

anemia; anemic see **ae**

angle an *angel* has wings

ångström abbr: **A**

anion def: negative ion

anneal; annealed; annealing

annihilate; annihilated; annihilation

anomaly pl *anomalies*

anonymous

ANSI abbr for American National Standards Institute

antarctic see **arctic**

ante- a prefix that means before; combines to form one word: *antecedent, anteroom*

ante meridiem def: before noon; abbr: **a.m.**; can also be written as *antemeridian*

antenna the proper plural in the technical sense is *antennas*; *antennae* should be limited to zoology; *antenna* has generally replaced the obsolescent *aerial*

anti- a prefix meaning opposite or contradictory to; generally combines to form one word: *antiaircraft, antiastigmatism, anticapacitance, anticoincidence, antisymmetric*; if combining word starts with i or is a proper noun, insert a hyphen: *anti-icing, anti-American*

antimeridian def: the opposite of meridian (of longitude); e.g. the antimeridian of 96°30′W is 83°30′E

anybody; any body *anybody* means any person; *any body* means any object: *anybody can attend; discard the batch if you find any body containing foreign matter*

anyone; any one *anyone* means any person; *any one* means any single item: *you may take anyone with you; you may take any one of the samples*

anyway; any way *anyway* means in any case or in any event; *any way* means in any manner: *the results may not be as good as you expect, but we want to keep them anyway; the work may be done in any way you wish*

AOL

apm advanced power manager

apparatus; apparatuses

apparent; apparently

appear(s); seem(s) use *appears* to describe a condition that can be seen: *the equipment appears to be new*; use *seems* to describe a condition that cannot be seen: *the temperature seems to be low*

appendix def: the part of a report that contains supporting data; pl: *appendixes*

applets def: JAVA computer programs

approximate(ly) abbr: **approx**; but *about* is a better word

aquarium(s)

arbitrary

arc; arced; arcing

archeology; archeologist; archeological

Archie

architect; architecture

arctic capitalize when referring to a specific area: *beyond the Arctic Circle*; otherwise use lc letters: *in the arctic*; never omit the first *c*

area the SI unit for area is the *hectare* (abbr: **ha**)

areal def: having area

around def: on all sides, surrounding, encircling

arrester, arrestor *arrester* pref

arteriosclerosis

article

artifact

artwork

as avoid using when the intended meaning is *since* or *because*; to write *he could not open his desk as he left his keys at home* is incorrect (replace *as* with *because*)

ASCII American standard code for information exchange

as per avoid using this hackneyed expression, except in specifications

asphalt *asfalt* also used, but less pref

assembly; assemblies abbr: **assy**

assure means to state with confidence that something has been or will be made certain; it is sometimes confused with *ensure* and *insure*, which it does not replace; see **ensure**

asthma; asthmatic

as well as avoid using when the meaning is *and*

asymmetric; asymmetrical

asynchronous

atmosphere abbr: **atm**

atomic weight abbr: **at. wt**

attenuator

atto def: 10^{-18}; abbr: **a**

audible; audibility

audio frequency abbr: **af** (pref) or **a-f**

audiovisual

audit; auditor

auger; augur an *auger* is a tool; *augur* means to sense something

aural def: that which is heard; avoid confusing with *oral*, which means that which is spoken

author; writer avoid referring to yourself as *the author* or *the writer*; use *I, me,* or *my*

authoritative

authorize; authorise *authorize* pref

auto- a prefix meaning self; combines to form one word: *autoalarm, autoconduction, autogyro, autoionization, autoloading, automation, automaton, autotransformer, autoworker*

automatic frequency control abbr: **afc** (pref) or **AFC**

automatic volume control abbr: **avc** (pref) or **AVC**

auxiliary abbr: **aux**; pl: **auxiliaries**

average see **mean**

averse def: reluctant; see **adverse**

avocation def: an interest or hobby; avoid confusing with *vocation*

ax; axe *axe* pref; pl: *axes*

axis the plural also is *axes*

azimuth abbr: **az**

B

bachelor of science abbr: B.S.

bacillus abbr: **bacilli**

back- as a prefix normally combines into one word: *backboard, backdate(d), backlog, backup*; but *back burner*

bacterium pl: **bacteria**

balance; remainder use *balance* to describe a state of equilibrium (as in *discontinuous permafrost is frozen soil delicately balanced between the frozen and unfrozen state*), or as an accounting term; use *remainder* when the meaning is the rest of: *the remainder of the shipment will be delivered next week*

balk(ed)/baulk(ed) *balk* pref

ball bearing

bandwidth

bare; bear *bare* means barren or exposed; *bear* means to withstand or to carry (or a wild animal)

barometer abbr: **bar.**

barrel; barrel(l)ed; barrel(l)ing *ll* pre; the abbr of *barrel(s)* is **bbl**

barretter

barring def: preventing, excepting

bases this is the plural of both *base* and *basis*

basically

basic input/output system abbr: **bios** (pref) or **BIOS**

baud; baud rate

baulk see **balk**

because; for use *because* when the clause it introduces identifies the cause of a result: *he could not open his desk because he left his keys at home*; use *for* when the clause introduces something less tangible: *he failed to complete the project on schedule, for reasons he preferred not to divulge*

becquerel def: a unit of activity of radionuclides (SI): abbr: **Bq**; other abbr: **PBq, TBq, GBq, kBq**; in SI, the *becquerel* replaces the *curie*

benefit; benefited; benefiting

beside; besides *beside* means alongside, at the side of; *besides* means as well as

between see **among**

bi- a prefix meaning two or twice; combines to form one word: *biangular, bicultural, bidirectional, bifilar, bilateral, bilingual, bimetallic, bizonal*

biannual(ly); biennial(ly) *biannual(ly)* means twice a year; *biennial(ly)* means every two years

bias; biased; biases; biasing

billion def: 10^9 (US); 10^{12} (Britain)

billion electron volts although the pref abbr is **GEV**, *beV* and *bev* are more commonly used

Bill of Materials abbr: **BOM**

bimonthly def: every two months

binary

binaural

bioelectronics

bionics def: application of biological techniques to electronic design

bios basic input/ouput system: a program that starts the computer and manages data flow between the operating system and the hard disk, video adapter, keyboard, mouse, printer, etc.

birdseye (view)

bit abbr: **kb**

Bitnet

bits per second abbr: **bps**

biweekly def: every two weeks

blow- as a prefix combines to form one word: *blowhole, blowoff, blowout*

blueprint

blur; blurred; blurring; blurry

board feet abbr: **fbm** (derived from *feet board measure*)

boiling point abbr: **bp**

boldface (type)

bookkeeper

boot *to boot* means the operating system is being loaded into the computer; also *reboot*

borderline

bps bits per second

brakedrum; brake lining; brakeshoe

brake horsepower; brake horsepower-hour abbr: **bhp, bhp-hr**

brand-new

break- when used as a prefix to form a compound noun or adj, *break* combines into one word: *breakaway, breakdown, breakup*; in the verb form it retains its single-word identity: *it was time to break up the meeting*

bridging

Brinell hardness number abbr: **Bhn**

British thermal unit abbr: **BTU**

budget; budgeted; budgeting

build- compounds as one word or in adj form, as in *corrosive buildup*; use two words in the v form, as *in to build up our resources, we have...*

buoy; buoyant

burned; burnt *burned* pref

bur(r) *burr* pref

buses; bused; busing; bus bar

business; businesslike; businessperson avoid using *businessman* or *businesswoman* unless referring to a specific male or female person

by- as a prefix, *by-* normally combines to form one word: *bylaw, byline, bypass, byproduct*

byte abbr: **kbyte** and **Mbyte** (pref), or **kB** and **MB**

B2B abbr for business-to-business (or **e-biz**); the exchange of products, services or information between businesses

B2G abbr for business-to-government; permits businesses and government agencies to use central Web sites to exchange information and do business with each other

C

cache memory

calendar; calender; colander a *calendar* is the arrangement of the days in a year; *calender* is the finish on paper or cloth; a *colander* is a sieve

caliber

calk/caulk *caulk,* pref

cal(l)iper *caliper* pref

calorie abbr: **cal**

calorimeter; colorimeter a *calorimeter* measures quantity of heat; a *colorimeter* measures color

cancel(l)ed; cancel(l)ing

candela def: unit of luminous intensity (replaces *candle*); abbr. **cd**; recommended abbr for candela per square foot and square metre are **cd/ft^2** and **cd/m^2**

candlepower; candlehour(s) abbr: **cp, c-hr**

candoluminescence

cannot one word pref; avoid using *can't* in technical writing

canvas; canvass *canvas* is a coarse cloth used for tents; *canvass* means to solicit

capacitor

capacity for never use *capacity to* or *capacity of*

capillary

capital letters abbr: **caps.**

car- as a prefix normally combines to form one word: *carload, carlot, carpool, carwash*

carburet(t)or *carburetor* pref; a third, seldom used spelling is *carburetter*; also: **carburetion**

carcino- as a prefix combines to form one word

cartilage

case- as a prefix normally combines to form one word: *casebook, caseharden, casework(er)*; exceptions: *case history, case study*

cassette

caster; castor use *caster* when the meaning is to swivel freely, and *castor* when referring to castor oil, etc.

catalyst; catalytic

cathode-ray tube abbr: **crt** (pref) or **CRT** (commonly used)

cation def: positive ion

caulk pref spelling; also *see* **calk**

CD-ROM

-ceed; -cede; -sede only one word ends in *-sede*: *supersede*; only three words end in *-ceed*: *exceed, proceed, succeed*; all others end in *-cede*: e.g. *precede, concede*

cell phone abbr: **cell**; called *mobile* in Br

Celsius abbr: **C**; see **temperature**

cement; concrete *cement* is the powder used to make concrete; *concrete* is the hard, finished product

central processing unit abbr: **cpu** (pref) or **CPU**

centerline abbr: (pref) or **CL**

center-to-center abbr: **c-c**

centi- def: 10^{-2}, as a prefix combines to form one word: *centiampere, centigram*; abbr:**c**; other abbr:

centigram	**cg**
centiliter	**cL**
centimeter	**cm**
centimeter-gram-second	**cgs**
centimeters per second	**cm/s**
square centimeter	**cm^2**

centigrade abbr: **C**; in SI, *centigrade* has been replaced by *Celsius*; see **temperature**

centri- a prefix meaning center; combines to form one word: *centrifugal, centripetal*

cga color/graphics adapter

chairperson avoid using *chairman* or *chairwoman*

chamfer

changeable; changeover (n and adj); **change over** (v)

chargeable

chassis both singular and plural are spelled the same

chat room a Web site, part of a Web site, or an online service in which users with a common interest can communicate in real time (see **real time**)

check- as a prefix combines to form one word: *checklist, checkoff, checkpoint* and *checkup* are one word as n or adj, but two words in v form

checksum def: a term used in computer technology

chlorophyll

chrominance

chromosome

chunk; chunking

cipher in Br also spelled *cypher*

circuit abbr: **cct**; also: **circuitous; circuit breaker**

cite def: to quote; see **site**

cleanup (n and adj); **clean up** (v)

climate avoid confusing *climate* with *weather*; *climate* is the average type of weather, determined over a number of years, experienced at a particular place; *weather* is the state of the atmospheric conditions at a specific place at a specific time

clockwise (turn) abbr: **CW**

cmc computer-mediated communication

cmos complementary metal-oxide conductor

co- as a prefix, *co-* generally means jointly or together; it usually combines to form one word: *coexist, coequal, cooperate, coordinate, coplanar* (*co-worker* is an exception); it is also used as the abbr for *complement of* (an arc or angle): *codeclination, colatitude*

coalesce; coalescent

coarse; course *coarse* means rough in texture or of poor quality; *course* implies movement or passage of time; *a coarse granular material; the technical writing course*

coaxial abbr: **coax.**

Cobol def. common business oriented language

coefficient abbr: **coef**

coerce; coercion

collaborate; collaborator avoid writing *collaborate together* (delete *together*)

collapsible

collateral

collide use *collide* to describe two moving objects that bang or crash into one another; use *drove into* or *bumped into* if one object is stationary

cologarithm abbr: **colog**

colon when a colon is inserted in the middle of a sentence to introduce an example or short statement, the first word following the colon is not capitalized; see guidelines on p. 325 for inserting a colon at, or omitting it from, the end of a sentence that introduces a list or subparagraphs; a hyphen should not be inserted after a colon

color/graphics adapter abbr: **cga** (pref) or **CGA**

colorimeter see **calorimeter**

column abbr: **col.**

combustible

comma a comma normally need not be used immediately before *and, but,* and *or*, but may be inserted if to do so will increase understanding or avoid ambiguity

commence in technical writing, replace *commence* with the more direct *begin* or *start*

commit; commitment; committed; committing

committee

communicate it is vague to write *I communicated the results to the client*; use a clearer verb: *I emailed/faxed/wrote/telephoned*

compare; comparable; comparison; comparative use *compared to* when suggesting a general likeness; use *compared with* when making a definite comparison

compatible; compatibility

complement; compliment *complement* means the balance required to make up a full quantity or a complete set; to *compliment* means to praise; *in a right angle, the complement of 60° is 30°; Mr. Perchanski complimented Janet Rudman for writing a good report*

complementary metal-oxide conductor abbr: **cmos** (pref) or **CMOS**

composed of; comprising; consists of all three terms mean "made up of" (specific items); if any one of these terms is followed by a list of items, it implies that the list is complete; if the list is not complete, the term should be replaced by *includes* or *including*

compound terms two or more words that combine to form a compound term are joined by a hyphen or are written as one word, depending on accepted usage and whether they form a verb, noun, or adjective; the trend is toward one-word compounds; (see page 345)

comprise; comprised; comprising to write *comprised of* is incorrect, because the verb comprise includes the preposition *of*

CompuServe

computer-mediated communication abbr: **cmc** (pref) or **CMC**

concrete the hard, rock-like substance used to make roads, bridges and buildings; avoid confusing with *cement*

concur; concurred; concurrent; concurring

condenser

conductor

config.sys a text file containing DOS commands that tells the operating system how the computer is set up

conform use *conform to* when the meaning is to abide by; use *conform with* when the meaning is to agree with

conscience; conscientious

conscious

consensus means a general agreement of opinion; hence to write *consensus of* *opinion* is incorrect; e.g. write: *the consensus was that a further series of tests would be necessary*

consistent with never *consistent of*

consists of; consisting of see **composed of**

contact *contact* should not be used as a verb when *email, write, visit, speak, fax,* or *telephone* better describes the action taken

continual; continuous *continual(ly)* means happens frequently but not all the time: *the generator is continually being overloaded* (is frequently overloaded); *continuous(ly)* means goes on and on without stopping: *the noise level is continuously at or above 100dB* (it never drops below 100 dB)

continue(d) abbr: **cont**

continuous wave abbr: **cw**

contra- as a prefix normally combines into a single word

contrast when used as a verb, *contrast* is followed by *with*; when used as a noun, it may be followed by either *to* or *with* (*with* pref)

control; controlled; controlling; controller

convenor

conversant with never *conversant of*

converter; convertible

conveyor

cooperate

coordinate; coordinator

copyright not *copywrite*

corollary

correlate

correspond *to correspond to* suggests a resemblance; *to correspond with* means to communicate in writing

corroborate

corrode; corrodible; corrosive

cosecant abbr: **csc** (pref) or **cosec**

cosine abbr: **cos**

cotangent abbr: **cot**

coulomb def: a quantity of electricity, electric charge (SI); abbr: **C**; other abbr: **kC, mC, μC, nC, pC, C/m^2**

counter- a prefix meaning opposite or reciprocal; combines to form one word: *counteract, counterbalance, counterflow, counterweight*

counterclockwise (turn) abbr: **CCW**

counterelectromotive force abbr: **cemf**; also known as *back emf*

counts per minute abbr: **cpm**

course see **coarse**

cpu central processing unit

criteria; criterion the singular is *criterion*, the plural is *criteria*: e.g. *one criterion; seven criteria*

criticism; criticize; critique

cross- as a prefix combines erratically: *cross-border, cross-check, cross-examine, crosshatch, crosstalk, cross-purpose, cross section*

cross-refer(ence) abbr: **x-ref**

crt cathode-ray tube

cryogenic

cryptic

crystal abbr: **xtal**

crystalline; crystallize

cubic abbr: **cu** or **3**; other abbr:

cubic centimeter(s)	**cm^3** (pref); **cc**
cubic decimeter(s)	**dm^3**
cubic foot (feet)	**ft^3** (pref); **cu ft**
cubic feet per minute	**cfm** (pref); **ft^3/min**
cubic feet per second	**cfs** (pref); **ft^3/sec**
cubic inch(es)	**in.^{3}** (pref); **cu in.**
cubic meter(s)	**m^3**
cubic millimeter(s)	**mm^3**
cubic yard(s)	**yd^3** (pref); **cu yd**

curb; kerb *curb* pref in US; *kerb* is Br

curie abbr: **Ci**; other abbr: **mCi, μCi**; in SI the *curie* is replaced by the *becquerel*

current; currant *current* refers to a flow (of water, electricity); a *currant* is a dried fruit

curriculum pl: *curriculums* (pref) or *curricula*

cursor

cw continuous wave

cyberspace

cycles per minute abbr: **cpm**

cycles per second abbr: **cps**; although occasionally used, this term has been replaced by **hertz**

cylinder; cylindrical abbr: **cyl**

D

daraf def: the unit of elastance

data def: gathered facts; although *data* is plural (derived from the singular *datum*, which is rarely used), it is more acceptable to use it as a singular noun: *when all the data has been received, the analysis will begin*

database

dateline

date(s) avoid vague statements such as "last month" and "next year" because they soon become indefinite; write a specific date, using day (in numerals), month (spelled out), and year (in numerals): *January 27, 2004* or *27 January 2004* (the latter form has no punctuation); to abbreviate, reduce month to first three letters and year to last two digits: *Jan 27, 04* or *27 Jan 04*; "th" is unnecessary after the "27"

day- as a prefix, generally combines to form one word: *daybook, daylight, daytime, daywork*

days days of the week are capitalized: *Monday, Tuesday*

dc direct current

de- a prefix that generally combines to form one word: *deaccentuate, deactivate, decentralize, decode, decompress, deemphasize, deenergize, deice, derate, destagger*; exceptions are *de-ionize* and *de-ice*

dead- as a prefix combines erratically: *deadbeat, dead center, dead end, deadline, deadweight, deadwood*

debug; debugged; debugging

decelerate def: to slow down; never use *deaccelerate*

decibel abbr: **dB**; the abbr for decibel referred to 1 mW is **dBm**

decimals for values less than unity (one), place a zero before the decimal point: *0.17, 0.0017*

decimate def: to reduce by one-tenth; can also mean to destroy much of

decimeter abbr: **dm**

declination abbr: **dec**

deductible

defective; deficient *defective* means unserviceable or damaged (generally lacking in quality); *deficient* means lacking in quantity (it is derived from *deficit*), and in the military sense incomplete: *a short circuit resulted in a defective transmitter; the installation was completed on schedule except for a deficient rotary coupler that will not be delivered until June 10*

defense; defensive

defer; deferred; deferring; deferrable; deference

definite; definitive *definite* means exact, precise; *definitive* means conclusive, fully evolved; e.g. *a definite price* is a firm price; *a definitive statement* concerns a topic that has been thoroughly considered and evaluated

defuse; diffuse *defuse* means to ease tension; *diffuse* means scattered

degree(s) abbr: **deg** (pref in narrative) or ° (following numerals); see **temperature**

demarcation

demi- a little-used prefix meaning half (generally replaced by *semi-*); combines to form one word: *demivolt*

demonstrate; demonstrator; demonstrable not *demonstratable*

depend; dependence; dependent (adj); **dependant** (n); **dependable**

deprecate; depreciate *deprecate* means to disapprove of; *depreciate* means to reduce the value of: *the use of "as per" in technical writing is deprecated; the vehicles depreciated by 50% the first year and 20% the second year*

depth

desiccant, desiccate(d)

desirable

desktop

desktop publishing abbr: **dtp** (pref) or **DTP**

desktop video conference abbr: **dtv** (pref) or **DTV**

deter; deterred; deterrence; deterrent; deterring

deteriorate

develop not *develope*

device; devise the noun is *device*, the verb is *devise*: *a unique device; he devised a new software program*

dext(e)rous *dexterous* pref

diagnose; diagnosis pl: **diagnoses**

dialog(ue) *dialogue* pref

dialysis

diameter abbr. **dia**

diaphragm

diazo

didn't never use this contraction in technical writing; use **did not**

die; dye; dying *die* means to end life; *dye* means a change of color

dielectric

diesel; diesel-electric

dietitian

differ use *differ from* to demonstrate a difference; use *differ with* to describe a difference of opinion

different *different from* is preferred; *different to* is sometimes used; *different than* should never be used

diffraction

diffusion; diffusible

digital library a collection of documents organized electronically on the Internet or CD-ROM

dilemma means to be faced with a choice between two unhappy alternatives; should not be used as a synonym for *difficulty*

diplex; duplex *diplex operation* means the simultaneous transmissions of two signals using a single feature, e.g. an antenna; *duplex operation* means that both ends can transmit and receive simultaneously

direct current abbr: **dc**

directly def: immediately; do not use when the meaning is as soon as

disassemble; dissemble *disassemble* means to take apart; *dissemble* means to conceal facts or put on a false appearance

disassociate see **dissociate**

disc; disk *disk* pref

discernible

discolor

discreet; discrete *discreet* means prudent or discerning: *his answer was discreet*; *discrete* means individually distinctive and separate: *discrete channels*; **discretion** is formed from *discreet*, not from *discrete*

disinterested; uninterested *disinterested* means unbiased, impartial; *uninterested* means not interested

disk cache

disk operating system abbr: **DOS**

dispatch; despatch *dispatch* pref

dispel; dispelled

disseminate

dissimilar

dissipate

dissociate; disassociate *dissociate* pref

distribute; distributor

don't; doesn't such contractions should not appear in technical writing

donut; doughnut for electronics/nucleonics, use *donut*

doppler capitalize only when referring to the Doppler principle

DOS disk operating system

double- as a prefix combines erratically: *double-barrelled, doublecheck, doublecross, double-duty, double entry, doublefaced, doubletalk*

down- as a prefix combines into one word: *downgrade, downrange, downtime, downwind*

dozen abbr: **doz**

drafting; draftsperson avoid writing *draftsman* or *draftswoman*

drawing(s) abbr: **dwg**

drier; dryer the adjective is always *drier*; the pref noun is *dryer*: *this material is drier; place the others back in the dryer*

drop; droppable; dropped; dropping

dtp desktop publishing

dtv desktop video conferencing

due to an overused expression; *because of* pref

duo- a prefix meaning two; combines to form one word: *duocone, duodiode, duophase*

duplex see **diplex**

duplicator

dye; dying

dynamic data exchange an operating system function which allows information to be shared between programs

dysfunction; not *disfunction*

E

each abbr: **ea**

east capitalize only if *east* is part of the name of a specific location: *East Africa*; otherwise use lc letters: *the east coast of Florida*; abbr: **E**; the abbr for *east-west* (control, movement) is **E-W**; *eastbound* and *eastward* are written as one word

e-Bay

e-biz abbr for business conducted electronically

eccentric; eccentricity

echo; echoes

e-commerce electronic commerce, or **EC**; the buying or selling of goods and services on the Internet; also called **e-business**

economic; economical use *economical* to describe economy (of funds, effort, time); use *economic* when writing about economics: *an economical operation* (it did not cost much); *an economic disaster* (it will have a major effect on the economy)

EDI Electronic Data Interchange; a standard format for exchanging business data

effect see **affect**

efficacy; efficiency *efficacy* means effectiveness, ability to do a job; *efficiency* is a measurement of capability, the ratio of work done to energy expended*: we hired a consultant to assess the efficacy of our training methods; the power house is to have a high-efficiency boiler*

e.g. def: *for example*; avoid confusing with **i.e.**; no comma is necessary after **e.g.**; may also be abbr **eg**

ega enhanced graphics adapter; abbr also **EGA**

eighth

electric(al) if in doubt, use *electric*; generally, *electric* means produces or carries electricity, whereas *electrical* means related to the generation or carrying of electricity; abbr: **elec**

electro- a prefix generally meaning pertaining to electricity; it normally combines to form one word: *electroacoustic, electroanalysis, electrodeposition, electromechanical, electroplate*; if the combining word starts with o, insert a hyphen: *electro-optics, electro-osmosis*

electromagnetic units abbr: **emu**

electromotive force abbr: **emf**

electronic(s) use *electronic* as an adjective, *electronics* as a noun: *electronic countermeasures; your career in electronics*

electronic mail abbr: **email**

electron volt(s) abbr: **eV** (pref) or **ev**

electrostatic units abbr: **esu**

elicit; illicit *elicit* means to obtain or identify; *illicit* means illegal

eligible; illegible *eligible* means "meets the required conditions"; also: *eligibility*; *illegible* means unreadable

ellipse; ellipsis; pl: **ellipses**

email

embarrass; embarrassed; embarrassing; embarrassment

embedded

embryo; embryos

emf abbr for electromotive force

emigrate; immigrate *emigrate* means to go away from; *immigrate* means to come into

emit; emitter; emittance; emission; emissivity

emu abbr for electromagnetic units

emulate; emulation

encase; incase *encase* pref

encipher

enclose; inclose *enclose* pref; *inclose* is used mainly as a legal term

enforce not *inforce*

engineer; engineered; engineering

enhanced graphics adapter abbr: **ega** (pref) or **EGA**

enquire; inquire *inquire* pref

enrol; enroll both are correct, but *enrol* pref; universal usage prefers *ll* for **enrolled** and **enrolling**

en route def: on the road, on the way; never use *on route*

ensure; insure; assure use *ensure* when the meaning is to make certain of: *the new oscilloscope will ensure accurate calibration*; use *insure* when the meaning is to protect against financial loss: *we insured all our drivers*; use *assure* when the meaning is to state with confidence that something has been or will be made certain: *he assured the meeting that production would increase by 8%*

entrepreneur; entrepreneurial

entrust; intrust *entrust* pref

envelop; envelope *envelop* is a verb that means to surround or cover completely; *envelope* is a noun that means a wrapper or a covering

environment; environmental; environmentally

EPROM def: abbr for erasable programmable read-only memory; can also be abbr as **eprom**

equi- a prefix that means equality; combines to form one word: *equiphase, equipotential, equisignal*

equilibrium; equilibriums

equip; equipped; equipping; equipment

equivalent abbr: **equiv**

erase; erasable

errata although *errata* is plural (from the singular *erratum*, which is seldom used), it can be used as a singular or plural noun: both *the errata are ready* and *the errata is complete* are acceptable

erratic

erroneous

escalator

esker

especially; specially *specially* pref when used as an adjective, as in: *a specially trained crew*; *especially* should introduce a phrase, as in: *they were well trained, especially the computer technicians*

esthetic see **ae**

esu abbr for electrostatic unit

et al. def: and others; now rarely used

et cetera def: and so forth, and so on; abbr: **etc.**; use with care in technical writing: *etc.* can create an impression of vagueness or unsureness: *the transmitters, etc., were tested* sounds much less definite than *the transmitter, modulator, and power supply were tested* (or, if to restate all the equipment is too repetitious, *the transmitting equipment was tested*)

euro European currency abbr:

everybody; every body *everybody* means every person, or all the persons; *every body* means every single body: *everybody was present; every body was examined for gun-powder scars*

everyone; every one *everyone* means every person, or all the persons; *every one* means every single item: *everyone is insured; every one had to be tested in a saline solution*

exa def: 10^{18}; abbr: **E**

exaggerate

exceed

excel; excelled; excellent; excelling

except def: to exclude; see **accept**

exhaust

exhibit; exhibitor

exorbitant

expedite; expediter (pref) or **expeditor**

explicit; implicit *explicit* means clearly stated, exact; *implicit* means implied (the meaning has to be inferred from the words): *the supervisor gave explicit instructions* (they were clear); *that the manager was angry was implicit in the words he used*

extemporaneous

extracurricular

extranet a private network that uses the Internet to securely share part of a business's information with users outside an organization

extraordinary

extremely high frequency abbr: **ehf**

eye- as a prefix, normally combines into one word: *eyeball, eyesight, eyewitness*; also: **eyed**

F

face- as a prefix normally combines to form one word: *facedown, facelift, faceplate*; exceptions: *face-saver, face-saving*

Fahrenheit abbr: **F**; see **temperature**

fail-safe

fallout one word as n; two words as v

familiarize; familiarization

farad def: a unit of electric capacitance; abbr: **F**; other abbr: **μF, nF, pF**

farfetched; far-out; far-reaching; farseeing; farsighted

farther; further *farther* means greater distance: *he traveled farther than the other salespeople*; *further* means a continuation of (as an adjective) or to advance (as a verb): *the promotion was a further step in her career plan,* and *to further his education, he took a part-time course in industrial drafting*

fasten; fastener

faultfinder; faultfinding

feasible; feasibility

February

feet; foot abbr: **ft**; other abbr: feet board measure

(board feet)	**fbm**
feet per minute	**fm**
feet per second	**fps**
foot-candle(s)	**fc** (pref); **ft-c**
foot-pound(s)	**fp** (pref); **ft-lb**
foot-pound-second (system)	**fps** system

femto def: 10^{-15}; abbr: **f**; other abbr:

femtoampere(s)	**fA**
femtovolt(s)	**fV**

ferri- a prefix meaning contains iron in the ferric state; combines to form one word: *ferricyanide, ferrimagnetic*

ferro- a prefix meaning contains iron in the ferrous state; combines to form one word: *ferroelectric, ferromagnetic, ferrometer*

ferrule; ferule a *ferrule* is a metal cap or lid; a *ferule* is a ruler

fewer; less use *fewer* to refer to items that can be counted: *fewer technicians than we predicted have been assigned to the project*; use *less* to refer to general quantities: *there was less water than predicted*

field- as a prefix normally does not combine into a single word or hyphenated form: *field glasses, field test, field trip*; but: *fieldwork(er)*

figure numbers in text, spell out the word *Figure* in full, or abbr it to **Fig.**: *the circuit diagram in Figure 26* and *for details, see Fig. 7*; use the abbreviated form beneath the illustration; always use numerals for the figure number

file transfer protocol abbr: **ftp** (pref) or **FTP**

final; finally; finalize

fire- as a prefix combines erratically: *firearm, fire alarm, firebreak, fire drill, fire escape, fire extinguisher, firefighter, firepower, fireproof, fire sale, fire wall*

firmware

first to write *the first two...* (or three, etc) is better than to write *the two first...*; never use *firstly*; as a prefix, *first-* combines erratically: *first-class, firsthand, first-rate*

fix in technical usage, *fix* means to firm up or establish as a permanent fact; avoid using it when the meaning is to repair

flameout; flameproof

flammable def: easily ignited; see **inflammable**

flatcar

flexible

flight- usually combines to form two words: *flight control, flight deck, flight plan*

flip-flop

floe floating ice

flotation this is the correct spelling for describing an item that floats: *flotation gear*

flowchart

fluid abbr: **fl**; the abbr for fluid ounces is **fl oz**

fluorescence; fluorescent

fluorine; fluoridation; fluorite; fluorocarbon; fluoroscope

flyer not *flier*

FM abbr for frequency modulation

focus; focused; focusing; focuses pl: *focuses* (pref) or *foci*

followup one word as n or adj; two words as v: **follow up**

foot- as a prefix normally combines into one word: *footbridge, footcandle, foothold, footnote, footwork*; also see **feet**

for see **because**

forceful; forcible use *forceful* to describe a person's character; use *forcible* to describe physical force

fore- def: that which goes before; as a prefix normally combines into one word: *foreclose, forefront, foreground, foreknowledge, foremost, foresee, forestall, forethought, forewarn*

forecast this spelling is correct for present and past tense

forego; forgo *forego* means to go before; *forgo* means to go without

foreman avoid using in a general sense, except when referring to a person specifically, as in *John Hayward, the foreman*; never use *forewoman* (a better word is *supervisor*)

foresee; foreseeable

forestall

foreword; forward a *foreword* is a preface or preamble to a book; *forward* means onward: *the scope is defined in the foreword to the book; he requested that we move the meeting date forward*

for example abbr: **e.g.** (pref) or **eg**

former; first use *former* to refer to the first of only two things; use *first* if there are more than two

formula pl: *formulas* (pref) or *formulae*

Fortran def: formula translation

forty def: 40; it is not spelled *fourty*

fourth def: 4th; it is not spelled *forth*

fractions when writing fractions that are less than unity, spell them out in descriptive narrative: *by the end of the heat run, nine-tenths of the installation had been completed*; for technical details use decimals rather than fractions, as in *a flat case 0.75 m wide by 0.060 m deep*, except when a quantity is normally stated as a fraction (such as $^3/_8$ *in. plywood*)

free- as a prefix normally combines into one word: *freehand, freehold, freestanding, freeway, freewheel*

free from use *free from* rather than *free of*: *he is free from prejudice*

free on board abbr: **fob** (pref), **f.o.b.** (commonly used), or **FOB**

frequency abbr: **freq**

frequency modulation abbr: **FM**

ftp file transfer protocol

fungus pl: **fungi**

further see **farther**

fuse as a verb, means join together or weld; as a noun, means a circuit protection device

fuselage

fuze def: a detonation initiation device

G

gadget; gadgetry

gallon gallons differ between US (3.785 dm^3) and Britain (4.546 dm^3); abbr: **gal**; other abbr:

gallons per day	**gpd**
gallons per hour	**gph**
gallons per minute	**gpm**
gallons per second	**gps**

gang; ganged; ganging

gas; gases; gassed; gassing; gasious; gassy

gauge or *gage*

gearbox; gearshift

geiger (counter)

gelatin(e) *gelatin* pref

genealogy; genealogist

geo- a prefix meaning of the earth; combines to form one word; *geocentric, geodesic, geomagnetic, geophysics*

giga def: 10^9, abbr: **G**; other abbr:

gigabecquerel(s)	**GBq**
gigahertz	**GHz**
gigajoule(s)	**GJ**
gigaohm(s)	**GΩ; Gohm**
gigapascal(s)	**GPa**
gigavolt(s)	**GV**

gimbal

glue; glueing; gluey

glycerin(e) *glycerin* pref

gnd abbr for *ground*

Gopher def: an Internet search tool

government capitalize when referring to a specific government either directly or by implication; use lc if the meaning is government generally: *the US Government; the Government specifications; no government would sanction such restrictions*

gram abbr: **g**; abbr for **gram-calorie** is **g-cal**

grammar; grammatical(ly)

grateful not *greatful*

gray def: absorbed dose of ionizing radiation (SI); abbr: **Gy**; other abbr: **mGy, µGy**; in SI, the *gray* replaces the *rad*

Greenwich mean time abbr: **GMT**

grey; gray def: a color; *gray* pref in US

grill(e) when the meaning is a loudspeaker covering or a grating, *grille* pref; when referring to cooking, use *grill*

ground (electrical) abbr: **gnd**; **groundcrew; ground floor**

guage wrongly spelled; the correct spelling is *gauge*

guarantee never *guaranty*; also **guarantor**

guesstimate

GUI a graphical (rather than textual) user interface to a computer; pronounced *goo-ee*

guideline

gyroscope abbr: **gyro**

H

half; halved; halves; halving as a prefix, *half* combines erratically; some common compounds are: *half-hourly, half-life, half-monthly, halftone, halfwave*; for others, consult your dictionary

hand- as a prefix normally combines to form one word: *handbill, handbook,* *handful, handfuls, handhold, handmade, handpicked, handset, handshake*

hangar; hanger a *hangar* is a large building for housing aircraft; a *hanger* is a supporting bracket

harass; harassed; harassment

hard- as a prefix normally combines into one word: *hardbound, hardhanded, hardhat, hardware*; exceptions: *hard-earned, hard-hitting*

haversine abbr: **hav**

H-beam

head- as a prefix normally combines into one word: *headfirst, headquarters, headset, headstart, headway*

heat- as a prefix, *heat* combines erratically; some typical compounds are: *heatresistant, heat-run, heatsink, heat-treat, heat wave*; for others, consult your dictionary

heavy-duty

hectare def: a large unit of area, used in surveying and agriculture; in SI, *hectare* replaces *acre*; abbr: **ha**

height (not *heighth*) abbr: **ht**; also: **heighten; heightfinder; heightfinding**

helix pl: *helixes* (pref) or *helices*

hemi- a prefix meaning half; combines to form one word: *hemisphere, hemitropic*

hemophilia; hemorrhage see **ae**

henry def: a unit of inductance; abbr: **H**; other abbr: **mH, µH, nH, pH**

here- whenever possible avoid using *here-* words that sound like legal terms, such as *hereby, herein, hereinafter, hereof*; they make a writer sound pompous; as a prefix, *here-* combines to form one word

hertz def: a unit of frequency measurement (similar to *cycles per second*, which it replaces); abbr: **Hz**; other abbr: **THz, GHz, MHz, KHz**

heterodyne

heterogeneous; homogeneous *heterogeneous* means of the opposite kind; *homogeneous* means of the same kind

hexadecimal

high- as a prefix either combines into one word or the two words are joined by a hyphen: *highhanded, highlight;* as compound adj: *high-frequency, high-power, high-priced, high-speed*

high frequency abbr: **hf**

high-pressure (as an adjective) abbr: **h-p**

high voltage abbr: **hv** (pref) or **HV**

hinge; hinged; hinging

homogeneous see **heterogenous**

hono(u)r *honor* pref

horizontal abbr: **hor**

horsepower abbr: **hp**; the abbr for horse-power-hour is **hp-hr**

hotkey

hovercraft

hour(s) abbr: **hr** or **h** (SI)

HTML hypertext markup language

http Hypertext Transfer Protocol; rules for exchanging files on the World Wide Web

hundred abbr: **C**

hundredweight def: 112 lb; abbr: **cwt**

hybrid

hydro- a prefix meaning of water; combines to form one word: *hydroacoustic, hydroelectric, hydromagnetic, hydrometer*

hyper- a prefix meaning over; combines to form one word; *hyperacidity, hypercritical*

hyperbola the plural is *hyperbolas* (pref) or *hyperbolae*

hyperbole def: an exaggerated statement

hyperbolic cosine, sine, tangent abbr: **cosh, sinh, tanh**

hyperlink

hypertext information organized into related chunks or units; the Web is a massive set of hypertext information

hypertext markup language abbr: **HTML**

hyphen in compound terms you may omit hyphens unlesss they need to be inserted to avoid ambiguity or to conform to accepted usage; e.g. *preemptive* is preferred without a hyphen, but *photo-offset* and *re-cover* (when the meaning is *to cover again*) both require one; refer to individual entries in this glossary

hypothesis pl: *hypotheses*

I

I-beam

ibid. def: Latin abbr for *ibidem*, meaning in the same place; used in footnoting, but becoming obsolete

icon

ID card

i.e. def: *that is*; avoid confusing with **e.g.**; no comma is necessary after *i.e.*; may also be abbr **ie**

ignition abbr: **ign**

ill- as a prefix combines into a hyphenated expression: *ill-advised, ill-defined, ill-timed*

illegible def: not readable; also see *eligible*

im- see **in-**

imbalance this term should be restricted for use in accounting and medical terminology; use *unbalance* in other technical fields

immalleable

immaterial

immeasurable

immigrate see **emigrate**

immittance

immovable

impasse

impel; impelled; impelling; impeller

imperceptible

impermeable

impinge; impinging

imply; infer speakers and writers can *imply* something; listeners and readers *infer* from what they hear or read: *in his closing remarks, Mr. Smith implied that further studies were in order; the technician inferred from the report that his work was better than expected*

impracticable; impractical *impracticable* means not feasible; *impractical* means not practical; a less-preferred alternative for impractical is *unpractical*

in; into *in* is a passive word; *into* implies action; *ride in the car; step into the car*

in-; im-; un- all three prefixes mean not; all combine to form one word: *ineligible, impossible, unintelligible*; if you are not sure whether you should use *in-*, *im-*, or *un-*, use *not*

inaccessible

inaccuracy

inadmissible

inadvertent

inadvisable; unadvisable *inadvisable* pref

inasmuch as a better word is *since*

inaudible

incalculable not *incalculatable*

incandescence; incandescent

incase *encase* pref

inch(es) abbr: **in.**; other abbr: inches per second **ips** (pref), **in./s**; inch-pound(s)**in.-lb**

inclose see **enclose**

includes; including abbr: **incl**; when followed by a list of items, *includes* implies that the list is not complete; if the list is complete, use *comprises* or *consists of*

incomparable

incompatible

incur; incurred; incurring

index pl: *indexes* pref, except in mathematics (where *indices* is common)

indicated horsepower abbr: **ihp**; the abbr for indicated horsepower-hour is **ihp-hr**

indifferent to never use *indifferent of*

indiscreet; indiscrete *indiscreet* means imprudent; *indiscrete* means not divided into separate parts

indispensable

indorse *endorse* pref

industrywide

ineligible

inequitable

inessential; unessential both are correct; *unessential* pref

inexhaustible

inexplicable

infallible

infer; inferred; inferring; inference also see **imply**

infinitesimal

inflammable def: easily ignited (derived from *inflame*); *flammable* is a better word: it prevents readers from mistakenly thinking the *in* of *inflammable* means *not*

inflexible

infrared

infrastructure

ingenious; ingenuous *ingenious* means clever, innovative; *ingenuous* means innocent, naive; *ingenuity* is a noun derived from ingenious

in-house

inoculate

inoperable not *inoperatable*

input/output abbr: **I/O**

inquire; enquire *inquire* pref

insanitary; unsanitary both are correct; *insanitary* pref

inseparable

inside diameter abbr: **ID**

in situ def: in the normal position

instal(l) *install, installed, installer, installing, installation* pref;

instantaneous

instrument

insure the pref def is to protect against financial loss; can also mean make certain of; see **ensure**

integer

integral; integrate; integrator

intelligence quotient abbr: **IQ**

intelligible

inter- a prefix meaning among or between; normally combines to form one word: *interact, intercarrier, interdependence, interdigital, interface, intermodulation, interoffice*

intermediate-pressure (as an adjective) abbr: **i-p**

intermittent

internal abbr: **int**

Internet a worldwide computer network of networks that allows users at one computer to access information from another computer

Internet relay chart abbr: **irc** or **IRC** (pref)

interrupt

into see **in**

intra- a prefix meaning within; normally combines to form one word: *intranet, intranuclear*; if combining word starts with *a*, insert a hyphen: *intra-atomic*

Intranet an organization's private Internet network, used to share company information and computing resources among employees

intrust *entrust* pref

I/O input/output

IQ

irc Internet relay chart

irrational

irregardless never use this expression; use *regardless*

irrelevant frequently misspelled as *irrevelant*

irreparable also see repairable

irreversible

irritate also see *aggravate*

ISO International Organization for Standardization

iso- a prefix meaning the same, of equal size; normally combines to form one word: *isoelectronic, isometric, isotropic*; if combining word starts with o, insert a hyphen: *iso-octane*

its; it's *its* means belonging to; *it's* is an abbr for it is: *the transmitter and its modulator; if the fault is not in the remote equipment, then it's most likely in master control*; in technical wirting *it's* should seldom be used: replace with *it is*

J

jackhammer

jobholder; jobseeker; job lot

joule def: a unit of energy, work, or quantity of heat (SI); abbr: **J**; other abbr: **TJ, GJ, MJ, kJ, mJ, J/m^3, J/K** (joule(s) per kelvin), **J/kg, J/mol** (joule(s) per mole)

journey; journeys

judg(e)ment *judgment* pref

judicial; judicious *judicial* means related to the law; *judicious* means sensible, discerning

K

kelvin def: the SI unit for thermodynamic temperature; abbr: **K**

kerb Br equivalent of *curb*

key- as a prefix normally combines to form one word: *keyboard, keypunch, keying, keynote, keystroke*; but *key word*

kilo def: 10^3, abbr: **k**; other abbr:

kiloampere(s)	**kA**
kilobecquerel(s)	**kBq**
kilobyte(s)	**kbyte** (pref) or **kb**
kilocalorie(s)	**kcal**
kilocoulomb(s)	**kC**
kilogram(s), (see **kilogram**)	**kg**
kilohertz	**kHz**
kilohm(s)	**kΩ**; **kohm**
kilojoule(s)	**kJ**
kiloliter(s)	**kL**
kilometer(s)	**km**
kilometers per hour	**km/h**
kilomole(s)	**kmol**
kilonewton(s)	**kN**
kilopascal(s)	**kPa**
kilosecond(s)	**ks** (pref); **ksec**
kilosiemens	**kS**
kilovolt(s)	**kV**
kilovolt-ampere(s)	**kVA**
kilovolt-ampere(s) reactive	**kVAr**
kilowatt(s)	**kW**
kilowatthour(s)	**kWh** (pref); **kw-hr**

avoid writing *kilo* in text as an abbr for *kilogram* or *kilometer*

kilogram def: the SI unit for mass; abbr: **kg**; other typical abbr: **Mg, g, mg, μg**; also:

kilogram-calorie(s)	**kg-cal**
kilogram(s) per meter	**kg/m**
kilogram(s) per square meter	**kg/m²**
kilogram(s) per cubic meter	**kg/m²**
kilogram meter(s) per second	**kg · m/s**

knockout as noun or adjective, one word

knot abbr: **kn**

know-how (n); **know how** (v)

knowledge; knowledgeable

L

label(l)ed; label(l)ing single *l* pref

laboratory abbr: **lab**

lacquer

lambert abbr: **L**; use the abbr **L** with care: it is also the SI abbr for *liter*

lampholder

laptop (computer)

large scale integration abbr: **LSI**

laser light amplification by stimulated emission of radiation

last; latest; latter *last* means final; *latest* means most recent; *latter* refers to the second of only two things (if more than two, use *last*); it is better to write *the last two* (or *three*, etc) than *the two last*

lath; lathe *a lath* is a strip of wood; a *lathe* is a machine

latitude abbr: **lat** or ∅

lay- as a prefix generally combines to form one word (as noun or adj): *layoff, layout, layover*

LCD liquid crystal display

lead; led as a n, *lead* is a metal; as a v, *lead* means to lead (someone); the past of the v is *led: a lead-filled pipe; he was asked to lead the project team; he led the project team*

learned; learnt *learned* pref

LED light emitting diode

leeway

left-hand(ed) abbr: **LH**

length the SI unit of length is the *meter*, expressed in multiples and submultiples of *kilometers* (**km**), *meters* (**m**), and *millimeters* (**mm**)

lengthy not *lengthly*

less see **fewer**

letter- as a prefix combines erratically: *letterhead, letter-perfect, letter writer*

letter of intent; letter of transmittal pl: *letters of intent, letters of transmittal*

liable to means under obligation to; avoid using as a synonym for *apt to* or *likely to*

liaison liaison is a noun; it is sometimes used uncomfortably as a verb: *liaise*

library

life- as a prefix combines erratically: *lifebelt, lifeboat, life cycle, lifeless, lifelong, life-size, lifespan, lifetime*

light- as a prefix *light-* generally combines to form one word: *lightface* (type), *lightweight*; but *light-year*; the past tense is *lighted*

light emitting diode abbr: **LED**

lightening; lightning *lightening* means to make lighter; *lightning* is an atmospheric discharge of electricity

likable not *likeable*

linear abbr: **lin**; the abbr for lineal foot is **lin ft**

lines of communication not *line of communications*

liquefy; liquefiers; liquefaction

liquid abbr: **liq**

liquid crystal display abbr: **LCD**

Listserv

litre; liter the SI spelling is **litre** (pref in Can. and Br), but in US *liter* is more common; abbr: **L**; other abbr: **kL, mL, μL**; the abbr for *liter(s) per day/hour/minute/second* are **L/d, L/h, L/m, L/s**

lock- as a prefix combines into a single word: *locknut, lockout* (n), *locksmith, lockstep, lockup, lockwasher;* but *lock out* (v)

locus pl: *loci*

logarithm abbr: (common) **log**; (natural) **ln**

logbook

logistic(s) use *logistic* as an adjective, *logistics* as a noun: *logistic control; the logistics of the move*

long- as a prefix normally combines into a single word or is hyphenated: *long-distance, longhand, longplaying,* *long-range, long-term, long-winded*; but *long shot*

longitude abbr: **long.** or λ

loophole

looseleaf

loran abbr for long-range air navigation system

lose; loose *lose* is a verb that refers to a loss; *loose* is an adjective or a noun that means free or not secured: *three loose nuts caused us to lose a wheel*

louver

low frequency abbr: **lf**

low-pressure (as an adjective) abbr: **l-p**

LSI large scale integration

lubricate; lubrication abbr: **lub**

lumbar; lumber *lumbar* is the lower back; *lumber* is wood

lumen def: a unit of luminous flux (SI); abbr: **lm**; other abbr:

lumen-hour(s)	**lm-h** (pref); **lm-hr**
lumens per square foot	**lm/ft^2**
lumens per square meter	**lm/m^2**
lumens per watt	**lm/W**
lumen-second(s)	**lm·s**
microlumen(s)	**μlm**
millilumen(s)	**mlm**

luminance; luminescence; luminosity; luminous

lux def: a unit of illuminance (SI); abbr: **lx**; other abbr: **klx**

M

Mach

macro- a prefix meaning very large; combines to form one word: *macroscopic; macroview*

magneto pl: *magnetos*; as a prefix normally combines to form one word: *magnetoelectronics, magnetohydrodynamics, magnetostriction*; if combining word starts with *o* or *io*, insert a hyphen: *magneto-optics*, *magneto-ionization*

magneton; magnetron a *magneton* is a unit of magnetic moment; a *magnetron* is an electronic device controlled by an external magnetic field

maintain; maintained; maintenance

majority use *majority* mainly to refer to a number, as in a *majority of 27*; avoid using it as a synonym for many or most; e.g. do not write *the majority of technicians* when the intended meaning is *most*

make- as a prefix normally combines to form one word as a n or adj: *makeshift, makeover, makeup*

malfunction

malleable

manage; managed; manageable; managing

manufacturer abbr: **mfr**

marketplace

mass see **kilogram**

mat; matt a *mat* is a covering; *matt* is a dull finish

material; materiel *material* is the substance or goods out of which an item is made; when used in the plural, it describes items of a like kind, such as *writing materials*; *materiel* are all the equipment and supplies necessary to support a project or undertaking (a term commonly used in military operational support)

maximum pl: *maximums* (pref) or *maxima*; abbr: **max**; like *minimize, maximize* can be used as a verb

maybe; may be *maybe* means "perhaps": *maybe there is a second supplier*; the verb form *may be* means "perhaps it will be" or "possibly there is": e.g. *there may be a second supplier*

mda monochrome display adapter; also **MDA**

mean; median the *mean* is the average of a number of quantities; the *median* is the midpoint of a sequence of numbers; e.g. in the sequence of five numbers 1, 2, 3, 7, 8, the mean is 4.2 and the median is 3

mean effective pressure abbr: **mep**

mean sea level abbr: **msl** (pref) or **MSL**

mediocre

medium when *medium* is used to mean substances, liquids, materials, or communication or advertising, the pref plural is *media*; in all other senses the pref plural is *mediums*

mega def: 10^6, abbr: **M**; other abbr:

megabyte(s)	**Mbyte** (pref) or **Mb**
megacoulomb(s)	**MC**
Megaelectronvolt(s)	**MeV**
megahertz	**MHz**
megajoule(s)	**MJ**
meganewton(s)	**MN**
megaohm(s	**MΩ**; Mohm
megapascal(s)	**MPa**
megavolt(s)	**MV**
megawatt(s)	**MW**

memorandum pl: *memorandums* (pref) or *memoranda*; abbr: **memo** (singular), **memos** (pl)

memory the electronic holding place for instructions and data that a computer's microprocessor can reach quickly

merit; merited; meriting

meteorology; metrology *meteorology* pertains to the weather; *metrology* pertains to weights, measures, and calibration

meter def: a measuring instrument (noun) or to measure out (verb)

metre; meter def: metric unit of length; the SI spelling is *metre,* but in US *meter* is more common; abbr: **m**; other typical abbr:

square meter(s)	**m^2**
cubic meter(s)	**m^3**
meters per second	**m/s**
newton-meter(s)	**N·m**
newtons per square meter	**N/m^2**
kilogram(s) per cubic meter	**kg/m^3**

metrication

micro def: 10^{-6}; abbr: **μ** (pref) or **u**; other abbr:

microampere(s)	**μA**
microcoulomb(s)	**μC**
microfarad(s)	**μF**

microgram(s)	**μg**
microgray(s)	**μGy**
microhenry(s)	**μH**
microhm(s)	**μΩ**; μohm
microlumen(s)	**μlm**
micromho(s)	**μmho**
micrometer(s)	**μm**
micromole(s)	**μmol**
micronewton(s)	**μN**
micropascal(s)	**μPa**
microsecond(s)	**μs** (pref); **μsec**
microsiemens	**μS**
microtesla(s)	**μT**
microvolt(s)	**μV**
microwatt(s)	**μW**

micro- as a prefix meaning very small, normally combines to form one word: *microammeter, microcomputer, micrometer, microprocessor, microswitch, microview, microwave*; but *micro-organism*

microchip a logic chip; often called an integrated circuit in computer circuitry

microphone abbr: **MIC** (pref) or **mike**

microprocessor a computer processor on a microchip; the "engine" that runs a computer; previously called the **CPU**

Microsoft disk operating system abbr: **MS-DOS**

mid- a prefix that means in the middle of; generally combines into one word: *midday, midpoint, midweek*; if used with a proper noun, insert a hyphen: *mid-Atlantic*

mile the word mile is generally understood to mean a statute mile of 5280 ft (1609 m), so the statement *I drove 326 miles* implies statute miles; when referring to the *nautical mile* (6080 ft; 1853 m), always identify it as such: *the flight distance was 4210 nautical miles* (or *4210 nmi*); abbr:

statute mile(s)	**mi**
nautical mile(s)	**nmi** (pref) or **n.m.**
miles per gallon	**mpg**
miles per hour	**mph**

mileage; milage *mileage* pref

milli def: 10^{-3}; abbr: **m**; other abbr:

milliampere(s)	**mA**
millicoulomb(s)	**mC**
millicurie(s)	**mCi**
millifarad(s)	**mF**
milligram(s)	**mg**
milligray(s)	**mGy**
millihenry(s)	**mH**
millijoule(s)	**mJ**
millikelvin(s)	**mK**
milliliter(s)	**mL**
millilumen(s)	**mlm**
millimeter(s)	**mm**
millimho(s)	**mmho**
millimole(s)	**mmol**
millinewton(s)	**mN**
milliohm(s)	**mΩ**; mohm
millipascal(s)	**mPa**
milliroentgen(s)	**mR**
millisecond(s)	**ms** (pref); **msec**
millisiemens	**mS**
millitesla(s)	**mT**
millivolt(s)	**mV**
milliwatt(s)	**mW**
milliweber(s)	**mWb**

milli- as a prefix, combines to form one word: *milliammeter, milligram, millimicron*

millibar def: a unit of pressure (= 100 Pa); abbr: **mbar**

mini- as a prefix combines to form one word: *minicomputer, minireport*

miniature; miniaturization

minimum pl: *minimums* (pref) or *minima*; abbr: **min**; also **minimize**

minority use mainly to refer to a number, as in *a minority by 2*; avoid using it as a synonym for several or a few; to write *a minority of the technicians* is incorrect when the intended meaning is *a few technicians*

minuscule not *miniscule*; def: very small

minute abbr:

time	**min**
angular measure	**′**

mis- a prefix meaning wrong(ly) or bad(ly); combines to form one word: *misalign, misfired, mismatched, misshapen*

miscellaneous

miscible

mnemonic

mole def: the SI unit for amount of substance; abbr: **mol**; other abbr: **kmol, mmol, μmol, mol/m³**

momentary; momentarily both mean *for a moment*, not *in a moment*

money- as a prefix normally combines to form one word: *moneymaking*,

monitor

mono- a prefix meaning one or single; combines to form one word: *monopulse, monorail, monoscope*

monochrome display adapter abbr: **mda** (pref) or **MDA**

months the months of the year are always capitalized: *January, February*, etc; if abbr, use only the first three letters: *Jan, Feb*, etc; the abbr for *month* is **mo**

mortice; mortise *mortise* pref

mosaic

most never use as a short form for *almost*; to say *most everyone is here* is incorrect

motherboard

movable; moveable *movable* pref

Mr.; Ms. address men as *M.* and women as *Ms.*; use *Miss* or *Mrs.* only if you know the person prefers to be so addressed; the period (punctuation) may be omitted after *Mr* and *Ms*

MS-DOS Microsoft disk operating system

msl abbr for mean sea level; also **MSL**

mucous

multi- a prefix meaning many; combines to form one word: *multiaddress, multicavity, multielectrode, multimedia, multistate*

municipal; municipality

myself often used wrongly; write "Peter and I…", not "Peter and myself"

N

NAND-gate

nano def: 10^{-9}; abbr: **n**; other abbr:

nanoampere(s)	**nA**
nanocoulomb(s)	**nC**
nanofarad(s)	**nF**
nanohenry(s)	**nH**
nanometer(s)	**nm**
nanosecond(s)	**ns** (pref); **nsec**
nanotesla(s)	**nT**
nanovolt(s)	**nV**
nanowatt(s)	**nW**

naphtha(lene)

national information infrastructure abbr: **NII**

nationwide

navigate; navigator; navigable

NB means note well, and is the abbr for *nota bene*; it is more common to use the word ***Note***

NC abbr for *normally closed* (contacts)

negative abbr: **neg**

negligible

nevertheless

newsgroup

newton def: a unit of force (SI); abbr: **N**; other abbr: **MN, kN, mN, μN, N·m** (newton meter), **N/m** (newtons per meter)

next write the *next two* (or *next three*, etc.) rather than *the two next* (etc.)

nickel

night never use *nite*; write *nighttime* as one word

NII national information infrastructure

nineteen; ninety; ninth all three are frequently misspelled

nitroglycerine

NO abbr for *normally open* (contacts)

No. abbr for **number;** avoid using # sign

nomenclature

nomogram; nomograph *nomogram* pref

non- as a prefix meaning not or negative, normally combines to form one word: *nonconductor, nondirectional, nonnegotiable, nonlinear, nonstop*; if combining word is a proper noun, insert a hyphen: *non-American*; avoid forming a new word with *non-* when a similar word that serves the same purpose already exists (i.e. you should not form *nonaudible* because *inaudible* already exists)

none when the meaning is "not one," treat as singular; when the meaning is "not any," treat as plural: *none* (not one) *was satisfactory; none* (not any) *of the receivers were repaired*

no one two words

NOR-gate

norm def: the average or normal (distribution, situation, or condition)

normalize

normally closed; normally open (contacts) abbr: **NC, NO**

normal to def: at right angles to

north abbr: **N**; other abbr:

northeast	**NE**
northwest	**NW**
north-south (control, movement)	**N-S**

northbound and *northward* are written as one word; for rule on capitalization, see **east**

notable

not applicable abbr: **N/A**

note well abbr: **NB** (derived from *nota bene*), but ***Note*** is more common

NOT-gate

notice; noticeable; notification

not to exceed an overworked phrase that should be used only in specifications; in all other cases use *not more than*

nth (harmonic, etc)

nuclear frequently misspelled

nucleus the plural is *nuclei* (pref) or *nucleuses*

null

number although **no.** would appear to be the logical abbr for number (and is pref), **No.** is much more common (the symbol # is no longer used as an abbr for number); the abbr *no.* or *No.* must always be followed by a quantity in numerals: it is incorrect to write *we have received a No. of shipments*; for the difference in usage between *amount* and *number*, see **amount**

numbers (in narrative) as a general rule, spell out up to and including nine, and use numerals for 10 and above; for specific rules, see pages 353 and 355

O

oblique; obliquity

obsolete; obsolescent

obstacle

obtain; secure use *obtain* when the meaning is simply to get; use *secure* when the meaning is to make safe or to take possession of (possibly after some difficulty); *we obtained four additional samples; we secured space in the prime display area*

occasional; occasionally

occur; occurred; occurrence; occurring

o'clock avoid using; see **time**

OCR optical character recognition

off- as a prefix either combines into one word, or a hyphen is inserted: *offset, offshoot, off-center(ed), off-scale, off-the-shelf*

offline

off of an awkward construction; omit the word *of*

ohm def: a unit of electric resistance; abbr: **Ω** or **ohm**; other abbr: **GΩ, Gohm, MW, Mohm, kW, kohm, mW, mohm, μW, uohm**; abbr for ohm-centimeter(s) is **ohm-cm**; *ohmmeter* has *mm*

oilfield; oil-filled; oilsands

omit; omitted; omission

omni- a prefix meaning all or in all ways; combines to form one word: *omnibearing, omnidirectional, omnirange*

on; onto *on* means positioned generally; *onto* implies action or movement: *the report is on Mr. Cord's desk; the speaker stepped onto the platform*

one- as a prefix mostly combines with a hyphen: *one-piece, one-sided, one-to-one, one-way*; but *oneself* and *onetime*

online

onward(s) *onward* pref

opaque; opacity

op. cit. def: Latin abbr for *opere citato*, meaning the work cited; used in footnoting, now obsolescent

operate; operator; operable not *operatable*

optical character recognition abbr: **OCR**

optimum pl: *optima* (pref) or *optimums*; also: **optimal**

oral def: spoken, avoid confusing with *aural*

orbit; orbital; orbited; orbiting

organize; organizer; organization

OR-gate

orient; orientation the noun form is *orientation*; the pref verb form is *orient, oriented, orienting*

orifice

origin; original; originally

oscillate

oscilloscope slang abbr: **scope**

ounce(s) abbr: **oz**; other abbr:

ounce-foot	**oz-ft**
ounce-inch	**oz-in.**

out- as a prefix normally combines to form one word: *outbreak, outcome, outdistance*; when *out-* is followed by *of*, insert hyphens if used as a compound adjective (as in *an out-of-date list*), but treat as separate words when used in place of a noun (as in *the printing schedule is out of phase*)

outside diameter abbr: **OD**

outward(s) *outward* pref

over- as a prefix meaning above or beyond, normally combines to form one word: *overbunching, overcurrent, overdriven, overexcited, overrun*; avoid using as a synonym for more than, particularly when referring to quantities: *more than 17 were serviceable* is better than *over 17 were serviceable*

overage means either too many or too old

overall an overworked word; as an adjective it often gives unnecessary additional emphasis (as in *overall impression*) and should be deleted; avoid using as a synonym for *altogether, average, general,* or *total*

oxyacetylene

P

pacemaker; pacesetter

page; pages abbr: **p.; pp.**

paid not *payed*, when the meaning is to spend

pair(s) abbr: **pr**

pamphlet

paper- as a prefix mostly combines to form one word: *paperback, paperbound, paperwork*; but *paper-covered, paper-thin*

parabola; parabolas; parabolic; paraboloid

paragraph(s) abbr: **para**

parallax

parallel; paralleled; paralleling; parallelism; parallelogram both *parallel to* and *parallel with* are correct

parameter; perimeter *parameter* means a guideline; *perimeter* means a border or edge

paraplegic

paraprofessional

parenthesis the pl is *parentheses*

parity

particles

partly; partially use *partly* when the meaning is "a part" or "in part"; use *partially* when the meaning is "to a certain extent," or when preference or bias is implied

parts per million abbr: **ppm**

part-time

pascal def: a unit of pressure (SI); abbr: **Pa**; other abbr: **Gpa, Mpa, kPa, mPa, μPa, pPa, Pa · s** (pascal second)

pass- as a prefix normally combines to form one word: *passbook, passkey, passport, password*

passed; past as a general rule, use *passed* as a verb and *past* as an adjective or noun: *the test equipment has passed quality control inspection; past experience has demonstrated a tendency to fail at low temperature; in the past…*

pay- as a prefix normally combines to form one word: *paycheck, payload, payroll*; but *pay day*

pcb printed circuit board

PCMCIA Personal Computer Memory Card International Association

pel

pendulum pl: *pendulums*

penultimate def: the next to last

people; persons *people* pref: *all the people were present*; use *persons* to refer only to small numbers of people: *one person was interviewed; seven people failed the test*

per in technical writing it is acceptable to use *per* to mean either *by, a,* or *an*, as in *per diem* (by the day) and *miles per hour;* avoid using *as per* in all writing

percent abbr: %; use % only after numerals: 42%; use *percent* after a spelled-out number: *about forty percent*; avoid using the expression *a percentage of* as a synonym for *a part of* or *a small part*; also: **percentage** and **percentile**

perceptible

peripheral

permafrost

permeable; permeameter; permeance

permissible

permit; permitted; permitting; permittivity; permit-holder

perpendicular abbr: **perp**

persevere; perseverance

persistent; persistence; persistency

personal; personnel *personal* means concerning one person; *personnel* means the members of a group, or the staff: *a personal affair; the personnel in the powerhouse*

Personal Computer Memory Card International Association abbr: **PCMCIA**

peta def: 10^{15}; abbr: **P**; other abbr: **PBq** (petabecquerel)

petrochemical

pharmacy; pharmacist; pharmaceutical

phase in the nonelectric sense, *phase* means a stage of transmission or development; it should not be used as a synonym for aspect; it is used correctly in *the second phase called for a detailed cost breakdown*

phase-in; phaseout but use two words for the verb forms: *to phase in, to phase out*

phenolic

phenomenon pl: *phenomena*

photo- as a prefix, normally combines to form one word: *photoelectric, photogrammetry, photoionization, photomultiplier*; if combining word starts with *o*, insert a hyphen: *photo-offset*

pico def: 10^{-12}, abbr: **p**; other abbr:

picoampere(s)	**pA**
picocoulomb(s)	**pC**
picofarad(s)	**pF**
picohenry(s)	**pH**
picosecond(s)	**ps** (pref); **psec**
picowatt(s)	**pW**

piecemeal; piecework

piezoelectric; piezo-oscillator

pilot; piloted; piloting

pipeline

pixel

plagiarism def: to copy without acknowledging the original source

plateau pl: *plateaus* (pref) or *plateaux*

platform the underlying operating system of a computer, on which application programs can run

plug; plugged; plugging

plumbbob; plumb line

p.m. def: after noon (post meridiem)

pneumatic

polarize; polarizing; polarization

poly- a prefix meaning many; combines to form one word: *polydirectional, polyethylene, polyphase*

polyvinyl chloride abbr: **pvc**

pop-up window

positive abbr: **pos**

post- a prefix meaning after or behind; mostly combines to form one word: *postacceleration, postdated, postgraduate, postpaid*; but *post-mortem, post office; post-secondary*

post meridiem def: after noon; abbr: **p.m.**; can also be written as *post-meridian* (less pref)

potentiometer abbr: **pot.**

pound(s) (weight) abbr: **lb**; other abbr:

pound-foot	**lb-ft**
pound-inch	**lb-in.**
pounds per square foot	**psf** (pref); **lb/ft²**
pounds per square inch	**psi** (pref); **lb/in.²**
pounds per square inch, absolute	**psia**

power factor abbr: **pf** or spell out

powerhouse; power line; powerpack

practicable; practical these words have similar meanings but different applications that sometimes are hard to differentiate; *practicable* means feasible to do: *it was difficult to find a practicable solution* (one that could reasonably be implemented); *practical* means handy, suitable, able to be carried out in practice: *a practical solution would be to combine the two departments*

practice

pre- a prefix meaning before or prior; normally combines to form one word: *preamplifier, predetermined, preemphasis, preignite, preset*; if combining word is a proper noun, insert a hyphen; *pre-Roman*

precede; proceed *precede* means to go before; *proceed* generally means carry on or continue: *a brief business meeting preceded the dinner* (the meeting occurred first); *after dinner, we proceeded with the annual presentation of awards*; see **proceed**

precedence; precedent *precedence* means priority (of position, time, etc): *the pressure test has precedence* (it must be done first); a *precedent* is an example that is or will be followed by others: *we may set a precedent if we grant his request* (others will expect similar treatment)

précis

predominate; predominant; predominantly

prefer; preferred; preference; preferable avoid overstating *preferable*, as in *more preferable* and *highly preferable*

prescribe; proscribe *prescribe* means to state as a rule or requirement; *proscribe* means to deny permission or forbid

presently use *presently* only to mean soon or shortly; never use it to mean *now* (use *at present* instead)

pressure-sensitive

prestigious

pretense

prevalent; prevalence

preventive; preventative *preventive* pref

previous def: earlier, that which went before; avoid writing *previous to* (use *before*); see **prior**

principal; principle as a noun, *principal* means (1) the first one in importance, the leader; or (2) a sum of money on which interest is paid: *one of the firm's principals is Martin Dawes; the invested principal of $10,000 earned $950 in interest last year*; as an adjective, *principal* means most important or chief: *the principal reason for selecting the Arrow microprocessor was its low capital cost*; *principle* means a strong guiding rule, a code of conduct, a fundamental or primary source (of information, etc): *his principles prevented him from taking advantage of the error*

printed circuit board abbr: **pcb**

printout (noun and adj form)

prior; previous use only as adjectives meaning earlier: *he had a prior appointment*, or *a previous commitment prevented Mr. Perchanski from attending the meeting*; write *before* rather than *prior to* or *previous to*

proceed; proceeding; procedure use *proceed to* when the meaning is to start something new; use *proceed with* when the meaning is to continue something that was started previously

processor

producible

prohibit use *prohibit from*; never *prohibit to*

prominent; prominence

proofread

propel; propelled; propelling; propellant (noun); **propellent** (adjective); **propeller**

prophecy; prophesy use *prophecy* only as a noun, *prophesy* only as a verb

proposition in its proper sense, *proposition* means a suggestion put forward for argument; it should not be used as a synonym for *plan, project*, or *proposal*

pro rata def: assign proportionally; sometimes used in the verb form as **prorate**: *I want you to prorate the cost over two years*

prospectus; prospectuses

protein

protocol a term used in information technology to describe a special set of rules used to make telecommunication connections

proved; proven use *proven* only as an adjective or in the legal sense; otherwise use *proved: he has been proven guilty; he proved his case*

psycho- as a prefix normally combines to form one word: *psychoanalysis, psychopathic, psychosis*; if combining word starts with *o*, insert a hyphen: *psycho-organic*

purge; purging

pursuant to avoid using this wordy expression

Q

quality control abbr: QC

quantity; quantitative the abbr of quantity is **qty**

quart abbr: **qt**

quasi- a prefix meaning seemingly or almost; insert a hyphen between the prefix and the combining word: *quasi-active, quasi-bistable, quasi-linear*

question mark insert a question mark after a direct question: *how many booklets will you require?*; omit the question mark when the question posed is really a demand: *may I have your decision by noon on Monday*

questionnaire

quick- as a prefix normally combines with a hyphen: *quick-acting, quick-freeze, quick-tempered*; exceptions: *quicklime, quicksilver*

quiescent; quiescence

R

rack-mounted

racon def: a radar beacon

radian def: a unit of angular measurement; abbr: **rad**

radiator

radio- as a prefix, combines to form one word: *radioactive, radiobiology, radioisotope, radioluminescence*; if combining word starts with *o*, omit one of the *o*'s: *radiology, radiopaque*; in other instances *radio* may be either combined or treated as a separate word, depending on accepted usage; typical examples are *radio compass, radio countermeasures, radio direction-finder, radio frequency* (as a noun), *radio-frequency* (as an adj), *radio range, radiosonde, radiotelephone*

radio frequency abbr: **rf**

radio frequency interference abbr: **rfi** (pref) or **RFI**

radius pl: *radii* (pref) or *radiuses*

radix pl: *radices* (pref) or *radixes*

rain- as a prefix normally combines to form one word: *raincoat, rainproof, rainwear*; exception: *rain check*

rally; rallied; rallying

RAM def: random access memory; where a computer stores the operating system, application programs and data in current use

ramdrive

R and D abbr for research and development

range; ranging; rangefinder; range marker

rare; rarely; rarity; rarefy; rarefaction

ratemeter

ratio; ratios

rational; rationale *rational* means reasonable, clear-sighted: *John had a rational explanation for the error*; *rationale* means an underlying reason: *Tricia explained the company's rationale for diversifying the product line*

re def: a Latin word meaning in the case of; avoid using *re* in technical writing, particularly as an abbr for *regarding, concerning, with reference to*

re- a prefix meaning to do again, to repeat; normally combines to form one word: *reactivate, rediscover, reemphasize, reentrant, reignition, reorganize, rerun, reset, reunite*; if the compound term forms an existing word that has a different meaning, insert a hyphen to identify it as a compound term, as in *re-cover* (to cover again)

reaction use *reaction* to describe chemical or mechanical processes, not as a synonym for *opinion* or *impression*

reactive kilovolt-ampere; reactive volt-ampere see **kilo** or **volt**

readability

read-only memory a protected computer storage for data that normally can only be read, not written to; abbr: **ROM**

readout (noun and adj form)

realize; realization

real time defines computer responsiveness in a human rather than a machine sense of time; insert a hyphen when two words are combined into a compound adj; also: **real-time chat; real-time transmission**

reboot reload the operating system; see **boot**

recede

receive; receiver; receiving; receivable

rechargeable

recipe; receipt often confused; *recipe* means cooking instructions; *receipt* means a written record that something has been received

recommend; recommendation

reconcile; reconcilable

reconnaissance

recover; re-cover *recover* means to get back, to regain; *re-cover* means to cover again

recur; recurred; recurring; recurrence these are the correct spellings; never write *reoccur* (etc.)

recycle; recyclable

reducible

reenforce; reinforce *reenforce* means to enforce again; *reinforce* means to strengthen: *Rick Davis reenforced his original instructions by circulating a second memorandum; The Artmo Building required 34,750 tons of reinforced concrete*

refer; referred; referring; referral; referee; reference

refuel; refueled

reiterate def: to say again

relaid; relayed *relaid* means laid again, like a carpet; *relayed* means to send on, as a message would be relayed from one person to another

remit; remitted; remitting; remitter; remittance

remodel; remodeled; remodeling see **model**

removable

rent-a-car

reoccur(rence) never use; see **recur**

repairable; reparable both words mean in need of repair and capable of being repaired; *reparable* also implies that the cost to repair the item has been taken into account and it is economically worthwhile to effect repairs

replaceable

reproducible

rescind

research and development abbr: **R and D**

reservoir

reset; resetting; resettability

resin; rosin these words have become almost synonymous, with a preference for *resin*; use *resin* to describe a gluey substance used in adhesives, and *rosin* to describe a solder flux-core

respective(ly) this overworked word is not needed in sentences that differentiate between two or more items; e.g. it should be deleted from a sentence such as: *pins 4, 5, and 7 are marked R, S, and V respectively*

restart

resume def: a personal biography; the correct spelling is *résumé* (with two accents), but the single accent (*résume*) or no accent has become standard

retrieve; retrieval

retro- a prefix meaning to take place before, or backward; normally combines to form one word: *retroactive, retrofit, retrogression*; if combining word starts with *o*, insert a hyphen: *retro-operative*

reverse; reverser; reversal; reversible

revolutions per minute; revolutions per second abbr: **rpm; rps**

rfi radio frequency interference

rheostat

rhombus pl: *rhombuses* (pref) or *rhombi*

rhythm; rhythmic; rhythmically

ricochet; ricocheted; ricocheting

right-hand(ed) abbr: **RH**

RISC

rivet; riveted; riveter; riveting

road- as a prefix normally combines to form one word: *roadblock, roadmap, roadside*

roentgen abbr: **R**

role; roll a *role* is a person's function or the part that he or she plays (in an organization, project, or play); a *roll*, as a technical noun, is a cylinder; as a verb, it means to rotate: *the technician's role was to make the samples roll toward the magnet*

rollover one word as a n or adj; two words as a v

ROM def: read-only memory

root mean square abbr: **rms**

rosin see **resin**

rotate; rotator; rotatable; rotary

ruggedize

run-off insert a hyphen when used as n or adj

rustproof; rust-resistant

S

salvageable

same avoid using *same* as a pronoun; to write *we have repaired your receiver and tested same* is awkward; instead, write *we have repaired and tested your receiver*

satellite

saturate; saturation; saturable

save; savable

sawtooth; saw-toothed

scalar; scaler *scalar* is a quantity that has magnitude only; *scaler* is a measuring device

scarce; scarcity

sceptic(al); skeptic(al) *skeptic(al)* pref

schedule

schematic although really an adjective (as in *schematic diagram*), in technical terminology *schematic* can be used as a noun (meaning a *schematic drawing*)

science; scientific(ally); scientist

screwdriver; screw-driven

seam-weld

seasonal; seasonable *seasonal* means affected by or dependent on the season; *seasonable* means appropriate or suited to the time of year; *a seasonal activity; seasonable weather*

seasons the seasons are not capitalized: *spring, summer, autumn* or *fall, winter*

secant abbr: **sec**

secede; secession

second as a prefix, *second-* combines erratically: *second-class, second-guess, secondhand, second-rate, second sight*; the abbr for *second* (time) is **sec**, and for *second* (angular measure) it is "

secure see **obtain**

-sede *supersede* is the only word to end with *-sede*; others end with *-cede* or *-ceed*

seem(s) see **appear(s)**

self- insert a hyphen when used as a prefix to form a compound term: *self-absorption, self-bias, self-excited, self-locking, self-setting*; but there are exceptions: *selfless, selfsame*

semi- a prefix meaning half; normally combines to form one word: *semiactive, semiannually* (every six months), *semiconductor, semimonthly* (half-monthly), *semiremote, semiweekly* (half-weekly); if combining word starts with *i*, insert a hyphen: *semi-idle, semi-immersed*

separate; separable; separator; separation all are frequently misspelled

sequence; sequential

serial input/output abbr: **SIO**

serial number abbr: **ser no.** or **S/N**

series-parallel

serrated

serviceable

serviceperson avoid using *serviceman* or *servicewoman*

servo- as a prefix, combines to form one word: *servoamplifier, servocontrol, servosystem*; as a noun, *servo* is an abbr for *servomotor* or *servomechanism*

sewage; sewerage *sewage* is waste matter; *sewerage* is the drainage system that carries away the waste matter

SGML abbr for standard generalized markup language

shall *shall* is rarely used in technical writing (*will* is pref), except in specifications when its use implies that the specified action is mandatory

short- as a prefix, may combine with a hyphen, as in *short-circuit, short-form* (report), *short-lived, short-term*; in some cases it may combine into one word, as in *shorthand* (writing), *shorthanded, shortcoming, shortsighted, shortwave* (n or adj)

sic a Latin word which means a quotation has been copied exactly, even though there was an error in the original; e.g. *the report stated: "Our participation will be an issential (sic) requirement."*

siemens def: a unit of electric conductance (SI); abbr: **S**; other abbr: **kS, mS, μS.**

signal-to-noise (ratio)

silhouette

silverplate; silver-plate use *silverplate* as a noun or adjective; *silver-plate* as a verb

similar not *similiar*

sine abbr: **sin**

singe; singeing the *e* must be retained to avoid confusion with *singing*

singlehanded

siphon not *syphon*

sirup; syrup *syrup* pref

site; sight; cite three words that often are misspelled; a *site* is a location: *the construction site*; *sight* implies the ability to see: *mud up to the axles became a familiar sight*; *cite* means quote: *I cite the May 17 progress report as a typical example of good writing*

skeptic(al) pref spelling

skil(l)ful *skillful* pref; note that this is contradictory to most *l* and *ll* situations listed in this glossary

slip- usually combines into a single word: *slippage, slipshod, slipstream*; but *slip ring(s)*

smelled; smelt *smelled* pref

smo(u)lder *smolder* pref

soft key; software

solder

solely

soluble

someone; some one *someone* is correct when the meaning is any one person; *some one* is seldom used

some time; sometimes *some time* means an indefinite time: *some time ago*; *sometimes* means occasionally: *he sometimes works until after midnight*

sound- combines irregularly: *sound-absorbent, sound-absorbing, sound-powered, soundproof, sound track, sound wave*

south abbr: **S**; other abbr:

southeast	**SE**
southwest	**SW**

southbound and *southward* are written as one word; for rule on capitalization, see **east**

space- as a prefix normally combines to form one word: *spacecraft, spaceflight*

spare(s) can be used as a noun meaning spare part(s)

specially see **especially**

specific gravity abbr: **sp gr**

specific heat abbr: **sp ht**

spectro- as a prefix, combines to form one word: *spectrometer, spectroscope*; if combining word starts with *o*, omit one *o*: *spectrology*

spectrum pl: *spectra* (pref) or *spectrums*

split infinitive to split an infinitive is to insert an adverb between the word *to* and a verb: *to really insist* is a split infinitive; although grammarians used to claim that you should never split an infinitive, they now suggest you may do so if rewriting would result in awkward construction, ambiguity, or extensive rewriting

spotweld

square abbr: **sq** or ^{2}; other abbr:

square foot/feet	**ft^2** (pref); **sq ft**
square inch(es)	**in.^{2}** (pref); **sq in.**
square metre(s)	**m^2**
square centimeter(s)	**cm^2**
square millimeter(s)	**mm^2**
curies per square meter	**Ci/m^2**
milliwatts per square metre	**mW/m^2**

standard generalized markup language abbr: **SGML**

standby; standoff; standstill all combine into one word when used as noun or adjective

standing-wave ratio abbr: **swr** or **SWR**

startup one word as n or adj; two words as a v: **start up**

state-of-the-art

stationary; stationery *stationary* means not moving: *the vehicle was stationary when the accident occurred*; *stationery* refers to writing materials: *the main item in the October stationery requisition was an order for one thousand writing pads*

statutory

stereo- as a prefix, combines to form one word: *stereometric, stereoscopic*; *stereo* can be used alone as a noun meaning multi-channel system

stimulus pl: *stimuli*

stocklist, stockpile

stop- as a prefix usually combines to form one word: *stopgap, stopnut, stopover* (when used as noun or adjective); but *stop payment, stop watch*

stoppage

straightened; straitened *straightened* means straight; *straitened* means restricted

strato- a prefix that combines to form one word: *stratocumulus, stratosphere*

stratum pl: *strata*

structural

stylus dictionaries list *styli* as the pref pl, but *styluses* is much more commonly used, and recommended

sub- a prefix generally meaning below, beneath, under; combines to form one word: *subassembly, subcarrier, subcommittee, subnormal, subpoint*

subparagraph abbr: **subpara;** abbr for *subsubparagraph* is **subsubpara**

subpixel

subtle; subtlety; subtly

succinct; succinctly

sufficient in technical writing, *enough* is a better word than *sufficient*

summarize

super- a prefix meaning greater or over; combines to form one word: *superabundant, superconductivity, superregeneration*

superhigh frequency abbr: **shf**

superimpose; superpose *superimpose* means to place or impose one thing generally on top of another; *superpose* means to lay or place exactly on top of, so as to be coincident with

supersede see **-sede**

supra- a prefix meaning above; normally combines to form one word: *supramolecular*; if combining word starts with *a*, insert a hyphen: *supra-auditory*

surfeit def: to have more than enough

surveillance

surveyor

susceptible

switch- *switchboard, switchbox, switchgear*

syllabus pl: *syllabuses* (pref) or *syllabi*

symmetry; symmetrical

symposium pl: *symposiums (pref) or symposia*

synchro as a prefix combines to form one word: *synchromesh, synchronize, synchronous, synchroscope*; *synchro* can also be used alone as a noun meaning synchronous motor

synonymous use *synonymous with*, not *synonymous to*

synopsis pl: *synopses*

synthesis pl: *syntheses*

synthetic

syphon *siphon* pref

syringe

syrup; syrupy

systemwide

T

tail- as a prefix normally combines to form one word: *tailboard, tailless, tailwind*; but *tail end, tail fin*

take- *takeoff; takeover; takeup*; as nouns and adjectives these terms all combine into a single word

tangent abbr: **tan**

tangible

tape deck

target; targeted

taxable; tax-exempt; taxpayer

TCI/IP abbr for transmission control protocol/Internet protocol; def: the basic communication language or protocol of the Internet

teamwork

technician

Teflon

tele- a prefix meaning at a distance; combines to form one word: *teleammeter, telemetry, telephony, teletype(writer)*

telecom; telecon *telecom* is the abbr for *telecommunication(s)*; *telecon* is the abbr for *telephone conversation*

television abbr: **TV**

Telnet

temperature abbr: **temp**; combinations are *temperature-compensating* and *temperature-controlled*; when recording temperatures, the abbr for *degree* (deg or °) may be omitted: *an operating temperature of 85C; the water boils at 100C or 212F*; the pref (SI) unit for temperature is the degree Celsius (°C)

tempered

template; templet both spellings are correct; *template* pref

temporary; temporarily

tenfold

tensile strength abbr: **ts**

tentative; tentatively

tenuous

tera def: 10^{12}; abbr: **T**; other abbr:

terabecquerel(s)	TBq
terahertz	THz
terajoule(s)	TJ
terawatt(s)	TW

terminus pl: *termini* (pref) or *terminuses*

tesla def: a unit of magnetic flux density, magnetic inductance (SI); abbr: **T**; other abbr: **mT, μT, nT**

that is abbr: **i.e.** (pref) or **ie**

their; there; they're these words are frequently misspelled, more through carelessness than as an outright error; *their* is a possessive, meaning belonging to them: *the staff took their holidays earlier than normal*; *there* means in that place: *there were 18 desks in the room*, or *put it there*; *they're* is a contraction of *they are* and should not appear in technical or business writing

there- as a prefix combines to form one word: *thereafter, thereby, therein, thereupon*

therefor(e) *therefore* pref

thermo- a prefix generally meaning heat; combines to form one word: *thermoammeter, thermocouple, thermoelectric, thermoplastic*

thermodynamic temperature the SI unit is the kelvin (abbr: **K**), expressed in degrees Celsius (°C)

thesis pl: *theses*

thousand abbr: **k**

thousand foot-pound(s) **kip-ft**

thousand pound(s) **kip**

three- when used as a prefix, a hyphen normally is inserted between the combining words: *three-dimensional, three-phase, three-ply, three-wire*; exceptions are *threefold* and *threesome*

threshold

through never use *thru*

tieing, tying *tying* pref

timber; timbre *timber* is wood; *timbre* means tonal quality

time always write time in numerals, if possible using the 24-hour clock: *08:17* or *8:17 a.m., 15:30* or *3:30 p.m.;* 24-hour times may be written as *20:45* (pref), *20:45 hr*, or *20:45 hours*; never use the term "o'clock" in technical writing: write *15:00* or *3 p.m.* rather than *3 o'clock*

time- typical combinations are *time base, time-card, time clock, time constant, time-consuming, time lag, timesaving, time-slot, timetable, time-wasting*

tinplate; tin-plate use *tinplate* as a noun, *tin-plate* as a verb or adjective

to; too; two frequently mispelled, most often through carelessness; *to* is a preposition that means in the direction of, against, before, or until; *too* means as well; *two* is the quantity 2

today; tonight; tomorrow never use *tonite*

tolerance abbr: **tol**

ton; tonne the US ton is 2000 lb and is known as a *short ton*; the Br ton is 2240 lb and is known as a *long ton*; the metric ton is 1000 kg (2204.6 lb) and is known as a *tonne (*abbr: **t**); other terms: *tonmile* and *tonnage*

toolbox; toolmaker; toolroom

top- top-heavy, top-loaded, top-up

torque; torqued; torquing

touch-tone (dialing)

toward(s) *toward* pref

traceable

trade- as a prefix combines most often into a single word; *trademark, tradeoff*; but *trade-in, trade name*, and *trade show*

trans- a prefix meaning over, across, or through; it normally combines to form one word: *transadmittance, transcontinental, transship*; if combining word is a proper noun, insert a hyphen; *trans-America* (an exception is *transatlantic* and *transpacific)*; *transonic* has only one *s*

transceiver def: a transmitter-receiver

transfer; transferred; transferring; transferable; transference

translator

transmission control protocol/Internet protocol abbr: TCI/IP

transmit; transmitted, transmitting; transmittal, transmitter; transmission

transverse; traverse *transverse* means to lie across; *traverse* means to track horizontally

tri- a prefix meaning three or every third; combines to form one word: *triangulation, tricolour, trilateral, tristimulus, triweekly*

triple- all compounds are hyphenated: *triple-acting, triple-spaced*

trouble-free; troubleshoot(ing)

truncated

tune; tunable; tuneup (noun or adjective)

turbo- a prefix meaning turbine-powered; combines to form one word: *turboelectric, turboprop*

turbulence; turbulent

turn- *turnaround* and *turnover* form one word when used as adjectives or nouns; *turnstile* and *turntable* always form one word; *turns-ratio* is hyphenated

two- when used as a prefix to form a compound term, a hyphen normally is inserted: *two-address, two-phase, two-ply, two-position, two-wire*; an exception is *twofold*

type- as a prefix normally combines into one word: *typeface, typeset(ting)*

tyre Br spelling of *tire* (on a car wheel)

U

UCD user-centered drive

ultimatum pl: ultimatums

ultra- a prefix meaning exceedingly; normally combines to form one word: *ultrasonic, ultrasound, ultraviolet*; if combining word starts with *a*, insert a hyphen: *ultra-audible, ultra-audion*

ultrahigh frequency abbr: **uhf** or UHF

un- a prefix generally meaning not or negative; normally combines to form one word: *uncontrolled, undamped, unethical, unnecessary*; if combining word is a proper noun, or if term combines to form an existing word that has a different meaning, insert a hyphen: *un-American, un-ionized* (meaning not ionized); if uncertain whether to use *un-, in-*, or *im-*, try using *not*

unadvisable; inadvisable both are correct; *inadvisable* pref

unbalance; imbalance for technical writing, *unbalance* pref; see **imbalance**

unbiased

under- a prefix meaning below or lower; combines to form one word: *underbunching, undercurrent, underexposed, underrated, undershoot, undersigned, underway*

underage means a shortage or deficit, or too young

unequal(l)ed *unequaled* pref; see **equal**

unessential; inessential *unessential* pref

unforeseen; unforeseeable

uni- a prefix meaning single or one only; combines to form one word: *uniaxial, unidirectional, unifilar, univalent*

uninterested def: not interested; avoid confusing with *disinterested*

unionized; un-ionized *unionized* refers to a group of people who belong to a union; *un-ionized* means not ionized

unique def: the one and only, without equal, incomparable; use with great care and never in any sense where a comparison is implied; you cannot write *this is the most unique design*; rewrite as *this design is unique*, or (if a comparison must be made) *this is the most unusual design*

universal resource locator abbr: **URL**

unmistakable

unnavigable

unparalleled

unpractical *impractical* pref

unsanitary

unserviceable abbr: **u/s**

unstable but *instability* is better than *unstability*

untraceable

unwieldy

up- as a prefix combines to form one word: *update, upend, upgrade, uprange, upswing;* but *up-to-date*

uppercase def: capital letters; abbr: **uc**

uppermost

URL universal resource locator

use; usable; usage; using; useful

Usenet

user-centered drive abbr: UCD

utilize; avoid using *utilizes* when *uses* or *employs* would be a better word

vacuum

valance; valence *valance* means a cover over a drapery track; *valence* is an electronic or nucleonic term, as in *valence electron*

valve-grind(ing)

vari- as a prefix meaning varied, combines into one word: *varicoloured, variform*

variance write *at variance with*, never *at variance from*

varimeter; varmeter def for both: a meter for measuring reactive power; *varimeter* pref

V-chip

vdisk virtual disk

vehicle, vehicular

vender; vendor *vendor* pref

ventilator

verbatim written exactly as originally said

versed sine abbr: **vers**

versus def: against; abbr: **vs**

vertex def: top; pl: *vertexes* (pref) or *vertices*; avoid confusing with *vortex*

very high frequency abbr: **vhf** or **VHF**

VGA video graphics locator

vice versa def: in reverse order

video- as a prefix normally combines to form one word: *videocast, videocassette, videotape*; the abbr for videocassette recorder is **VCR** or **vcr**

video frequency abbr: **vf** or **VF**

video graphics array abbr: **VGA**

viewfinder; viewpoint

virtual disk abbr: **vdisk** (pref) or **VDISK**

visor; vizor *visor* pref

viz def: namely; this term is seldom used in technical writing

vocation; avocation *vocation* is a trade or calling; *avocation* means an interest or hobby

voice-over; voiceprint

volatile memory

volt def: electric potential or potential difference; abbr: **V**; other abbr: **MV, kV, mV, μV, nV**; also

volt-ampere(s)	VA
volt-ampere(s), reactive	VAr
volts, alternating current	Vac
volts, direct current	Vdc
volts, direct current, working	Vdcw
volts per meter	V/m

volt- combines into one word: *voltammeter, voltohmyst*

volume abbr: **vol**

vortex def: spiral; pl: *vortexes* (pref) or *vortices*; avoid confusing with *vertex*

VU-meter

W

WAIS wide area information server

waive; waiver; waver *waive* and *waiver* mean to forgo one's claim or give up one's right; *waver* means to hesitate, to be irresolute

walkie-talkie

war- as a prefix, combines to form one word: *warfare, wartime*

warranty

waste; wastage

water- combines irregularly: *water-cool(ed), water cooler, waterflow, water level, waterline, waterproof, water-soluble, watertight*

watt def: a unit of power, or radiant flux (SI); abbr: **W**; other abbr: **TW, GW, MW, kW, mW, μW, nW, pW, W/m²**; the abbr for *watt-hour(s)* is **Wh** (pref) or **W-hr**; as a prefix, *watt-* forms *watthourmeter* and *wattmeter*

wave- normally combines to form one word: *waveband, waveform, wavefront, waveguide, wavemeter, waveshape*; exceptions are *wave angle* and *waveswept*

wavelength abbr: λ

waver see **waiver**

wear and tear *no* hyphens

weather use only as a noun; never write *weather conditions*; avoid confusing with *climate* and *whether*

weatherproof

Web slang abbr for World Wide Web

weber def: a unit of magnetic flux (SI); abbr: **Wb**; other abbr: **mWb**

Web site

Wednesday often misspelled

week(s) abbr: **wk**

weekend

weight abbr: **wt**

well- as a prefix normally combines with a hyphen: *well-adjusted, well-defined, well-timed*

west abbr: **W**; *westbound* and *westward* are written as one word; for rule on capitalization, see **east**

where- as a prefix combines to form one word: *whereas, wherein*; when combining word starts with *e*, omit one *e*: *wherever*

whether; weather *whether* means if; *weather* has to do with rain, snow, sunshine, etc.

while; whilst *while* pref

whoever

wholly not *wholely*

wide; width abbr: **wd**

wide area information server abbr: **WAIS**

wideband; widespread

wirecutter(s); wire-cutting; wirewound

withheld; withhold

word processor; word processing abbr: **WP**; as an adj, insert a hyphen: *the word-processing software*

words per minute abbr: **wpm**

work- as a prefix usually combines to form one word: *workbench, workflow, workforce, workload, workshop*; but *work station*

working volts, dc abbr: **Vdcw**

worldwide

World Wide Web abbr: **www** or **The Web**; all the resources and users on the Internet that use the Hypertext Transfer Protocol (http)

wrap; wrapped; wrapping; wraparound

writeoff; writeup both combine into one word when used as noun or adjective

writer see **author**

writing only one *t*

www World Wide Web

wysiwyg def: What You See Is What You Get

X

x- *x-axis, X-band, x-particle, x-radiation, x-ray*

Xerox

X-Y recorder

Y

y- *Y-antenna, y-axis, Y-connected, Y-network, Y-signal*

Yahoo

yard(s) abbr: **yd**

yardstick

year(s) abbr: **yr**; typical combinations are *year-end* and *year-round*

yocto def: 10^{-24}; abbr: **y**

yotta def: 10^{24}; abbr: **Y**

your; you're *your* means belonging to or originating from you: *I have examined your prototype analyzer*; *you're* is a contraction of *you are* and should not appear in technical or business writing

Z

z-axis

zepto def: 10^{-21}; abbr: **z**

zero pl: *zeros* (pref) or *zeroes*; typical combinations are *zero-access, zero-adjust, zero-beat, zero-hour, zero level, zero-set, zero reader*

zetta def: 10^{21}; abbr: **Z**

zip code

zoology; zoological; zoologist

Index

MARKING CONTROL CHART

This control chart will show you which aspects of your writing need attention and, as time progresses, whether you have successfully corrected your most predominant faults. As each assignment is returned to you, count up the errors indicated as marginal notations by your instructor and enter them on the chart. For instance, if on assignment 1 your instructor enters "F" and "U" once, and "S" three times, in the margin, you are being told you have used the wrong format (F), your work is untidy (U), and you have three spelling errors (S). In column 1 of the chart enter "1" in the squares opposite F and U, and "3" in the square opposite "S".

ASSIGNMENT NUMBER

1	2	3	4	5	6	7	8	9	10	11	12	13	14	15

A – Awkward construction

B – Brevity overdone; too few details

C – Continuity weak; paragraph lacks coherence/unity

D – Development inadequate; support your argument

E – Error! Check your facts, data, information

F – Format incorrect

G – Grammar fault

H – Heavy going; dull; uninteresting

I – Illogical or irrelevant (correct or omit)

J – Jumpy — too many short sentences, reads like primary reader

K – King-size paragraph or sentence (shorten it)

L – Low Information Content words or phrase (delete)

M – Missing words or information

N – No! Never do this; never use slang, contractions, unexplained abbreviations, etc.

O – Organization poor

P – Punctuation error, or punctuation missing

Q – Query: what does this mean? not understood; can't read your writing

R – Repetition

S – Spelling error

T – Tone wrong

U – Untidy, messy, or careless work (improve "presentation")

V – Vague; ambiguous; not clear enough

W – Wishy-washy; weak argument; unconvincing

X – X-out (delete, omit) this unnecessary statement

Y – Yak! Yak! Yak! — too wordy; too many generalities

Z – Lacks continuity; needs better transitions

– Numbers wrongly presented

// – Use parallel construction